Molecular Modeling and Simulation

Applications and Perspectives

Series editor

Edward Maginn, University of Notre Dame, Notre Dame, IN, USA

This series aims at providing a comprehensive collection of works on developments in molecular modeling and simulation, particularly as applied to the various research fields of engineering. The Series covers a broad range of topics related to modeling matter at the atomistic level. The series provides timely and detailed treatment of advanced methods and their application in a broad range of interrelated fields such as biomedical and biochemical engineering, chemical engineering, chemistry, molecular biology, mechanical engineering and materials science. The Series accepts both edited and authored works, including textbooks, monographs, reference works, and professional books. The series welcomes manuscripts concerned with developments in and applications of molecular modeling and simulation to contemporary research in myriad technological fields, including, but not limited to:

- New Materials Development
- Process Engineering
- Fuel Engineering
- Combustion
- Polymer Engineering
- Biomechanics
- Biomaterials
- Fluid Flow and Modeling
- Nano and Micro Fluidics
- Nano and Micro Technology
- Thin Films
- Phase Equilibria
- Transport Properties
- Computational Biology

More information about this series at http://www.springer.com/series/13829

Jianzhong Wu
Editor

Variational Methods in Molecular Modeling

Springer

Editor
Jianzhong Wu
Department of Chemical and Environmental
Engineering and Department of Mathematics
University of California
Riverside, CA
USA

ISSN 2364-5083 ISSN 2364-5091 (electronic)
Molecular Modeling and Simulation
ISBN 978-981-10-2500-6 ISBN 978-981-10-2502-0 (eBook)
DOI 10.1007/978-981-10-2502-0

Library of Congress Control Number: 2016955927

Printed on acid-free paper

This Springer imprint is published by Springer Nature
The registered company is Springer Nature Singapore Pte Ltd.
The registered company address is: 152 Beach Road, #22-06/08 Gateway East, Singapore 189721, Singapore

Series Editor's Preface

The series *Molecular Modeling and Simulation—Application and Perspectives* seeks to publish a comprehensive collection of volumes highlighting the most important and groundbreaking developments in molecular modeling and simulation. The goal is to publish volumes where leading researchers can describe the latest advances in their field in a comprehensive and nuanced manner that makes the material both accessible to those outside the field while at the same time being useful to other experts. The series encourages authors to expand their treatment of their topic in ways that are impossible to achieve in normal journal articles. With this second volume in the series "Variational Methods in Molecular Modeling," Editor Jianzhong Wu has assembled an outstanding collection of contributions from the top people in the field of variational methods. The volume starts with a pedagogical introduction to the topic that should be of great interest to students desiring to learn about these methods. In the subsequent nine chapters, the authors provide an overview of variational methods for the particular topic of their chapter and follow this with examples that illustrate the application of these methods. The volume closes with an appendix treating the calculus of variations.

I am deeply grateful to Prof. Jianzhong Wu of the University of California, Riverside, for his willingness to take on this project and for his wisdom and effort in putting together such an outstanding volume. I am confident this volume will play an important role in the future application of these methods to the field of molecular modeling.

Edward Maginn
University of Notre Dame

Preface

Calculus of variations is a branch of mathematical analysis that deals with functionals, i.e., algebraic relations mapping functions into real numbers. The original ideas were established by Leonhard Euler in 1733, and since then, variational methods have found widespread applications in science and engineering. A key objective in the calculus of variations is to identify a specific function that minimizes (or maximizes) a given functional. The mathematical procedure is naturally applicable to statistical thermodynamics as demonstrated in the groundbreaking works of J. Willard Gibbs. Today, the maximum entropy principle (MaxEnt), a cornerstone of the so-called Bayesian statistics, is broadly used not only in equilibrium as well as non-equilibrium statistical mechanics but also in pattern recognition and image processing, risk analysis, urban and regional planning, and business financing, just to name a few from a large class of probabilistic problems. Calculus of variations is also useful in both classical and quantum mechanics as shown in the pioneering works of Joseph-Louis Lagrange and in the variational principle for determining the ground states of quantum systems.

This monograph is an exposition of recent applications of variational methods in molecular modeling for thermodynamic systems. While variational principles have been routinely used in both Lagrangian mechanics and the Kohn–Sham density functional theory, their applications to complex molecular systems are rarely discussed in the conventional texts of molecular modeling and statistical mechanics. Instead of describing molecular motions and electronic structures, this book is mostly concerned with the formulation and application of free-energy functionals that connect thermodynamic variables with potential fields or the ensemble averaged atomic, molecular, or particle distributions. Prime examples include classical density functional theory for simple as well as complex fluids, self-consistent-field theories for ionic mixtures and polymer blends, phase-field methods for phase separations and interfacial phenomena, and the Ginzburg-Landau-type theories for molecular self-assembly and order-disorder transitions. In addition, this book covers the applications of variational methods to describing time-dependent phenomena and to solving quantum many-body problems.

To introduce these fascinating topics to a broad audience, each chapter in this book provides a pedagogical overview of variational methods for specific subjects, with the key theoretical results illustrated with tutorial examples. With emphasis placed on physical understanding rather than on rigorous mathematical derivations, this monograph should be accessible to graduate students and researchers in the broad areas of applied mathematics, condensed matter physics, materials science and engineering, chemistry, and chemical and biomolecular engineering without specific training in the calculus of variations.

I am tremendously grateful to all contributors of this monograph for their dedicated work and cooperation in finishing their writings in a timely manner. Preparation of pedagogical materials is not most rewarding in today's academic environment, yet it is very time-consuming to summarize the previous research in particular publications from others. Therefore, I feel especially lucky to have contributions to this book from a cohort of very distinguished authors. I also want to thank all reviewers of this monograph for their careful examination of individual chapters and professional services: Jaydeep P. Bardhan, Northeastern University; Daniel Borgis, École Normale Supérieure; Joachim Dzubiella, Humboldt University; Jian Jiang, California Institute of Technology; Isamu Kusaka, Ohio State University; Yu Liu, East China University of Science and Technology; Umberto M.B. Marconi, University of Camerino; Friederike Schmid, University of Mainz; Cyrus Umrigar, Cornell University; Qiang (David) Wang, Colorado State University; Rik Wensink, University of Paris-Sud XI; Zhenli Xu, Shanghai Jiao Tong University; and Pingwen Zhang, Peking University. Last, but not least, I would like to thank Prof. Edward Maginn, the Chief Editor of this book series, for inviting me to prepare this monograph and Mr. Praveen Kumar, the Springer Project Coordinator, for his considerable help and patience to put things together.

Riverside, CA, USA Jianzhong Wu

Contents

Editor and Contributors

About the Editor

Dr. Jianzhong Wu is a professor of Chemical Engineering and a cooperating faculty member of Mathematics Department at the University of California, Riverside. His research is focused on the development and application of statistical mechanical methods, in particular density functional theory, for predicting the microscopic structure and physiochemical properties of confined fluids, soft materials, and biological systems.

Contributors

Jan Forsman Theoretical Chemistry, Chemical Centre, Lund, Sweden

Alexander H. Lewis Mathematical Institute, University of Oxford, Oxford, UK

Leo Lue Department of Chemical and Process Engineering, University of Strathclyde, James Weir Building, Glasgow, UK

Hartmut Löwen Institut für Theoretische Physik II: Weiche Materie, Heinrich-Heine-Universität Düsseldorf, Düsseldorf, Germany

Apala Majumdar Department of Mathematical Sciences, University of Bath, Claverton Down, Bath, UK

Brenda Rubenstein Lawrence Livermore National Laboratory, Quantum Simulations Group, Livermore, CA, USA; Department of Chemistry, Brown University, Providence, RI, USA

An-Chang Shi Department of Physics and Astronomy, McMaster University, Hamilton, ON, Canada

Zhen-Gang Wang Division of Chemistry and Chemical Engineering, California Institute of Technology, Pasadena, CA, USA

Guo-Wei Wei Department of Mathematics, Michigan State University, East Lansing, MI, USA

Clifford E. Woodward School of Physical Environmental and Physical Sciences University College, University of New South Wales, Kensington, Australia

Jianzhong Wu Department of Chemical and Environmental Engineering and Department of Mathematics, University of California, Riverside, CA, USA

Yongcheng Zhou Department of Mathematics, Colorado State University, Fort Collins, CO, USA

Variational Methods in Statistical Thermodynamics—A Pedagogical Introduction

Zhen-Gang Wang

1 Introduction

In this chapter, we provide a pedagogical introduction to variational methods in statistical thermodynamics. This chapter is written primarily for graduate students who have had a first course in statistical mechanics. It is hoped that this chapter provides a useful and insightful introduction to both the conceptual and practical aspects of the variational methods that can be helpful in their research in statistical mechanics.

Variational principle is at the very heart of thermodynamics. Its root is the second law, usually stated in terms of the entropy, which is an inequality—the only fundamental physical law that takes the form of an inequality rather than equality [1]. The statistical mechanical correspondence of the variational method is the maximum-term method in the evaluation of the partition function, which forms the basis for minimization of the free energy. We thus start with a brief review of the variational nature of thermodynamics and its corresponding statistical mechanical origin.

Except for a limited number of special cases, the partition function cannot be evaluated exactly. Thus in constructing the free energy of a system of interest, we usually have to make approximations. One of the simplest and most useful approximations is the mean-field approximation, which in essence reduces an intractable many-body problem to a single-body problem in an effective external field, which is then determined self-consistently. While for simple systems, the mean-field approximation can often be constructed intuitively, more systematic derivations are based on variational methods. We present two common variational methods for approximating the partition function (or equivalently the free energy)—the Gibbs-Bogoliubov-Feynman (GBF) variational bound, and the steepest-descent method. By way of a toy example in the evaluation of an integral, we illustrate the use of these two methods. We then

Z.-G. Wang (✉)
Division of Chemistry and Chemical Engineering, California Institute of Technology, Pasadena, CA 91125, USA
e-mail: zgw@caltech.edu

J. Wu (ed.), *Variational Methods in Molecular Modeling*,
Molecular Modeling and Simulation, DOI 10.1007/978-981-10-2502-0_1

use the variational methods to derive the mean-field solution for the Ising model and the Poisson-Boltzmann theory for electrolyte solutions. Finally, we show that the GBF method provides a natural way to include fluctuation effects in weakly correlated systems.

2 The Variational Nature of Thermodynamics

The second law of thermodynamics is commonly expressed by the Clausius inequality:

$$dS \geq \frac{\delta Q}{T} \tag{2.1}$$

where S is the entropy of the system, Q the heat into the system and T the absolute temperature. In the most general case the temperature refers to that of the heat bath. For an adiabatic process, the right hand side of the expression is zero, and Eq. 2.1 reduces to the well-known expression

$$dS \geq 0 \tag{2.2}$$

This is often termed the principle of increasing entropy. If we consider the system and surrounding as a closed, isolated system, then this principle states that the entropy cannot decrease—it increases for an irreversible (nonequilbrium) process and reaches a maximum at equilibrium.

The principle of increasing entropy implies a variational principle. Imagine we start an isolated system in some nonequilibrium state, then the second law tells us that the state of the system will evolve in such a way as to increase its entropy. However, if the system is to finally attain a well defined final equilibrium, as postulated in thermodynamics, then the entropy will asymptotically reach a final value—the maximum value—and cease to increase. Since the entropy is at its maximum in this final state, its first differential with respect to any real or virtual change in the macro states of the system (for example, the density distribution in a gas in the absence of external fields) must vanish, while its second differential will be negative. Thus mathematically, the condition of equilibrium for an isolated system is expressed as:

$$\frac{\delta S}{\delta X} = 0 \tag{2.3}$$

and

$$\frac{\delta^2 S}{\delta X^2} < 0 \tag{2.4}$$

where we use the generic notation X to denote some unconstrained macrostate variable [2]. The second derivative is usually unnecessary (it rarely is explicitly evaluated), as the problem statement usually makes it obvious whether the extreme is a maximum or minimum. Henceforth we will focus on the first derivative. For simplicity, we consider here only one variable, but generalization to multivariables is straightforward.

Isolated systems are not the most convenient systems to work with. In most cases, the system of interest is in thermal contact with a heat bath. Heat bath or reservoir is an idealized conceptual construct in thermodynamics. Its size is considered sufficiently large (in fact infinite) and is always in its own internal equilibrium. These two attributes imply that any finite change in its extensive variables, such as energy, volume, particle number, are infinitesimal processes for the bath, so that it is always maintained at internal equilibrium and its intensive properties, such as temperature, pressure, or chemical potential, remain unchanged.

For an isothermal process where the system is kept in thermal equilibrium with a thermal bath (thus having the same temperature as the thermal bath), making use of the first law $dE = \delta Q + \delta W$, the Clausius inequality becomes

$$d(TS) \geq dE - \delta W \tag{2.5}$$

or

$$d(E - TS) \leq \delta W \tag{2.6}$$

(We use E rather than U as the notation for the energy of the system because E is the more common notation in statistical mechanics.) Equation 2.6 naturally leads to the definition of the Helmholtz free energy:

$$F = E - TS \tag{2.7}$$

In terms of the Helmholtz free energy, the second law now becomes

$$dF \leq \delta W \tag{2.8}$$

In the special case of no work, we thus have

$$dF \leq 0 \tag{2.9}$$

Therefore, for an isothermal system, the Helmholtz free energy will decrease for a spontaneous process and will reach a minimum at equilibrium. Following similar arguments as for entropy, the equilibrium condition in terms of the Helmholtz free energy is

$$\frac{\delta F}{\delta X} = 0 \tag{2.10}$$

We can obtain Eq. 2.10 using an alternative approach, which serves to illustrate the use of the variational condition on the entropy, Eq. 2.3. Recall that the maximum entropy condition is for an isolated system. Therefore, to make use of Eq. 2.3, we

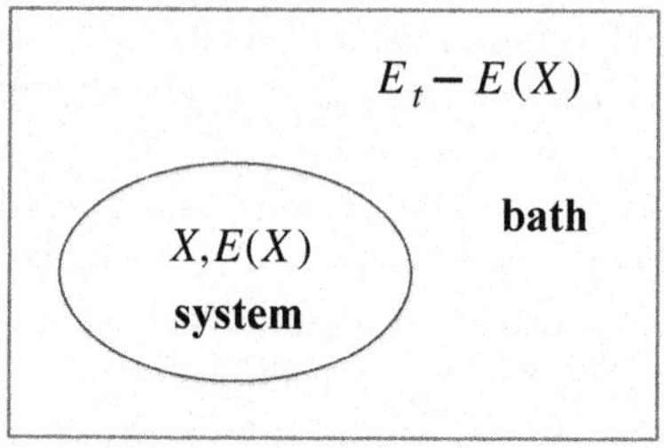

Fig. 1 Isothermal system (enclosed within the *oval*) in contact with the thermal bath. The system and the bath can be considered an isolated "super" system whose boundary is indicated by the *rectangle*. The bath is assumed to be much larger in extent and so $E_t \gg E(X)$

consider a "super system" consisting of the system of interest and the bath; see Fig. 1 for illustration. Let the total energy of the super system be E_t, the energy of the system of interest be E. We use X to denote the unconstrained internal macrostate variable for the system. In general, the system energy depends on X, and we write $E(X)$ to account for this dependence. The total entropy of the super system is then

$$S_t\left[E(X), X; E_t\right] = S\left[E(X), X\right] + S_b\left[E_t - E(X)\right] \tag{2.11}$$

Because the bath is much larger than the system, we may Taylor expand the last term around E_t. Doing so yields,

$$S_t\left[E(X), X; E_t\right] = S\left[E(X), X\right] + S_b\left(E_t\right) - \left(\frac{\partial S_b}{\partial E_b}\right)_{E_b=E_t} E(X) \tag{2.12}$$

We recognize that the derivative $(\partial S_b/\partial E_b)$ is nothing but the inverse of the temperature of the bath, so

$$S_t\left[E(X), X; E_t\right] = S\left[E(X), X\right] - \frac{E(X)}{T_b} + S_b\left(E_t\right) = -\frac{F(X)}{T_b} + S_b\left(E_t\right) \tag{2.13}$$

where $F(X) = E(X) - T_b S\left[E(X), X\right]$ is just the Helmholtz free energy of the system at the isothermal condition when the system temperature is kept at the bath temperature. Since the last term is an immaterial constant independent of the system variables, the condition of maximum in S_t for the super system is equivalent to the condition of minimum in F for the system.

Variational conditions using other thermodynamic potentials can be derived following similar arguments. Of particular interest is the variational condition using the grand potential, which for a single-component system, is defined as

$$W\left[N(X), X\right] = F\left[N(X), X\right] - \mu_b N(X) \tag{2.14}$$

where μ_b is the chemical potential of the particle reservoir (bath). X, for example, can be the density distribution of the particles. The grand potential is convenient for treating systems with spatial inhomogeneity as well as for describing phase transitions, because the temperature and chemical potential of the system are set by the reservoir (i.e., do not need to be solved for), so equality of chemical potential is automatically satisfied.

3 The Variational Origin of Statistical Thermodynamics

The central task of statistical thermodynamics is the computation of the partition function in a given ensemble, from which we obtain the appropriate thermodynamic potential (free energy) and all other equilibrium thermodynamic properties. The most common ensemble is the canonical ensemble at fixed volume, particle number and temperature (set by the thermal bath). Symbolically, we write the canonical partition function as:

$$Q = \int d\Gamma \exp\left[-\beta H(\Gamma)\right] \tag{3.1}$$

where Γ is a collective symbol to denote the microstates of the system, H is the Hamiltonian, and $\beta = (kT)^{-1}$ with T being the temperature of the bath. $\int d\Gamma$ is a short-hand notation for summing over the microstates. For example, in the case of classical fluids, it denotes summing over all the particle momenta and positions. For the Ising model, it corresponds to summing over all spin states.

From the partition function, we obtain the free energy as

$$F = -kT\ln Q = -kT\ln\left\{\int d\Gamma \exp\left[-\beta H(\Gamma)\right]\right\} \tag{3.2}$$

To see the connection to variational principle, we perform the summation over the microstates in two steps. First, we group all the microstates that correspond to a particular value of the macrostate X, and then we sum over the value of the macrostate X. Mathematically, this is accomplished by inserting the following identity $\int dX\delta\left[X - \hat{X}(\Gamma)\right] = 1$ into the partition function

$$Q = \int d\Gamma \int dX\delta\left[X - \hat{X}(\Gamma)\right]\exp\left[-\beta H(\Gamma)\right] = \int dX \int d\Gamma\delta\left[X - \hat{X}(\Gamma)\right]\exp\left[-\beta H(\Gamma)\right] \tag{3.3}$$

Where $\hat{X}(\Gamma)$ is the microscopic denition of X in terms of the microstate Γ. The inner integral over Γ yields a constrained partition function

$$Q(X) = \int d\Gamma\delta\left[X - \hat{X}(\Gamma)\right]\exp\left[-\beta H(\Gamma)\right] \tag{3.4}$$

from which we can define a constrained free energy

$$F(X) = -kT \ln Q(X) \tag{3.5}$$

Thus, the partition function now becomes

$$Q = \int dX \exp\left[-\beta F(X)\right] \tag{3.6}$$

$F(X)$ defines a free energy surface in the macrostate variable X; it is a constrained free energy for a fixed value of the internal variable X. In general $F(X)$ is some highly nonlinear function of X and as such the integral cannot be evaluated exactly. Here we use Laplace's method [3], by noting that the integral will be dominated by values of X around the maximum of the integrand, or the minimum of the free energy $F(X)$. Denoting by X^* the value of X at the minimum, it is obtained from

$$\frac{\partial F(X)}{\partial X} = 0 \tag{3.7}$$

Expanding $F(X)$ around $F(X^*)$ to quadratic order, we get

$$F(X) = F(X^*) + \frac{1}{2}F^{(2)}(X - X^*)^2 \tag{3.8}$$

where $F^{(2)}$ is the second derivative evaluated at $X = X^*$. Performing the straightforward Gaussian integral, we obtain

$$Q = \sqrt{\frac{2\pi}{F^{(2)}}} \exp\left[-\beta F(X^*)\right] \tag{3.9}$$

The equilibrium free energy is then

$$F = -kT \ln Q = F(X^*) + \frac{1}{2}kT \ln \frac{F^{(2)}}{2\pi} \tag{3.10}$$

Because the free energy is extensive, i.e., $O(N)$, but the second term is at most $O(\ln N)$, for large N, the second term can be safely ignored and we have simply

$$F = -kT \ln Q = F(X^*) \tag{3.11}$$

This is just the maximum-term method for evaluating the partition function [4] and is practically exact for thermodynamically large systems. We thus see that the maximum-term method, or equivalently, the minimization of the constrained free energy $F(X)$ with respect to the variable X, is the origin of variational principle in statistical thermodynamics [2]. We will thus use the term variational free energy synonymously with the constrained free energy.

As an illustration of the use of the maximum-term method, we take the energy E as the macrovariable. Thus we have,

$$Q(X) = \int d\Gamma \delta [E - H(\Gamma)] \exp [-\beta H(\Gamma)] = \Omega (E) e^{-\beta E} \tag{3.12}$$

where $\Omega (E)$ is the degeneracy, i.e., the microcanonical partition function. The variational free energy is then

$$F(E) = -kT \ln Q(E) = E - kT \ln \Omega (E) \tag{3.13}$$

The variational condition Eq. 3.7 becomes

$$1 - kT \frac{\partial \ln \Omega (E)}{\partial E} = 0 \tag{3.14}$$

i.e.,

$$\frac{\partial \ln \Omega (E)}{\partial E} = \beta \tag{3.15}$$

Note that the left hand side of this equation is the microcanonical definition of β for the system. Thus this variational condition has the simple interpretation that the value of the energy that minimizes the free energy of an isothermal system is such that it results in a temperature of the system that equals the temperature of the thermal bath. This is just the condition for thermal equilibrium!

4 The Method of Steepest Descent

While the free energy minimization principle and the maximum-term method are exact, analytical expressions for the exact variational free energy are seldom available. Therefore, in practice variational methods are most often used to construct approximate theories for interacting many-body systems. Two variational methods are widely used in the literature: the method of steepest descent (also called the saddle-point, or stationary-phase method) and the Gibbs-Bogoliubov-Feynman variational bound. We start with the method of steepest descent in this section.

The method of steepest descent is a generalization of Laplace's method [3] introduced in the last section in our discussion of the maximum-term method. We consider the following integral

$$I = \int dx q(x) \exp [-\alpha h(x)] \tag{4.1}$$

where x is a real variable and $h(x)$ and $q(x)$ are analytic functions of x which may in general be complex, and α is a large positive parameter. We put a negative sign in the exponential to make apparent the analogy with the Boltzmann weight. Since α

is large, the dominant contribution to the integral comes from the neighborhood of the stationary point of $h(x)$. The integration is often extended to the complex plane, in which case the stationary point becomes a saddle point [3]; hence the method is also called the saddle-point method. We will use these two terms interchangeably. For many examples in statistical mechanics, the saddle point occurs for imaginary values of x. Extending the function $h(x)$ to $h(z)$, and expanding $h(z)$ about $z = z^*$ to quadratic order, and performing the resulting Gaussian integral along the steepest-descent direction, we obtain

$$I = \sqrt{\frac{2\pi}{\alpha h^{(2)}(z^*)}} \exp\left[-\alpha h(z^*)\right]\left[q(z^*) + O\left(\alpha^{-1}\right)\right] \tag{4.2}$$

where $h^{(2)}(z^*)$ is the second derivative in the steepest-descent direction [3].

Since we usually work more with the free energy than with the partition function, we define $f = -\ln I$. f is then given by

$$f = \alpha h(z^*) + \frac{1}{2}\ln\frac{\alpha h^{(2)}(z^*)}{2\pi} - \ln q(z^*) + O\left(\alpha^{-1}\right) \tag{4.3}$$

If we keep only the leading $O(\alpha)$ term, f can be further approximated as

$$f \approx \alpha h(z^*) \tag{4.4}$$

To explain the saddle-point method and illustrate its use, we consider the following integral:

$$I = \int_{-\infty}^{\infty} dx \exp\left[-\alpha\left(\frac{1}{2}x^2 - ikx\right)\right] \tag{4.5}$$

This integral can, of course, be evaluated exactly by simply completing the square in the exponential, but we use it here to show the key ideas in the saddle-point method. Although the variable x is real, the integrand is complex, and it is convenient to extend the integration on the real axis to the complex domain z by using the Cauchy theorem. We note that there is a stationary point at $z^* = ik$, which is purely imaginary. This suggests that we make a closed contour as shown in Fig. 2, with the direction of the segments indicated by the red dash arrow. Since there are no residues in the region enclosed by the contour, by the Cauchy theorem we have

$$I_{C1} + I_{C2} + I_{C3} + I_{C4} = 0 \tag{4.6}$$

where I_{C1} is just the original integral I. It can be easily shown that the stationary point $z^* = ik$ is a saddle point: it is a maximum (for the integrand) with respect to variations in x but a minimum with respect to variations in y. Thus I_{C3} runs through the saddle point in the direction of the steepest descent. By taking the limit $b \to \infty$, the integration along $C2$ and $C4$ vanish (because of the vanishing of the integrand).

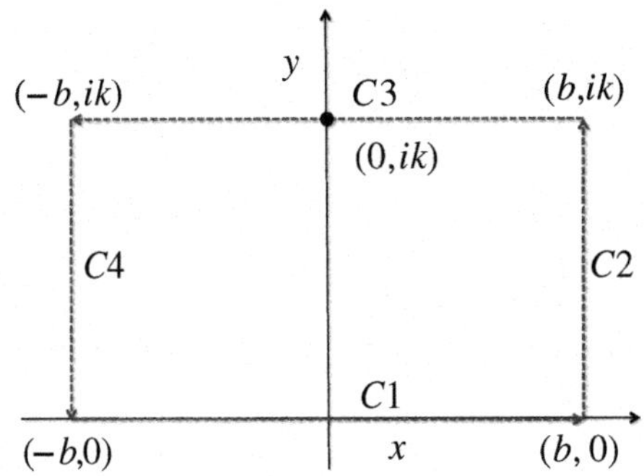

Fig. 2 Integration contour in the complex plane used to evaluate integral Eq. 4.5. The saddle point is indicated by the *black dot* located at (0, ik), and $b \to \infty$

We thus have

$$I = -I_{C3} = \int_{-\infty+ik}^{\infty+ik} dz \exp\left[-\alpha\left(\frac{1}{2}z^2 - ikz\right)\right] \tag{4.7}$$

Completing the square in the exponential, we get

$$I = \mathrm{e}^{-\frac{1}{2}\alpha k^2} \int_{-\infty+ik}^{\infty+ik} dz \exp\left[-\frac{1}{2}\alpha\,(z-ik)^2\right] \tag{4.8}$$

Along the $C3$ contour, $z = x + ik$. Thus integrating along x, i.e., the steepest-descent direction, we obtain

$$I = \mathrm{e}^{-\frac{1}{2}\alpha k^2} \int_{-\infty}^{\infty} dx \exp\left[-\frac{1}{2}\alpha x^2\right] = \sqrt{\frac{2\pi}{\alpha}}\mathrm{e}^{-\frac{1}{2}\alpha k^2} \tag{4.9}$$

As expected, this result is just Eq. 4.2 without the higher-order corrections.

The real power of the saddle-point method, of course, is to evaluate integrals that cannot be performed exactly. As an example, consider the following integral,

$$I = \int_{-\infty}^{\infty} dx \exp\left[-h(x)\right] \tag{4.10}$$

with $h(x) = \frac{1}{2}x^2 - ikx - \lambda(\mathrm{e}^{-ix} + \mathrm{e}^{ix})$ where $\lambda > 0$. We choose this example, because it shares similar mathematical form as the functional integral we introduce in our derivation of the Poisson-Boltzmann equation for the electrolyte solution. Clearly, for nonvanishing values of λ, the integral cannot be evaluated in closed form. We thus obtain an approximate solution using the saddle-point method. Extending the variable to the complex domain, we find the saddle-point condition to be given by

$$z^* = ik + i\lambda(\mathrm{e}^{iz^*} - \mathrm{e}^{-iz^*}) \tag{4.11}$$

Inspection of this equation suggests that the saddle point is located on the imaginary axis. We thus denote the saddle-point value as $z^* = iy^*$, so that the above equation becomes

$$y^* = k - \lambda(e^{y^*} - e^{-y^*}) = k - 2\lambda \sinh y^* \tag{4.12}$$

which yields a real solution for y^*. (It can be checked that the real solution is unique; this can be done, for example, by graphing the functions on the two sides of the equation.) The value of the function at the saddle point, upon using Eq. 4.12, is

$$h(y^*) = \frac{1}{2}y^{*2} + 2\lambda\left(y^* \sinh y^* - \cosh y^*\right) \tag{4.13}$$

It can be easily seen that the steepest-descent direction around the saddle point is in the direction of x, with a second derivative given by

$$h^{(2)}(y^*) = 1 + 2\lambda \cosh y^* \tag{4.14}$$

The approximate value of the integral is then

$$I \approx \left(\frac{2\pi}{1 + 2\lambda \cosh y^*}\right)^{1/2} \exp\left[2\lambda\left(\cosh y^* - y^* \sinh y^*\right) - \frac{1}{2}y^{*2}\right] \tag{4.15}$$

5 The Gibbs-Bogoliubov-Feynman Variational Principle

We start with the mathematical inequality

$$e^x \geq 1 + x \tag{5.1}$$

for any real number x. If x is a stochastic variable, we may write

$$e^x = e^{\langle x\rangle + x - \langle x\rangle} \geq e^{\langle x\rangle}\left(1 + x - \langle x\rangle\right) \tag{5.2}$$

where the angular brackets denotes the average over the distribution of x. Taking the average of both sides in the above expression, we get

$$\langle e^x\rangle \geq e^{\langle x\rangle} \tag{5.3}$$

Now consider the partition function Eq. 3.1. In general, the partition function cannot be evaluated exactly. The idea of the Gibbs-Bogoliubov-Feynman variational method is to evaluate it approximately using a reference Hamiltonian $H_R(\Gamma)$ for which the partition function can be obtained exactly. Let the partition function for this reference Hamiltonian be Q_R, so

$$Q_R = \int d\Gamma \exp[-\beta H_R(\Gamma)] \tag{5.4}$$

Next, we write

$$Q = \int d\Gamma \exp[-\beta H_R(\Gamma) - \beta H(\Gamma) + \beta H_R(\Gamma)] \tag{5.5}$$

Multiplying and dividing by Q_R, and noting

$$\langle A \rangle_R = Q_R^{-1} \int d\Gamma A(\Gamma) \exp[-\beta H_R(\Gamma)] \tag{5.6}$$

where A is any variable that depends on the microstate Γ, we obtain

$$Q = Q_R \langle \exp[-\beta H + \beta H_R] \rangle_R \geq Q_R \exp[-\beta \langle H \rangle_R + \beta \langle H_R \rangle_R] \tag{5.7}$$

Taking the logarithm and multiplying by $-kT$, we obtain

$$F \leq F_R + \langle H \rangle_R - \langle H_R \rangle_R \tag{5.8}$$

where F_R is the free energy for the reference system. This is the Gibbs-Bogoliubov-Feynman inequality [5–7], which allows to estimate the closest upper bound that can be achieved *for a given choice of the reference system*. Noting further that,

$$F_R = \langle H_R \rangle_R - TS_R \tag{5.9}$$

Equation 5.8 can alternatively be written as

$$F \leq \langle H \rangle_R - TS_R \equiv F_{var} \tag{5.10}$$

Therefore, to best approximate the true free energy F, we choose a reference Hamiltonian that makes F_{var} a minimum.

6 A Toy Example

In order to illustrate the use of the steepest-descent and GBF methods and compare between the two, we consider the following integral,

$$I = \int_{-\infty}^{\infty} dx \exp[-h(x)] \tag{6.1}$$

in which $h(x) = \frac{1}{2}\epsilon x^2 + \frac{1}{4!}ux^4 - kx$ is a purely real function with $u > 0$. (Note the parameter α is unnecessary as it can be absorbed by a rescaling of the variable x.) This function has a similar mathematical form to the ϕ^4 field theory widely used in the study of critical phenomena [8], with k playing the role of the external field. We will be interested in the value of the integral as a function of k as well as the mean and variance for the variable x. Without loss of generality we can set $u = 1$ (this can always be achieved by a redefinition of ϵ and k.) For large and positive ϵ, the quartic term is unimportant. As ϵ decreases towards zero; the effect of the quartic term becomes more pronounced. Since the integral has the mathematical structure of a partition function, we refer to h as the "Hamiltonian".

As the function is real with a real minimum, the steepest-descent approximation reduces to the Laplace approximation. The minimum condition is given by

$$\epsilon x^* + \frac{1}{6}x^{*3} = k \tag{6.2}$$

The value of the "Hamiltonian" at the minimum is

$$h(x^*) = -\frac{1}{2}\epsilon x^{*2} - \frac{1}{8}x^{*4} \tag{6.3}$$

The second derivative is given by

$$h^{(2)}(x^*) = \epsilon + \frac{1}{2}x^{*2} \tag{6.4}$$

The approximate value of the "free energy" corresponding to the integral is

$$f = -\ln I \approx -\frac{1}{2}\epsilon x^{*2} - \frac{1}{8}x^{*4} + \frac{1}{2}\ln\frac{\epsilon + \frac{1}{2}x^{*2}}{2\pi} \tag{6.5}$$

As the nonlinear effect is strongest for $\epsilon = 0$, we investigate this special case. One easily gets in this case $x^* = 6^{1/3}k^{1/3}$, and the "free energy" is then

$$f = -\frac{3}{4}6^{1/3}k^{4/3} + \frac{1}{2}\ln\frac{6^{2/3}k^{2/3}}{4\pi} \tag{6.6}$$

Note that the free energy becomes logarithmically divergent as $k \to 0$, while the original integral is clearly convergent. This is indication of the breaking down of the approximation.

Within the steepest-descent approximation, the average is taken to be the saddle-point value,

$$\langle x \rangle = x^* \tag{6.7}$$

(upon ignoring higher-order corrections in an expansion in α^{-1}; see Eqs. 4.1–4.4), and the variance of x is given by

$$\langle(\Delta x)^2\rangle = \frac{1}{h^{(2)}(x^*)} \tag{6.8}$$

which follows from the use of linear response theory

$$\langle(\Delta x)^2\rangle = \frac{\partial\langle x\rangle}{\partial k} \tag{6.9}$$

We now evaluate the integral Eq. 6.1 using the GBF bound. To this end, we introduce a two-parameter reference "Hamiltonian",

$$h_R = \frac{1}{2}a(x - x^*)^2 \tag{6.10}$$

where x^* and a are the variational parameters. The GBF bound now reads:

$$f = -\ln I \leq f_R + \langle h\rangle_R - \langle h_R\rangle_R \tag{6.11}$$

where the average $\langle\cdots\rangle_R$ here means

$$\langle\cdots\rangle_R = \left(\frac{a}{2\pi}\right)^{1/2} \int_{-\infty}^{\infty} dx\,(\cdots)\exp\left[-\frac{1}{2}a(x - x^*)^2\right] \tag{6.12}$$

For the reference "Hamiltonian" Eq. 6.10, one easily finds

$$\langle h_R\rangle_R = \frac{1}{2} \tag{6.13}$$

$$f_R = \frac{1}{2}\ln\left(\frac{a}{2\pi}\right) \tag{6.14}$$

$$\langle h\rangle_R = \frac{1}{2}\epsilon x^{*2} + \frac{1}{2}\epsilon a^{-1} + \frac{1}{4!}\left(x^{*4} + 6a^{-1}x^{*2} + 3a^{-2}\right) - kx^* \tag{6.15}$$

The parameters x^* and a are obtained by minimization of the right hand side of Eq. 6.11 using Eqs. 6.13–6.15, yielding, respectively

$$\epsilon x^* + \frac{1}{3!}x^{*3} + \frac{1}{2}a^{-1}x^* - k = 0 \tag{6.16}$$

and

$$2a^2 - 2\epsilon a - ax^{*2} - 1 = 0 \tag{6.17}$$

The positive root of Eq. 6.17 is given by

$$a = \frac{1}{4}\left[2\epsilon + x^{*2} + \sqrt{8 + (2\epsilon + x^{*2})^2}\right] \tag{6.18}$$

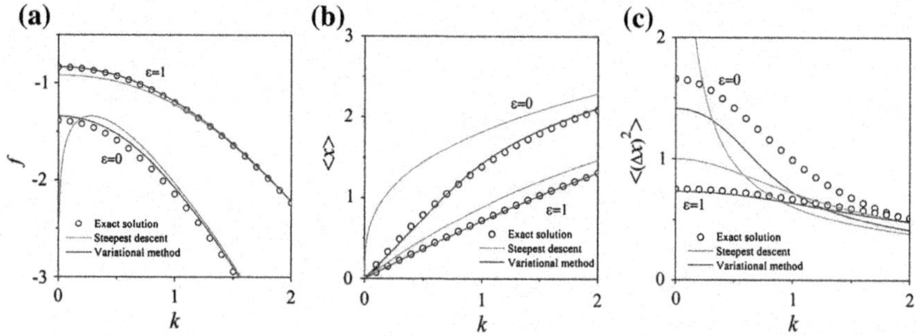

Fig. 3 Results obtained from the steepest-descent method (*blue curve*) and the GBF variational method (*red curve*), compared to the exact numerical results (*black circles*), for the example given by Eq. 6.1. From *left* to *right* the "free energy", the mean, and the variance

The minimized "free energy" after simplification using Eqs. 6.16 and 6.17 is:

$$f = \frac{1}{2}\ln\left(\frac{a}{2\pi}\right) - \frac{1}{4} + \frac{1}{4}\epsilon a^{-1} - \frac{1}{2}\epsilon x^{*2} - \frac{1}{8}x^{*4} - \frac{3}{8}a^{-1}x^{*2} \tag{6.19}$$

By the Gaussian ansatz, the mean and variance are given respectively by

$$\langle x \rangle = x^* \tag{6.20}$$

$$\langle (\Delta x)^2 \rangle = a^{-1} \tag{6.21}$$

In Fig. 3, we plot the "free energy", the mean and the variance as a function of k for two values of the parameter ϵ. For comparison, we include the results from the steepest-descent method (blue curve), from the GBF method (red curve), and from exact numerical solution (black circles). In all cases, the variational method provides a far closer agreement with the numerical results than the simple steepest descent. For the mean, which in essence is an equation of state, the agreement is excellent even for $\epsilon = 0$, and very good agreement is also obtained for the "free energy" under the "worst" condition $\epsilon = 0$ and $k = 0$. Note also that the GBF variational free energy is always higher than the true free energy, though no such restrictions apply for the steepest-descent method. The result for the fluctuation from the GBF method is quite good for $\epsilon = 1$; it is less accurate quantitatively for $\epsilon = 0$, but still captures the qualitative behavior. In contrast, the steepest-descent method yields qualitatively incorrect behavior for $\epsilon = 0$. Finally we note that both approximations become increasingly more accurate at large k and/or ϵ.

Those who are familiar with the statistical field theory will recognize the steepest-descent and the GBF treatments as the respective analogue of the random-phase approximation (RPA) [9, 10] and the self-consistent Hartree approximation [11]. (We note, however, that the term RPA has different meanings in the different communities. For example, in liquid-state theory, it usually refers to approximating the direct correlation function by the pair interaction potential in the closure for the Ornstein-Zernike equation [12]. In many-body and condensed matter physics, RPA is considered synonymous with the self-consistent Hartree approximation [13]. In this chapter, we will use the term RPA to refer to a simple Gaussian approximation around the saddle-point, or equivalently, linear response by perturbation around the saddle-point [9, 10].)

7 Mean-Field Solution for the Interacting Ising Model

The Ising model is the best known model for a wide range of phase transitions, including ferromagnetism, liquid-vapor transition, and phase separation in binary mixtures. In its most common form, the model consists of N "spins" on a lattice with coordination number z in an external field h. Each spin can be either in an up or down state, which is denoted by the spin variable $s_i = \pm 1$ with $+$ for the up state and $-$ for the down state. The spins interact with each other through nearest neighbor pairs in such a way as to favor the same orientation. The Hamiltonian for the system is

$$H = -\frac{1}{2}J{\sum_{i,j}}' s_i s_j - \sum_i h_i s_i \tag{7.1}$$

where $J > 0$ is the coupling constant, h_i is the external field, here allowed to be spatially dependent for generality, and the prime on the first sum restricts i and j to be nearest neighbors of each other. Henceforth, to economize notation, we set the energy unit to be kT and the entropy unit to be k.

The partition function of the system is then

$$Z = \sum_{\{s\}} \exp(-H) \tag{7.2}$$

where $\{s\}$ is a collective notation for all the spin states. It can be easily shown [4] that the model can be applied to describe a lattice-gas fluid in a grand canonical ensemble by introducing the occupation variable $\sigma = (1+s)/2$ and identifying the nearest-neighbor attraction energy $\epsilon = -4J$ and chemical potential $\mu = -4Jz + 2h$.

The mean-field solution of the Ising model in typical textbooks is usually derived using a Braggs-William random mixing approximation [4], which amounts to ignoring the correlations due to interaction and treating the distribution of the spins as completely random. Alternatively, one may invoke a local self-consistent field argument by replacing the local instantaneous field at location i, $h_i + J{\sum_j}' s_j$ with the average $h_i + J{\sum_j}' m_j$ where m_i is the average of s_i. The local average of m_i is in turn obtained through the self-consistency condition

$$m_i = \frac{\sum_{s_i} s_i \exp\left[(h_i + J\sum_j{}' m_j)s_i\right]}{\sum_{s_i} \exp\left[(h_i + J\sum_j{}' m_j)s_i\right]} = \tanh(h_i + J\sum_j{}' m_j) \tag{7.3}$$

For spatially uniform system, $m_i = m$, and the above equation simplifies to:

$$m = \tanh(h + Jzm) \tag{7.4}$$

where z is the lattice coordination number. This is the well-known mean-field equation of state relating the magnetization per spin to the external field. From Eq. 7.4, we find the familiar mean-field critical point at $h_c = 0$ and $J_c z = 1$

We now derive the mean-field solution using a variational approach. We make the reference Hamiltonian an effective non-interacting one, with spins in an effective one-body field $h_{i,R}$

$$H_R = -\sum_i h_{i,R} s_i \tag{7.5}$$

The partition function for H_R can be trivially worked out to be

$$Z_R = 2^N \prod_i \cosh\left(h_{i,R}\right) \tag{7.6}$$

from which we obtain the free energy of the reference system

$$F_R = -\sum_i \ln\cosh\left(h_{i,R}\right) - N\ln 2 \tag{7.7}$$

$\langle H\rangle_R$ and $\langle H_R\rangle_R$ can be straightforwardly evaluated to be

$$\langle H\rangle_R = -\frac{1}{2}J\sum_{i,j}{}' m_i m_j - \sum_i h_i m_i \tag{7.8}$$

and

$$\langle H_R\rangle_R = -\sum_i h_{i,R} m_i \tag{7.9}$$

where m_i is given by

$$m_i = \langle s_i\rangle_R = -\frac{\partial F_R}{\partial h_{i,R}} = \tanh\left(h_{i,R}\right) \tag{7.10}$$

The variational free energy $F[h_{i,R}]$ is obtained by combining Eqs. 7.7–7.9, and the effective field $h_{i,R}$ is obtained from

$$\frac{\partial F}{\partial h_{i,R}} = 0 \tag{7.11}$$

which, along with Eq. 7.10, gives

$$h_{i,R} = h_i + J\sum_j{}' m_j \tag{7.12}$$

Substituting this back in Eq. 7.10, we obtain the self-consistent Eq. 7.3.

Although m_i is defined through Eq. 7.10, it can be shown that within the variational free energy framework, it is just the average of the spin variable at i. To demonstrate this, we start with

$$\langle s_i \rangle = -\frac{\partial F}{\partial h_i} \tag{7.13}$$

By construction, the only place where h_i enters directly in the variational free energy is through $\langle H \rangle_R$. Furthermore, the dependence of $h_{i,R}$ on h_i does not contribute because of the variational condition Eq. 7.11. We thus have

$$\langle s_i \rangle = -\frac{\partial \langle H \rangle_R}{\partial h_i} = m_i \equiv \langle s_i \rangle_R \tag{7.14}$$

This conclusion that the ensemble average is equal to the average in the reference system is a general one within the variational approximation.

The variational condition Eq. 7.11 can alternatively be cast in a different form by directly using the local magnetization m_i. Since Eq. 7.10 establishes a one-to-one monotonic relationship between the variational parameter $h_{i,R}$ and m_i, we may use m_i itself as the variational parameter. Because the reference system consists of uncoupled spins, each having two states, its entropy is simply

$$S_R = -\sum_i \left[p(s_i = 1)\ln p(s_i = 1) + p(s_i = -1)\ln p(s_i = -1)\right] \tag{7.15}$$

Noting that $m_i = p(s_i = 1) - p(s_i = -1) = 2p(s_i = 1) - 1$, the above equation can be written as

$$\begin{aligned} S_R &= -\sum_i \left[\frac{1+m_i}{2}\ln\frac{1+m_i}{2} + \frac{1-m_i}{2}\ln\frac{1-m_i}{2}\right] \\ &= N\ln 2 - \frac{1}{2}\sum_i [(1+m_i)\ln(1+m_i) + (1-m_i)\ln(1-m_i)] \end{aligned} \tag{7.16}$$

where N is the total number of spins. $N\ln 2$ in the second line is simply the entropy of a completely unbiased system of N independent spins, whereas the remaining terms account for the decrease in entropy due to ordering.

With Eq. 7.16, the variational free energy is

$$F_{var} = -\frac{1}{2}J\sum_{i,j}{}' m_i m_j - \sum_i h_i m_i$$
$$+ \frac{1}{2}\sum_i [(1+m_i)\ln(1+m_i) + (1-m_i)\ln(1-m_i)] - N\ln 2 \quad (7.17)$$

The minimized free energy is obtained from the variational condition

$$\frac{\partial F_{var}}{\partial m_i} = 0 \quad (7.18)$$

which yields the mean-field equation of state, Eq. 7.10.

The variational free energy can be more systematically obtained using a combination of the identity transformation and the saddle-point method. To this end, we insert the identity $\int dm_i \delta(m_i - s_i)$ for each spin, to write the Boltzmann weight $\exp(-H)$ as

$$\exp[-H(\{s_i\})] = \int \prod_i dm_i \delta(m_i - s_i) \exp[-H(\{m_i\})]$$
$$= \frac{1}{(2\pi)^N} \int \prod_i dm_i d\lambda_i \exp\left[-H(\{m_i\}) + i\sum_i \lambda_i (m_i - s_i)\right] \quad (7.19)$$

where the second line follows from making use of the Fourier representation of the delta function:

$$\delta(m_i - s_i) = \frac{1}{2\pi}\int d\lambda_i \exp[i\lambda_i (m_i - s_i)] \quad (7.20)$$

The identify transformation has turned the problem of interacting spins into one of independent spins in fluctuating "external" field $i\lambda$. The summation over the spin variable s_i can now be performed trivially to yield the partition function

$$Z = \frac{1}{(2\pi)^N}\int \prod_i dm_i d\lambda_i \exp\left[-H(\{m_i\}) + i\sum_i \lambda_i m_i + \sum_i \ln\cosh(i\lambda_i)\right] \quad (7.21)$$

We now make the saddle-point approximation on this multidimensional integral, by taking the stationary point of the exponent with respect to m_i and λ_i; this yields,

$$i\lambda_i = \frac{\partial H}{\partial m_i} = -J\sum_j{}' m_j - h_i \quad (7.22)$$

and

$$im_i = -i\tanh(i\lambda_i) \quad (7.23)$$

respectively. Clearly from Eq. 7.22, the saddle point lies on the imaginary axis of λ, so we denote the saddle-point value of λ as $\lambda^* = i\eta$, and the above equations become

$$\eta_i = J \sum_j m_j + h_i \tag{7.24}$$

and

$$m_i = \tanh \eta_i \tag{7.25}$$

which are the same as Eqs. 7.10 and 7.12, with the identification $\eta_i = h_{i,R}$

A closer examination of Eq. 7.21 reveals that the three terms in the exponent correspond respectively to the $\langle H \rangle_R$, $\langle H_R \rangle_R$ and F_R terms in the GBF variational approach. This is no accident, because the identity transformation essentially turns the problem of interacting spins into one of independent spins in an effective fluctuating field, and the saddle-point condition is the condition for finding the optimal value for the effective field.

8 The Poisson-Boltzmann Equation

The Poisson-Boltzmann equation is a mean-field equation for the mean-electrostatic potential generated by some fixed external charge distribution in the presence of small mobile ions. The mobile ions are treated as point particles whose spatial distribution is given by the Boltzmann weight with the energy being the electrostatic energy of an ion in the mean electrostatic potential. To simplify notation, we scale the energy by kT, charge by the elementary charge e, and define a dimensionless permittivity $\epsilon = \varepsilon_r \varepsilon_0 kT/e^2$, where ε_0 is the vacuum permittivity and ε_r the dielectric constant (allowed to be spatially dependent). For simplicity, we consider the electrolyte to be a monovalent 1:1 salt. The mean electrostatic potential ψ in the presence of a fixed external charge distribution ρ_{ex} is then described by

$$-\nabla \cdot (\epsilon \nabla \psi) = \rho_{ex} - 2c^b \sinh \psi \tag{8.1}$$

where c^b is the bulk salt concentration and the electrostatic potential is defined such that $\psi = 0$ in the bulk far away from the fixed charge.

While the Poisson-Boltzmann equation can be constructed quite intuitively by combining the Poisson equation for electrostatics with the Boltzmann distribution for the small mobile ions, which are treated as ideal gas particles in the electrostatic potential, here we derive it using variational approaches. The systematic derivation is necessary for calculating the free energy of the system. We first provide a simple, thermodynamic derivation and then a more systematic one using field theoretical techniques.

First, we note that the electrostatic energy of a system is given by

$$E_c = \frac{1}{2}\int d\mathbf{r}\int d\mathbf{r}'\rho(\mathbf{r})C\left(\mathbf{r},\mathbf{r}'\right)\rho(\mathbf{r}') \tag{8.2}$$

where $\rho(\mathbf{r})$ is the charge density and $C\left(\mathbf{r},\mathbf{r}'\right)$ is the Coulomb operator given by the solution of the Poisson equation

$$-\nabla\cdot\epsilon\nabla C\left(\mathbf{r},\mathbf{r}'\right) = \delta\left(\mathbf{r}-\mathbf{r}'\right) \tag{8.3}$$

For spatially uniform dielectric permittivity, $C\left(\mathbf{r},\mathbf{r}'\right) = 1/\left(4\pi\epsilon|\mathbf{r}-\mathbf{r}'|\right)$.

Equation 8.2 is a quadratic form with the vector $\rho(\mathbf{r})$ and matrix operator $C\left(\mathbf{r},\mathbf{r}'\right)$. Noting the following identity for a quadratic form,

$$\frac{1}{2}\mathbf{x}\cdot\mathbf{A}\cdot\mathbf{x} = -\min_{\mathbf{y}}\left(\frac{1}{2}\mathbf{y}\cdot\mathbf{A}^{-1}\cdot\mathbf{y}-\mathbf{x}\cdot\mathbf{y}\right) = \max_{\mathbf{y}}\left(-\frac{1}{2}\mathbf{y}\cdot\mathbf{A}^{-1}\cdot\mathbf{y}+\mathbf{x}\cdot\mathbf{y}\right) \tag{8.4}$$

where $\mathbf{x}$, and $\mathbf{y}$ are vectors, $\mathbf{A}$ is a matrix operator and $\mathbf{A}^{-1}$ is its inverse, and noting that the inverse operator of $C\left(\mathbf{r},\mathbf{r}'\right)$ is $-\nabla_{\mathbf{r}}\cdot\delta\left(\mathbf{r}-\mathbf{r}'\right)\epsilon\left(\mathbf{r}'\right)\nabla_{\mathbf{r}'}$, we can introduce a variational "Hamiltonian"

$$\begin{aligned} H_{var} &= \int d\mathbf{r}\left[\rho\psi+\frac{1}{2}\psi\nabla\cdot\epsilon\nabla\psi\right] \\ &= \int d\mathbf{r}\left[\rho\psi-\frac{\epsilon}{2}\left(\nabla\psi\right)^2\right] \end{aligned} \tag{8.5}$$

The second line of the equation follows from integration by parts and applying the divergence theorem (at the system boundary, taken to be at infinity). Noting the following result for functional derivative

$$\frac{\delta}{\delta\psi}\int d\mathbf{r}f(\psi,\nabla\psi) = \frac{\partial f}{\partial\psi}-\nabla\cdot\frac{\partial f}{\partial\nabla\psi} \tag{8.6}$$

it can be easily seen that maximizing the Hamiltonian Eq. 8.5 leads to the Poisson equation.

Now the total charge density is the sum of the fixed external charge plus the charges due to the mobile ions

$$\rho\left(\mathbf{r}\right) = \rho_{ex}\left(\mathbf{r}\right)+c_+\left(\mathbf{r}\right)-c_-\left(\mathbf{r}\right) \tag{8.7}$$

where c_+ and c_- are respectively the concentration of the cations and anions. We now construct the variational free energy for an open system with chemical potential μ_+ and μ_- for the cations and anions, respectively. Adding in the translational entropy of the ions, we have

$$F_{var} = \int d\mathbf{r} \left[(\rho_{ex} + c_+ - c_-)\,\psi - \frac{\epsilon}{2}\,(\nabla\psi)^2 \right] + \int d\mathbf{r} \left[c_+ \ln(c_+ v_+) - c_+ + c_- \ln(c_- v_-) - c_- - \mu_+ c_+ - \mu_- c_- \right] \quad (8.8)$$

v_+ and v_- are volume scales which can be taken as the cube of the thermal de Broglie wavelength. However, the exact form is immaterial and will be absorbed into the definition of the chemical potentials.

Taking the functional derivatives of this variational free energy with respect to c_+, c_- and ψ and setting the derivatives to zero, we obtain

$$c_\pm = \frac{1}{v_\pm} \exp\left(\mu_\pm \mp \psi\right) \quad (8.9)$$

and

$$-\nabla \cdot (\epsilon \nabla \psi) = \rho_{ex} + c_+ - c_- \quad (8.10)$$

Setting the electrostatic potential in the uniform bulk to be zero, Eq. 8.9 can be written as

$$c_\pm = c_\pm^b e^{\mp\psi} \quad (8.11)$$

where $c_\pm^b = v_\pm^{-1} e^{\mu_\pm}$ is the bulk concentration of the cations/anions. Because of electroneutrality, $c_+^b = c_-^b \equiv c^b$. Substitution of Eq. 8.11 into Eq. 8.10 yields the PB equation, Eq. 8.1.

We now provide a statistical mechanical derivation using field theoretical techniques. To begin, we write the particle density operator for the ions

$$\hat{c}_\pm(\mathbf{r}) = \sum_{i=1}^{n_\pm} \delta(\mathbf{r} - \mathbf{r}_i) \quad (8.12)$$

the total charge density operator is then

$$\hat{\rho}(\mathbf{r}) = \rho_{ex}(\mathbf{r}) + \hat{c}_+(\mathbf{r}) - \hat{c}_-(\mathbf{r}) \quad (8.13)$$

The total Coulomb energy of the system is

$$H = \frac{1}{2} \int d\mathbf{r} d\mathbf{r}' \hat{\rho}(\mathbf{r}) C(\mathbf{r}, \mathbf{r}') \hat{\rho}(\mathbf{r}') \quad (8.14)$$

where $C(\mathbf{r}, \mathbf{r}')$ is the Coulomb operator given by Eq. 8.3. Note that the above two equations are identical in form to the corresponding Eqs. 8.2 and 8.7. However, here the concentration and charge density refer to the instantaneous particle configuration whereas in Eqs. 8.2 and 8.7, they are the thermally averaged quantities. Because of the discrete nature of the ions, the interaction energy Eq. 8.3 contains the self energy of the ions, which is infinite for the point-charge model. At the mean-field (saddle-point) level, the self-energy does not show up. It will show up when we consider fluctuations, in which case a finite charge distribution on the ion is required to render the self energy finite.

We start with the canonical partition function of the system

$$Q = \frac{1}{n_+!n_-!v_+^{n_+}v_-^{n_-}} \int \prod_{i=1}^{n_+} d\mathbf{r}_i \prod_{j=1}^{n_-} d\mathbf{r}_j \exp(-H) \tag{8.15}$$

Note that the Boltzmann weight with the Hamiltonian given by Eq. 8.14 is a Gaussian form in ρ with $\mathbf{r}$ as (the continuous) index. It is straightforward to show that a multivariable Gaussian function can be written as

$$e^{-\frac{1}{2}\sum_{i,j} x_i A_{ij} x_j} = \frac{1}{\sqrt{\det A}} \int_{-\infty}^{\infty} \cdots \int_{-\infty}^{\infty} e^{-\frac{1}{2}\sum_{i,j} y_i A_{ij}^{-1} y_j + i\sum_j x_j y_j} \frac{dy_1}{\sqrt{2\pi}} \cdots \frac{dy_N}{\sqrt{2\pi}} \tag{8.16}$$

This representation of a Gaussian function by an integral of another (shifted) Gaussian function is the the Gauss [8] or Hubbard-Stratonovich [13] transformation. It has the effect of decoupling the variables x_i and x_j at the expense of a new coupling between x_i and y_i. This transformation can be generalized to variables with continuous indices. Thus we perform the Gauss-Hubbard-Stratonovich transformation to decouple the quadratic interaction in Eq. 8.14 by introducing a field variable $\xi(\mathbf{r})$. This gives

$$Q = \frac{1}{n_+!n_-!v_+^{n_+}v_-^{n_-}} \frac{1}{Z_C} \int D\xi \int \prod_{i=1}^{n_+} d\mathbf{r}_i \prod_{j=1}^{n_-} d\mathbf{r}_j \exp\left\{-\int d\mathbf{r}\left[\frac{1}{2}\epsilon(\nabla\xi)^2 + i\hat{\rho}\xi\right]\right\} \tag{8.17}$$

where Z_C is a normalization factor given by

$$Z_C = \int D\xi \exp\left[-\frac{1}{2}\int d\mathbf{r}\epsilon(\nabla\xi)^2\right] = (\det C)^{1/2} \tag{8.18}$$

Now the partition function has the form of uncoupled n_+ cations and n_- anions in an "external" fluctuating field $i\xi$. Writing the partition function for each cation and anion respectively as q_+ and q_-, where

$$q_\pm = \int d\mathbf{r} \exp[\mp i\xi(\mathbf{r})] \tag{8.19}$$

The partition function is now transformed to

$$Q = \frac{1}{Z_C} \int D\xi \frac{q_+^{n_+} q_-^{n_-}}{n_+!n_-!v_+^{n_+}v_-^{n_-}} \exp\left\{-\int d\mathbf{r}\left[\frac{1}{2}\epsilon(\nabla\xi)^2 + i\rho_{ex}\xi\right]\right\} \tag{8.20}$$

It is convenient to work in the grand canonical ensemble with chemical potentials μ_+ and μ_- for the cations and anions, respectively. The grand partition function is

$$\Xi = \sum_{n_+=0}^{\infty} \sum_{n_-=0}^{\infty} Q(n_+, n_-) e^{n_+\mu_+} e^{n_-\mu_-} \tag{8.21}$$

The summation over the number of the mobile ions can be easily performed and we obtain

$$\Xi = \frac{1}{Z_C} \int D\xi \exp\{-L[\xi]\} \tag{8.22}$$

where the "action" L is

$$L = \int d\mathbf{r} \left[\frac{1}{2} \epsilon (\nabla \xi)^2 + i\rho_{ex}\xi - \lambda_+ e^{-i\xi} - \lambda_- e^{i\xi} \right] \tag{8.23}$$

In Eq. 8.23, $\lambda_\pm$ is the fugacity of the ions defined as

$$\lambda_\pm = \frac{e^{\mu_\pm}}{v_\pm} \tag{8.24}$$

We now evaluate the partition function using the saddle-point method. We first locate the saddle point by setting

$$\frac{\delta L}{\delta \xi} = 0 \tag{8.25}$$

which results in

$$\nabla \cdot (\epsilon \nabla \xi) = i\rho_{ex} + i\lambda_+ e^{-i\xi} - i\lambda_- e^{i\xi} \tag{8.26}$$

It should be clear that the saddle-point value for ξ is purely imaginary. We denote the value at the saddle point as ξ^* and define $\psi = i\xi^*$, so the above equation becomes

$$-\nabla \cdot (\epsilon \nabla \psi) = \rho_{ex} + \lambda_+ e^{-\psi} - \lambda_- e^{\psi} \tag{8.27}$$

This is the Poisson-Boltzmann equation upon identifying the fugacity $\lambda_\pm$ with the bulk concentration $c_\pm^b$.

Using Eq. 8.27 in Eq. 8.23, we obtain the "action" at the saddle point as

$$L^* = -\int d\mathbf{r} \left[c_+(\mathbf{r}) + c_-(\mathbf{r}) \right] + \frac{1}{2} \int d\mathbf{r} \psi(\mathbf{r}) \left[\rho_{ex}(\mathbf{r}) - c_+(\mathbf{r}) + c_-(\mathbf{r}) \right] \tag{8.28}$$

To obtain the full expression for the partition function at the saddle-point level, we expand L to second order in the steepest-descent direction $\chi = \xi - \xi^* = \xi + i\psi$, to get

$$L = L^*[\{\psi(\mathbf{r})\}] + \frac{1}{2} \int d\mathbf{r} \left[\epsilon (\nabla \chi)^2 + \left(\lambda_+ e^{-\psi} + \lambda_- e^{\psi} \right) \chi^2 \right] \tag{8.29}$$

Performing the Gaussian integral over χ, we obtain the following result for the partition function

$$\Xi = \left(\frac{\det G}{\det C}\right)^{1/2} \exp\left\{-L^*[\{\psi(\mathbf{r})\}]\right\} \tag{8.30}$$

where G is the Green's function satisfying

$$-\nabla \cdot \left[\epsilon \nabla G(\mathbf{r}, \mathbf{r}')\right] + \left(\lambda_+ e^{-\psi} + \lambda_- e^{\psi}\right) G(\mathbf{r}, \mathbf{r}') = \delta(\mathbf{r} - \mathbf{r}') \tag{8.31}$$

In most field theoretical formulations, the saddle-point approximation does not include factors involving the determinant of C and G; these terms can be shown in a systematic loop expansion to correspond to the one-loop corrections [8, 14, 15]. Thus, ignoring these one-loop corrections, the saddle-point approximation gives simply

$$\Xi \approx \exp\left\{-L^*[\{\psi(\mathbf{r})\}]\right\} \tag{8.32}$$

At this level of the saddle-point approximation, the concentration of the cation/anion is

$$c_\pm(\mathbf{r}) = \frac{\delta \ln \Xi}{\delta \mu_\pm} = \lambda_\pm e^{\mp\psi} \tag{8.33}$$

and upon noting the charge neutrality in the bulk, the concentration is

$$c_\pm(\mathbf{r}) = c^b e^{\mp\psi} \tag{8.34}$$

9 Fluctuations

In the toy example discussed in Sect. 6, we see how fluctuations are included in the variational treatment. In the steepest-descent approximation, fluctuations are determined by the local curvature around the saddle-point, whereas in the GBF method, they are reflected through the width of the variational Gaussian Boltzmann weight. Once a physical model for a many-body interacting system is transformed into a field-theoretical representation, the mathematical structure is similar to the toy-model integral, and fluctuations can be studied in a similar manner. The numerical examples shown in Fig. 3 clearly demonstrate the superiority of the GBF method in capturing the fluctuation effects. Therefore, in this section we illustrate the use of the GBF method in studying fluctuation effects for an electrolyte solution in the presence of a fixed charge distribution ρ_{ex} discussed in the previous section.

From Eq. 8.30, we see that the partition function contains contributions from local fluctuations around the saddle-point (approximated at the Gaussian level). The Green's function $G(\mathbf{r}, \mathbf{r}')$ is the correlation function for the fluctuating part of the field χ (i.e., deviation from the saddle-point value),

$$G(\mathbf{r}, \mathbf{r}') = \langle \chi(\mathbf{r})\chi(\mathbf{r}')\rangle \tag{9.1}$$

However, in the saddle-point approximation, the correlation is unidirectionally determined by the saddle-point value ψ from Eq. 8.31, i.e., the correlation does not affect the solution for ψ, which is obtained by ignoring the fluctuations. Alternatively, the same correlation function can be obtained from linear response of the saddle-point free energy to a small, spatially varying perturbation—this is just the random-phase approximation [9, 10].

To include fluctuations in a self-consistent manner, we employ the GBF variational method in approximating the partition function Eq. 8.22 [16]. We use a Gaussian reference action of the form [17, 18]

$$L_R[\xi] = \frac{1}{2}\int d\mathbf{r}d\mathbf{r}'[\xi(\mathbf{r}) + i\psi(\mathbf{r})]G^{-1}(\mathbf{r}, \mathbf{r}')[\xi(\mathbf{r}') + i\psi(\mathbf{r}')] = \frac{1}{2}\int d\mathbf{r}d\mathbf{r}'\chi(\mathbf{r})G^{-1}(\mathbf{r}, \mathbf{r}')\chi(\mathbf{r}') \tag{9.2}$$

where the average electrostatic field ψ and the Green's function G are taken to be variational parameters. G^{-1} is the functional inverse of the Green's function defined through

$$\int d\mathbf{r}''G^{-1}(\mathbf{r}, \mathbf{r}'')G(\mathbf{r}'', \mathbf{r}') = \delta(\mathbf{r} - \mathbf{r}') \tag{9.3}$$

Following the GBF principle, we construct the following variational grand free energy

$$\begin{aligned} W &= W_R + \langle L[\xi] - L_R[\xi]\rangle \\ &= -\frac{1}{2}\ln\left(\frac{\det G}{\det C}\right) - \frac{1}{2}\int d\mathbf{r}d\mathbf{r}'G^{-1}(\mathbf{r}, \mathbf{r}')G(\mathbf{r}, \mathbf{r}') \\ &- \frac{1}{2}\int d\mathbf{r}d\mathbf{r}'\left[\delta(\mathbf{r}' - \mathbf{r})\epsilon(\nabla\psi)^2 - \nabla_{\mathbf{r}}\cdot\left[\epsilon(\mathbf{r})\nabla_{\mathbf{r}'}\delta(\mathbf{r} - \mathbf{r}')\right]G(\mathbf{r}, \mathbf{r}')\right] \\ &+ \int d\mathbf{r}\left[\rho_{ex}\psi - \lambda_+ e^{-\psi}\langle e^{-i\chi}\rangle - \lambda_- e^{\psi}\langle e^{i\chi}\rangle\right] \end{aligned} \tag{9.4}$$

Here the reference free energy W_R is $W_R = -\ln \Xi_R$ and the reference partition function is in turn given by

$$\Xi_R = \frac{1}{Z_C}\int D\xi \exp\{-L_R[\xi]\} = \frac{(\det G)^{1/2}}{(\det C)^{1/2}} \tag{9.5}$$

The average $\langle \cdots \rangle$ is to be understood as $\langle \cdots \rangle_R$, i.e., average taken in the reference ensemble with action L_R (cf. discussions surrounding Eq. 7.14 in Sect. 7.) The second term in the first line of the above expression is from $-\langle L_R[\xi]\rangle$ upon using Eq. 9.1. The terms in the second and third lines of the expression arise from $\langle L[\xi]\rangle$ upon writing the field $\xi = i\psi + \chi$. We have made use of $\langle\xi\rangle = \xi^* = i\psi$ and the following relation:

$$\int d\mathbf{r}d\mathbf{r}'\delta(\mathbf{r}' - \mathbf{r})\epsilon\langle(\nabla\chi)^2\rangle = \int d\mathbf{r}d\mathbf{r}'\nabla_{\mathbf{r}} \cdot \left[\epsilon(\mathbf{r})\nabla_{\mathbf{r}'}\delta(\mathbf{r} - \mathbf{r}')\right] G(\mathbf{r}, \mathbf{r}') \qquad (9.6)$$

For more details of the derivation leading to Eq. 9.4, readers are kindly asked to consult Ref. [19]. Because the distribution of χ is Gaussian by ansatz, the average in Eq. 9.4 is

$$\langle e^{\mp i\chi}\rangle = \exp\left[-\frac{1}{2}\, G(\mathbf{r}, \mathbf{r})\right] \qquad (9.7)$$

We recognize $(1/2)G(\mathbf{r}, \mathbf{r})$ as the self energy of a point-charge [20], which is divergent. To eliminate the divergence, we introduce a finite charge distribution $h_\pm(\mathbf{r} - \mathbf{r}')$ [19] for an ion located at $\mathbf{r}'$, to replace the localized point-charge distribution $\delta(\mathbf{r} - \mathbf{r}')$. (For spherical ions, we have $h_\pm(\mathbf{r} - \mathbf{r}') = h_\pm(|\mathbf{r} - \mathbf{r}'|)$.) Then instead of $\langle e^{\mp i\chi}\rangle$, we now have

$$\langle e^{\mp i\hat{h}_\pm\chi}\rangle \qquad (9.8)$$

where we have used the short-hand notation $\hat{h}_\pm\chi$ to represent the local spatial averaging of χ by the charge distribution function on the ion:

$$\hat{h}_\pm\chi = \int d\mathbf{r}' h_\pm(\mathbf{r}' - \mathbf{r})\chi(\mathbf{r}') \qquad (9.9)$$

The Gaussian averaging then yields,

$$\langle e^{\mp i\hat{h}_\pm\chi}\rangle = e^{-u_\pm(\mathbf{r})} \qquad (9.10)$$

where the self energy of an ion at location $\mathbf{r}$ is now given by

$$u_\pm(\mathbf{r}) = \frac{1}{2}\int d\mathbf{r}'d\mathbf{r}'' h_\pm(\mathbf{r}' - \mathbf{r})G(\mathbf{r}', \mathbf{r}'')h_\pm(\mathbf{r}'' - \mathbf{r}) \qquad (9.11)$$

and is finite as long as the distribution $h_\pm(\mathbf{r}' - \mathbf{r})$ has a finite width. The same spatial averaging of the mean electrostatic potential ψ by the charge distribution is not necessary, since the width of the distribution is typically much smaller than the characteristic length scale for the variation of ψ, thus

$$\hat{h}_\pm\psi = \int d\mathbf{r}' h_\pm(\mathbf{r}' - \mathbf{r})\psi(\mathbf{r}') \approx \int d\mathbf{r}'\delta(\mathbf{r}' - \mathbf{r})\psi(\mathbf{r}') = \psi(\mathbf{r}) \qquad (9.12)$$

Setting the variation of W to zero with respect to ψ and G, yields, respectively,

$$-\nabla\cdot(\epsilon\nabla\psi)=\rho_{ex}+\lambda_{+}\mathrm{e}^{-\psi-u_{+}}-\lambda_{-}\mathrm{e}^{\psi-u_{-}} \tag{9.13}$$

and

$$-\nabla\cdot\left[\epsilon\nabla G(\mathbf{r},\mathbf{r}')\right]+2I(\mathbf{r})G(\mathbf{r},\mathbf{r}')=\delta(\mathbf{r}-\mathbf{r}') \tag{9.14}$$

where $I(\mathbf{r})$ is the local ionic strength,

$$I(\mathbf{r})=\frac{1}{2}\left(\lambda_{+}\mathrm{e}^{-\psi-u_{+}}+\lambda_{-}\mathrm{e}^{\psi-u_{-}}\right) \tag{9.15}$$

The density of the ions is obtained from:

$$c_{\pm}(\mathbf{r})=-\frac{\delta W}{\delta\mu_{\pm}}=\lambda_{\pm}\exp\left[\mp\psi(\mathbf{r})-u_{\pm}(\mathbf{r})\right] \tag{9.16}$$

We note that although Eq. 9.14 has the same form as Eq. 8.31, $G(\mathbf{r},\mathbf{r}')$ here is not determined unidirectionally by ψ, but rather feeds back on ψ through the self energy $u_{\pm}$. Therefore, in the GBF variational treatment $G(\mathbf{r},\mathbf{r}')$ and ψ are solved self-consistently.

Making use of Eqs. 9.13 and 9.14, we arrive at the expression for the equilibrium grand free energy,

$$W=-\int d\mathbf{r}\left[c_{+}(\mathbf{r})+c_{-}(\mathbf{r})\right]+\frac{1}{2}\int d\mathbf{r}\psi(\mathbf{r})\left[\rho_{ex}(\mathbf{r})-c_{+}(\mathbf{r})+c_{-}(\mathbf{r})\right]$$
$$+\frac{1}{2}\ln\left(\frac{\det C}{\det G}\right)-\int d\mathbf{r}I(\mathbf{r})G(\mathbf{r},\mathbf{r}) \tag{9.17}$$

The term involving the logarithm of the ratio of the two determinants can be converted to a charging integral following Ref. [21] (see Appendix B of that reference). The final form of the equilibrium free energy is

$$W=-\int d\mathbf{r}\left[c_{+}(\mathbf{r})+c_{-}(\mathbf{r})\right]+\frac{1}{2}\int d\mathbf{r}\psi(\mathbf{r})\left[\rho_{ex}(\mathbf{r})-c_{+}(\mathbf{r})+c_{-}(\mathbf{r})\right]$$
$$+\int d\mathbf{r}I(\mathbf{r})\left[G(\mathbf{r},\mathbf{r};\eta)-G(\mathbf{r},\mathbf{r})\right] \tag{9.18}$$

where η is a charging variable and $G(\mathbf{r},\mathbf{r};\eta)$ is the same-point Green's function obtained from solving Eq. 9.14 with the ionic strength $I(\mathbf{r})$ replaced by $\eta I(\mathbf{r})$. Although the same-point Green's function is divergent, the divergences from $G(\mathbf{r},\mathbf{r};\eta)$ exactly cancels that from $G(\mathbf{r},\mathbf{r})$, rendering the free energy finite even for point charges.

Comparing the free energy expression Eq. 9.18 with the saddle-point free energy Eq. 8.28, we are tempted to regard the charging term in Eq. 9.18 as the fluctuation correction. However, because the Green's function feeds back to the mean electrostatic potential and the ion concentration through the self-energy (see Eqs. 9.13 and 9.16), these properties also include fluctuation effects. Therefore, we should consider Eq. 9.18 in its entirety as the fluctuation-corrected free energy.

The appearance of the self-energy in the variational treatment of fluctuation can have qualitatively different effects than captured by the simple PB theory. For example, near a dielectric interface, the presence of image charge interaction—a form of self energy—can result in charge inversion [21] and attraction between like-charged plates [22].

10 Summary and Perspective

In this chapter, we have presented a pedagogical introduction to the variational methods in statistical thermodynamics, from the perspective of the second law and the maximum-term method in the evaluation of the partition function. This point of view implies that there exists an appropriate free energy that is a function or functional of a set of macrostate variables, and that the equilibrium state (including metastable state) corresponds to a particular macrostate that makes the free energy a minimum. Provided the free energy function or functional is known, by whatever means, application of the variational method is straightforward and yields the equilibrium state and its corresponding free energy.

However, the main utility of the variational methods in statistical thermodynamics is in their use to construct approximate theories for interacting systems. We have shown, using the examples of the Ising model and electrolyte solutions, that the variational methods provide a systematic means for constructing the mean-field theory. Furthermore, we have shown that when the model is transformed into a field-theoretical description, both the steepest-descent and the GBF methods can be used to study the leading order fluctuations, with the former corresponding to the random-phase approximation and the latter corresponding to a self-consistent Hartree approximation. In essence, we have extended the macrostate variables to include the variational parameters, such as the order parameters and the correlation functions, which can be considered generalized state variables. In doing so, we sometimes have to extend the parameter space from real to complex, analogous to the steepest-descent method in the approximate evaluation of integrals. It is in this sense that we view the variational methods as being rooted in the second law of thermodynamics or maximum-term method in statistical thermodynamics.

This particular view point we have taken on variational principle has dictated the choice of the methods we have included for discussion in this chapter; the two methods we have presented, the steepest-descent and the GBF, both involve the minimization (or more generally taking the stationary point) of a free-energy-like quantity. There are other variational methods that are not based on such a procedure. One such method is based on a perturbative calculation of the statistical average of the property of interest using a reference Hamiltonian (or Boltzmann weight) and demanding that the first correction term vanish [23]. Finally, even for the same variational method, there can be different levels of approximations, depending on the desired structure of the theory. For example, in the example provided in Sect. 9, the Green's function is treated as a parameter functional which is determined by solving Eq. 9.14; this will in general require numerical solutions. If instead we desire a more analytical solution, we may assume some parametrized functional form (e.g., screened Coulomb), in which case variation results in algebraic equations for the determination of the parameters in the assumed functional form.

References

1. The uncertainty principle also takes the form of an inequality, but it can be deduced from the fundamental equations of quantum mechanics, e.g., the Schrödinger equation
2. H.B. Callen, *Thermodynamics and an Introduction to Thermostatistics*, 2nd edn. (Wiley, New York, 1985)
3. A.H. Nayfeh, *Introduction to Perturbation Techniques* (Wiley, New York, 1981)
4. T.L. Hill, *Introduction to Statistical Thermodynamics* (Dover, New York, 1986)
5. A. Ishihara, J. Phys. A Gen. Phys. **1**, 539 (1968)
6. R.P. Feynman, *Statistical Mechanics: A Set of Lectures* (Addison- Wesley, Redwood, 1972)
7. H. Kleinert, *Path Integrals in Quantum Mechanics, Statistics, and Polymer Physics*, 2nd edn. (World Scientific, Singapore, 1995)
8. D. Amit, *Field Theory, Critical Phenomena and Renormalization Group*, 2nd edn. (World Scientific, Singapore, 1984)
9. P.-G. de Gennes, *Scaling Concepts in Polymer Physics* (Cornell University Press, Ithaca, 1979)
10. G.H. Fredrickson, *The Equilibrium Theory of Inhomogeneous Polymers* (Oxford University Press, New York, 2005)
11. A.L. Fetter, J.D. Walecka, *Quantum Theory for Many-Particle Systems* (Dover, Mineola, 2003)
12. J.-P. Hansen, I.R. McDonald, *Theory of Simple Liquids*, 2nd edn. (Academic Press, London, 1990)
13. P.M. Chaikin, T.C. Lubensky, *Principles of Condensed Matter Physics* (Cambridge University Press, Cambridge, 2000)
14. R.D. Coalson, A. Duncan, J. Chem. Phys. **97**, 5653 (1992)
15. R.R. Netz, H. Orland, Euro. Phys. J. E **1**, 203 (2000)
16. Because of the complex nature of the action in Eq. 8.23, the GBF bound does not hold, i.e., the true free energy is not always approached from above. D. Frydel has derived an set of exact relations based on the dual representation of the system—the original physical representation and the field-transformed representation. In the case of a two-parameter Gaussian reference, it is shown that the first two relations yield results identical to those by taking the stationary point of the variational action. See: D. Frydel, Eur. J. Phys.**36**, 065050 (2015)
17. R.R. Netz, H. Orland, Eur. Phys. J. E **11**, 310 (2003)
18. M.M. Hatlo, R.A. Curtis, L. Lue, J. Chem. Phys. **128**, 164717 (2008)
19. Z.-G. Wang, Phys. Rev. E **81**, 021501 (2010)
20. This can be understood by noting that the electrostatic energy of a charged body is $\frac{1}{2}\int d\mathbf{r}\rho(\mathbf{r})\psi(\mathbf{r})$ and the electrostatic potential is in turn $\psi(\mathbf{r}) = \int d\mathbf{r}' G(\mathbf{r}, \mathbf{r}')\rho(\mathbf{r}')$. For a point charge at $\mathbf{r}_1$, we then obtain the energy as $\frac{1}{2}G(\mathbf{r}_1, \mathbf{r}_1)$
21. R. Wang, Z.-G. Wang, J. Chem. Phys. **142**, 104705 (2015)
22. R. Wang, Z.-G. Wang, J. Chem. Phys. **139**, 124702 (2013)
23. M. Doi, S.F. Edwards, *The Theory of Polymer Dynamics* (Oxford University Press, New York, 1986)

Square-Gradient Model for Inhomogeneous Systems: From Simple Fluids to Microemulsions, Polymer Blends and Electronic Structure

Jianzhong Wu

1 Introduction

Statistical mechanics is concerned with the properties of many-body systems, i.e., systems containing a large number of either quantum or classical particles [1, 2]. Common examples of such particles include electrons and photons for quantum systems, or diverse microscopic objects for classical systems (e.g., atoms, inert-gas molecules, the repeating units of a polymer or macromolecule, colloidal particles, and globular proteins). To describe the inter-particle interactions, we may divide the elementary quantum particles into bosons and fermions, depending on the symmetric/antisymmetric nature of their wave functions. Such distinction is unnecessary for classical particles because the Newton's equation, often in the context of a semi-empirical potential, is used to describe the particle dynamics.

Regardless of the physical nature of particles, a number of common mathematical procedures may be taken to utilize statistical-mechanical principles to describe a wide variety of phenomena arising from many-body interactions. One of the best-known procedure is Monte Carlo (MC) simulation [3], applicable to calculating the configurational properties of virtually any thermodynamic system. The square-gradient approximation (SGA) discussed in this chapter represents another common but computationally much more efficient procedure. The basic ideas of SGA were introduced by van der Waals to describe the interfacial properties of coexisting vapor and liquid phases over a hundred years ago [4]. Because of its simplicity, SGA remains a popular choice for predicting the microscopic structure and thermodynamic properties of diverse inhomogeneous systems. Similar procedures have been extensively used, for example, to describe phase transitions in condensed-matter systems such as macroscopic phase separations, fluid wetting at solid surfaces, and formation of microemulsions or polymeric mesoscopic phases. In a slightly different context, the

J. Wu (✉)
Department of Chemical and Environmental Engineering and Department of Mathematics, University of California, Riverside, CA 92521, USA
e-mail: jwu@engr.ucr.edu

J. Wu (ed.), *Variational Methods in Molecular Modeling*,
Molecular Modeling and Simulation, DOI 10.1007/978-981-10-2502-0_2

gradient expansion method is also commonly used in theoretical descriptions of electronic properties for both chemical systems and materials.

In contrast to simulation methods, the universal applicability of SGA to various simple and complex fluids and electronic systems is rarely discussed as a common theme. Because the gradient expansion method was often introduced from the perspectives of seemingly unrelated physical phenomena with utterly different practical applications, SGA was "rediscovered" for a number of times and often named after re-inventors from different subfields of condensed matter physics. This chapter intends to establish a generic linkage among several incarnations of SGA. For pedagogy, our discussion begins with some basic concepts from statistical mechanics applicable to both quantum and classical systems. While for simplicity our discussion is mostly focused on systems containing only one type of particles, it should straightforward to extend similar ideas (and equations) to multicomponent systems. In addition to pedagogical purposes, this chapter intends to forge a common ground for better communication among different subfields of statistical mechanics and to facilitate future cross-field developments.

2 Statistical Mechanics

For a many-body system of practical concern, the dynamic and energetic properties are inevitably related to the particle positions or the spatial distributions of the microscopic constituents. A quantity of fundamental importance is thus the one-body particle density profile, viz. the average local number density of individual particles. For a system containing N indistinguishable particles of spherical shape, the instantaneous particle distribution may be specified by a summation of the Dirac-δ functions:

$$\hat{\rho}(\mathbf{r}) = \sum_{i=1}^{N} \delta(\mathbf{r} - \mathbf{r}_i) \tag{1}$$

where $\mathbf{r}_i$ represents the position to locate the center of mass for particle i. The one-body density profile is defined as an ensemble average of the instantaneous density

$$\rho(\mathbf{r}) =< \hat{\rho}(\mathbf{r}) >= \sum_{i=1}^{N} < \delta(\mathbf{r} - \mathbf{r}_i) > \tag{2}$$

where the angle brackets $< \cdots >$ denote an ensemble average, which is determined by the probability of the many-body system in different microstates $\{\nu\}$

$$< \cdots >= \sum_{\nu} p_\nu(\cdots). \tag{3}$$

The Dirac-δ function specifies the probability density distribution for a particle located at a specific position. Accordingly, the density profile $\rho(\mathbf{r})$ reflects the microscopic structure of the many-body system.

The grand canonical ensemble provides a convenient starting point to describe the properties of a many-particle system in terms of the one-body density profiles. For a one-component system at absolute temperature T, chemical potential μ, and volume V, the microstate probability is given by

$$p_\nu = \exp\left[\beta\left(\mu N_\nu - E_\nu\right)\right]/\Xi \tag{4}$$

where Ξ represents the grand partition function

$$\Xi \equiv \sum_\nu \exp\left[\beta\left(\mu N_\nu - E_\nu\right)\right]. \tag{5}$$

In Eqs. (4) and (5), $\beta = 1/(k_B T)$, k_B is the Boltzmann constant, N_ν and E_ν stand for the number of particles and the total energy at microstate ν, respectively. For an electronic system at 0 K, the microstate probability is defined in terms of the multi-body wave function $\Psi(\mathbf{r}^N, s^N)$. In that case, the one-body density profile can be written as

$$\rho(\mathbf{r}) = \int \sum_{s^N} \sum_{i=1}^{N} \delta(\mathbf{r} - \mathbf{r}_i)\Psi^*(\mathbf{r}^N, s^N)\Psi(\mathbf{r}^N, s^N)\mathrm{d}\mathbf{x}^N \tag{6}$$

where $\mathbf{r}^N = (\mathbf{r}_1, \mathbf{r}_2, \ldots, \mathbf{r}_N)$, $s^N = (s_1, s_s, \ldots, s_N)$ stands for the spin coordinates, and Ψ^* represents the complex conjugate of the multi-body wave fn function.

In principle, all equilibrium properties of the system can be derived from the grand potential

$$\Omega \equiv -k_B T \ln \Xi. \tag{7}$$

Taking two important thermodynamic variables as an example, we can calculate entropy from a partial derivative of the grand potential with respect to temperature

$$S \equiv -k_B \sum_\nu p_\nu \ln p_\nu = -\left[\frac{\partial \Omega}{\partial T}\right]_{\mu,V}, \tag{8}$$

and the internal energy is related to the grand potential by

$$U \equiv < K_\nu + \Gamma_\nu + \int d\mathbf{r}\hat{\rho}_\nu(\mathbf{r})\varphi^{\text{ext}}(\mathbf{r}) > = \left[\frac{\partial \beta\Omega}{\partial \beta}\right]_{\mu,V} + \mu N \tag{9}$$

where K_ν and Γ_ν stand for the kinetic and potential energies of the particles at microstate ν, respectively, $\varphi^{\text{ext}}(\mathbf{r})$ is the external potential for each particle, and $N = < N_\nu >$ denotes the average number of particles in the system.

From the grand potential, we can also derive a hierarchy of correlation functions. For example, the first derivative of the grand potential with respect to the one-body potential yields the one-body density profile

$$\frac{\delta\Omega}{\delta u(\mathbf{r})} = \rho(\mathbf{r}) \tag{10}$$

where $u(\mathbf{r}) \equiv \varphi^{\mathrm{ext}}(\mathbf{r}) - \mu$, and a second derivative leads to the density-density correlation function $\chi(\mathbf{r}, \mathbf{r}')$,

$$-\frac{\delta^2\Omega}{\delta^2 u(\mathbf{r})} = -\frac{\delta\rho(\mathbf{r})}{\delta u(\mathbf{r}')} = \beta\left\langle[\hat{\rho}(\mathbf{r}) - \rho(\mathbf{r})][\hat{\rho}(\mathbf{r}') - \rho(\mathbf{r}')]\right\rangle \equiv \beta\chi(\mathbf{r}, \mathbf{r}'). \tag{11}$$

Although the basic ideas of statistical mechanics are rather intuitive, the complexity in the dynamics of many particles makes direct evaluation of the grand partition function virtually impossible except for a few highly idealized systems. In MC simulation, the ensemble average is instead evaluated using some stochastic processes to sample the microstates with an electronic computer. Thanks to rapid advances in computing technology and algorithm developments, modern applications of statistical mechanics often hinge on simulation methods. Alternatively, the microstates can be generated following the dynamics of individual particles as in molecular dynamics (MD) simulations. While the numerical procedures for both MC and MD are formally exact and rather straightforward to implement, enumeration of the microscopic states of a many-body system is not only computationally demanding but also unhelpful for capturing the essential features of physical phenomena. Molecular simulation generates a large volume of often unrevealingly information. By contrast, theoretical methods are able to capture the universal principles underlying diverse physical phenomena and permit fast calculation of structural and thermodynamic properties without explicit consideration of the microscopic details.

3 Density Functional Theory (DFT)

Density functional theory (DFT) provides a generic mathematic framework to establish quantitative connections between thermodynamic properties of a many-body system and the underlying one-body density profiles. The central idea can be best introduced in terms of the Hohenberg-Kohn-Mermin (HKM) theorem [5, 6], which was established first in the context of inhomogeneous electrons at 0 K. The HKM theorem was later generalized to thermodynamic systems of both quantum and classical particles [7].

For a one-component system of identical particles, the HKM theorem asserts that the grand potential can be determined by minimization of the density functional

$$\Omega[\rho(\mathbf{r})] \equiv F[\rho(\mathbf{r})] + \int \rho(\mathbf{r})u(\mathbf{r})d\mathbf{r} \tag{12}$$

where $F[\rho(\mathbf{r})]$ stands for the *intrinsic* Helmholtz energy. Here by intrinsic we mean that the quantity is independent of the system external potential. Formally, $F[\rho(\mathbf{r})]$ is defined by the microstate probability and the intrinsic energy of the particles

$$F[\rho(\mathbf{r})] = \sum_{\nu} p_{\nu}(k_B T \ln p_{\nu} + K_{\nu} + \Gamma_{\nu}). \tag{13}$$

According to the HKM theorem, the one-body external potential is a unique functional of the one-body density profile. As a result, both the microstate probability p_{ν} and, subsequently, the intrinsic Helmholtz energy are unique functionals of the one-body density profile $\rho(\mathbf{r})$.

For a specific system, the one-body potential $u(\mathbf{r})$is fixed. Minimization of the grand potential functional with respect to the one-body potential leads to

$$\frac{\delta\Omega}{\delta\rho(\mathbf{r})} = \frac{\delta F}{\delta\rho(\mathbf{r})} + u(\mathbf{r}) = 0. \tag{14}$$

Equation (14) is known as the Euler-Lagrange equation. With an explicit expression for the intrinsic Helmholtz energy, Eq. (14) allows us to solve the one-body density profile, which serves as the starting point to predict other thermodynamic and structural properties of the system.

It is worth noting that DFT is formally exact and applicable to both quantum and classical systems. In other words, DFT represents a generic mathematical framework in statistical mechanics. The same procedure is similarly applicable to quantum and classical systems including electronic systems at zero temperature. At $T = 0$ K, the expression for the grand potential and the Euler-Lagrange equation remain the same but the thermodynamic entropy vanishes. In that case, the intrinsic Helmholtz energy becomes an *internal* energy, depending only on the kinetic and potential energy of the particles

$$F[\rho(\mathbf{r})] = \sum_{\nu} p_{\nu}(K_{\nu} + \Gamma_{\nu}) =< K_{\nu} + \Gamma_{\nu} > . \tag{15}$$

For electronic systems, the excitation energy, typically on the order of a few electron volts, is much higher than the thermal energy at room temperature (1 eV $\sim$ 40 k_BT). As a result, the entropy effects are relatively unimportant for the electronic properties.

While DFT is emerging as one of the most predominant approaches for the theoretical description of inhomogeneous quantum and classical systems, one noticeable caveat is that the HKM theorem does not provide any specific knowledge on the intrinsic Helmholtz energy. Nevertheless, analytical expressions are readily available for the density functional in the absence of inter-particle interactions. The ideal-gas systems provide a useful reference to formulate the *excess* intrinsic Helmholtz energy due to inter-particle interactions. Although exact results are no more attainable for most practical systems, excellent approximations can be established using analytical tools from both quantum and statistical mechanics [8, 9].

From a mathematical perspective, approximate methods in statistical mechanics are mostly based on perturbation expansions with respect to either the inter-particle energy or the local density inhomogeneity. For systems with a pairwise additive potential $\Gamma(\mathbf{r}_1, \mathbf{r}_2)$, the total potential energy at each microstate may be written as

$$\Gamma_\nu = \frac{1}{2}\int d\mathbf{r}_1 \int d\mathbf{r}_2 \sum_{i\neq j} \delta(\mathbf{r}_1 - \mathbf{r}_i)\Gamma(\mathbf{r}_1, \mathbf{r}_2)\delta(\mathbf{r}_2 - \mathbf{r}_j) \tag{16}$$

where a factor of 2 accounts for the fact that each pair potential involves two interacting particles. Using Eqs. (7) and (16), we may show that a functional derivative of the grand potential with respect to the pair potential leads to the two-body density distribution function, $\rho^{(2)}(\mathbf{r}_1, \mathbf{r}_2)$,

$$\frac{\delta\Omega}{\delta\Gamma(\mathbf{r}_1, \mathbf{r}_2)/2} = <\sum_{i\neq j} \delta(\mathbf{r}_1 - \mathbf{r}_i)\delta(\mathbf{r}_2 - \mathbf{r}_j) > \equiv \rho^{(2)}(\mathbf{r}_1, \mathbf{r}_2). \tag{17}$$

Equation (17) can be used to evaluate the difference between the intrinsic Helmholtz energy of a real system and that of an ideal system (ID).

At fixed temperature T and one-body potential $u(\mathbf{r})$, a functional integration of Eq. (17) with respective to the pair potential gives [10]

$$F[\rho(\mathbf{r})] = F^{ID}[\rho(\mathbf{r})] + \frac{1}{2}\int_0^1 d\lambda \int d\mathbf{r}_1 \int d\mathbf{r}_2 \rho^{(2)}(\mathbf{r}_1, \mathbf{r}_2, \lambda)\Gamma(|\mathbf{r}_1 - \mathbf{r}_2|) \tag{18}$$

where $F^{ID}[\rho(\mathbf{r})]$ represents the intrinsic Helmholtz energy of the non-interacting system, and $\rho^{(2)}(\mathbf{r}_1, \mathbf{r}_2, \lambda)$ stands for the two-body density correlation function of the system under consideration but with a reduced pair potential, $\Gamma_\lambda(r) = \lambda\Gamma(r)$, where $0 \leq \lambda \leq 1$. In writing Eq. (18), we assume that the inter-particle potential between spherical particles depends on the center-to-center distance, i.e., $\Gamma(\mathbf{r}_1, \mathbf{r}_2) = \Gamma(|\mathbf{r}_1 - \mathbf{r}_2|)$. The functional integration corresponds to the reversible work to add the inter-particle potential to non-interacting ideal particles [11]. For electronic systems, Eq. (18) is commonly known as the adiabatic connection [8].

For most systems of practical interest, the two-body correlation functions are extremely complicated, depending not only on two positions but also on the local density profile as well as all variables defining the thermodynamic state. For easy understanding, it is convenient to express the two-body density correlation function in terms of the radial distribution function (RDF)

$$g(\mathbf{r}_1, \mathbf{r}_2) \equiv \rho^{(2)}(\mathbf{r}_1, \mathbf{r}_2)/[\rho(\mathbf{r}_1)\rho(\mathbf{r}_2)] \tag{19}$$

or the total correlation function (TCF)

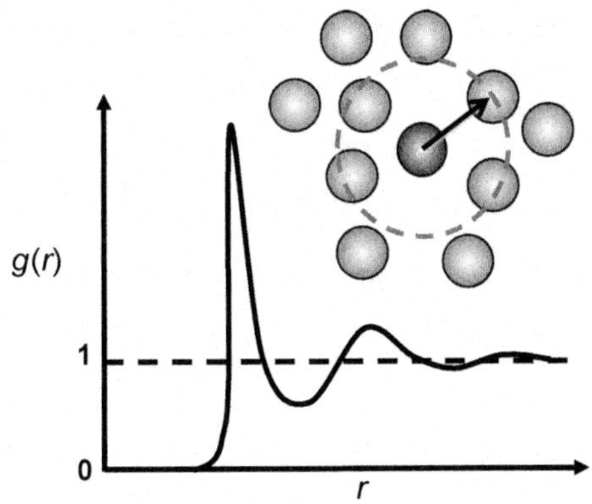

Fig. 1 A schematic representation of the radial distribution function (RDF). For a uniform system of spherical particles, $g(\mathbf{r}_1, \mathbf{r}_2)$ is a function of the distance $r = |\mathbf{r}_1 - \mathbf{r}_2|$. For an inhomogeneous system, however, RDF depends on the position and relative orientation. For both uniform and inhomogeneous systems, RDF vanishes at small separation and approaches unity at large distance owing to the short-range repulsion and the rapid decay of long-range inter-particle interactions

$$h(\mathbf{r}_1, \mathbf{r}_2) \equiv g(\mathbf{r}_1, \mathbf{r}_2) - 1. \tag{20}$$

Intuitively, RDF represents the probability of finding a particle given that the position of another particle is fixed at the origin. As shown schematically in Fig. 1, RDF vanishes at small separation due to the inter-particle repulsion and approaches unity at large distance when the particle densities become uncorrelated. In the mean-field approximation, it is commonly assumed $g(\mathbf{r}_1, \mathbf{r}_2) = 1$ or $h(\mathbf{r}_1, \mathbf{r}_2) = 0$, i.e., the total correlation function is completely neglected.

4 Square-Gradient Approximation (SGA)

For systems with a nearly uniform one-body density profile, the intrinsic Helmholtz energy may be approximated by a functional Taylor expansion with respect to that of a uniform system with an average density ρ_0:

$$F[\rho(\mathbf{r})] = F(\rho_0) + \mu \int \Delta\rho(\mathbf{r})d\mathbf{r} + \frac{1}{2}\int d\mathbf{r}_1 \int d\mathbf{r}_2 \Delta\rho(\mathbf{r}_1)\Delta\rho(\mathbf{r}_2)K(\mathbf{r}_1, \mathbf{r}_2) + \cdots \tag{21}$$

where $\Delta\rho(\mathbf{r}) \equiv \rho(\mathbf{r}) - \rho_0$. In the density expansion above, the reference system has the temperature and the particle chemical potential the same as those corresponding to the real system.

In Eq. (21), $K(\mathbf{r}_1, \mathbf{r}_2)$ has the units of energy and is referred to as the vertex function. Because the first-order functional derivative of the intrinsic Helmholtz energy with respect to $\rho(\mathbf{r})$ results in the one-body potential, the vertex function corresponds to the second-order functional derivative. For a uniform system, it depends only on the distance between $\mathbf{r}_1$ and $\mathbf{r}_2$, i.e.,

$$K(\mathbf{r}_1, \mathbf{r}_2) = -\left.\frac{\delta u(\mathbf{r}_1)}{\delta \rho(\mathbf{r}_2)}\right|_{\rho(\mathbf{r})=\rho_0} = K(|\mathbf{r}_1 - \mathbf{r}_2|). \tag{22}$$

In writing Eq. (22), we have utilized the Euler-Lagrange equation (i.e., Eq. (14)). Equation (22) suggests that the vertex function specifies the variational of the local one-body potential in response to the change in the particle density at another position.

In comparison to the exact expression given in Eqs. (18) and (21) has a major advantage because the vertex function depends on the distance between positions $\mathbf{r}_1$ and $\mathbf{r}_2$. For systems with a slow varying one-particle density $\rho(\mathbf{r})$, $\Delta\rho(\mathbf{r})$ is small and the functional Taylor expansion for the intrinsic Helmholtz energy may be truncated after the quadratic term. Similarly, the local density may be expressed as a truncated Taylor series

$$\rho(\mathbf{r}_2) = \rho(\mathbf{r}_1) + (\mathbf{r}_1 - \mathbf{r}_2)\cdot\nabla\rho(\mathbf{r}_1) + \frac{1}{2}(\mathbf{r}_1 - \mathbf{r}_2)(\mathbf{r}_1 - \mathbf{r}_2) : \nabla\nabla\rho(\mathbf{r}_1) + O(\nabla^3\rho) \tag{23}$$

where $\nabla\rho(\mathbf{r})$ denotes the density gradient, and symbol " :" is the scalar product of two tensors. As detailed in Appendix, the gradient expansions lead to a simple expression for the intrinsic Helmholtz energy of inhomogeneous systems:

$$F = \int d\mathbf{r} \left\{ f_0[\rho(\mathbf{r})] + \frac{\kappa}{2}|\nabla\rho(\mathbf{r})|^2 \right\} \tag{24}$$

where $f_0(\rho)$ represents the Helmholtz energy density of the uniform system at system temperature T and local density $\rho(\mathbf{r})$, and κ is called the influence parameter. The first term on the right side of Eq. (24) corresponds to the local density approximation (LDA) for the intrinsic Helmholtz energy, and the gradient term accounts for a correction to the intrinsic Helmholtz energy due to the local density inhomogeneity.

For a uniform system, $\nabla\rho(\mathbf{r}) = 0$, and Eq. (24) reduces the bulk Helmholtz energy. Because the correction to the local density approximation is a quadratic function of the density gradient, Eq. (24) is referred to as the square-gradient approximation (SGA). Alternatively, the mathematical form is also known as the Ginzburg–Landau theory or the Landau expansion. Similar methods are used extensively to describe structure formation in inhomogeneous systems and phase transitions [12].

As shown in Appendix, the influence parameter is related to the vertex function of the uniform system $K_0(r)$

$$\kappa = -\frac{2\pi}{3}\int_0^\infty r^4 K_0(r)dr = \lim_{q\to 0}\,[\tilde{K}_0(q) - \tilde{K}_0(0)]/q^2 \tag{25}$$

where $\tilde{K}_0(q)$ represents the 3-dimensional (3D) Fourier transform of $K_0(r)$

$$\tilde{K}_0(q) = \tilde{K}_0(\mathbf{q}) \equiv \int K_0(r)e^{-i\mathbf{q}\cdot\mathbf{r}}d\mathbf{r} = \int \frac{\sin(qr)}{qr}K_0(r)d\mathbf{r}. \tag{26}$$

Alternatively, it may be expressed in terms of the density-density correlation function

$$\beta\kappa = \lim_{q\to 0}\,[1/\tilde{\chi}_0(q) - 1/\tilde{\chi}_0(0)]/q^2 \tag{27}$$

where $\tilde{\chi}_0(q)$ corresponds to the 3D Fourier transform of the density–density correlation

$$\chi(\mathbf{r},\mathbf{r}') \equiv \langle[\hat{\rho}(\mathbf{r}) - \langle\hat{\rho}(\mathbf{r})\rangle][\hat{\rho}(\mathbf{r}') - \langle\hat{\rho}(\mathbf{r}')\rangle]\rangle\,. \tag{28}$$

In deriving the influence parameter, we assume that the vertex function is independent of the local density. Accordingly, κ is determined from the correlation functions of the uniform reference system. For better numerical performance, however, κ is often evaluated from $K_0(r)$ of a uniform system at the local density.

SGA requires as an input the local Helmholtz energy density and the vertex function (or the density–density correlation function) of the corresponding uniform system. On the one hand, the Helmholtz energy density is typically provided by an equation of state or an excess free-energy model for corresponding bulk systems. Alternatively, the local Helmholtz energy density may be obtained from an empirical correlation based on simulation results. For classical systems, the correlation functions can be solved from the integral-equation theories (e.g., the Ornstein-Zernike equation) or from mean-field approximations and analytical functions derived for ideal systems (e.g., correlation functions for Gaussian chains). Because gradient expansions are applied to both the intrinsic Helmholtz energy and the one-body density profile, we expect SGA to perform well for inhomogeneous systems with near uniform density profiles. With the influence parameter evaluated from the local densities (or locally averaged densities [13], however, SGA may also be used for highly inhomogeneous systems.

5 Simple Fluids

A simple fluid consists of argon-like molecules. In addition to noble gases, other examples of simple fluids include a large number of nonpolar gases of low molecular weight such as methane and, from a thermodynamics perspective, colloidal dispersions and aqueous solutions of globular proteins.

For a simple fluid, the bulk Helmholtz energy can be readily derived from an equation of state. Taking the van der Waals theory as an example, the reduced Helmholtz energy density per volume is given by

$$\beta f_0(\rho) = \rho \left[\ln\left(\frac{\rho\Lambda^3}{1-b\rho}\right) - 1 - \beta\rho a \right] \tag{29}$$

where a and b are van der Waals' parameters, Λ represents the thermal wavelength. More accurate expressions for the Helmholtz energy density are available [14].

The vertex function of a simple fluid may be expressed in terms of the direct correlation function (DCF), $c(\mathbf{r}, \mathbf{r}')$,

$$\beta K(\mathbf{r}, \mathbf{r}') = \frac{\delta(\mathbf{r} - \mathbf{r}')}{\rho(\mathbf{r})} - c(\mathbf{r}, \mathbf{r}'). \tag{30}$$

Mathematically, $c(\mathbf{r}, \mathbf{r}')$ corresponds to the second-order functional derivatives of the excess Helmholtz energy, $F^{\text{ex}} \equiv F - F^{ID}$

$$c(\mathbf{r}, \mathbf{r}') \equiv -\frac{\delta^2 \beta F^{\text{ex}}}{\delta\rho(\mathbf{r})\delta\rho(\mathbf{r}')}. \tag{31}$$

Without inter-particle interactions, the Helmholtz energy of the ideal system is exactly known

$$F^{ID} = k_B T \int \rho(\mathbf{r})\{\ln[\rho(\mathbf{r})\Lambda^3] - 1\} d\mathbf{r}. \tag{32}$$

As shown in Appendix, DCF and RDF are related through the Ornstein-Zernike equation, which provides a basis for numerical solutions of the correlation functions. Analytical expressions of $c_0(r)$ are also available for a number of simple fluids over a broad range of thermodynamic conditions [15].

Figure 2 shows schematically the direct correlation function for a uniform fluid. While there is an apparent connection between the direct correlation function and the reduced pair potential at large distance, it is important to recognize that $c_0(r)$ depends not only on the distance but also on thermodynamic parameters defining the equilibrium state. At large distance, the direct correlation function approaches to the reduced pair potential as

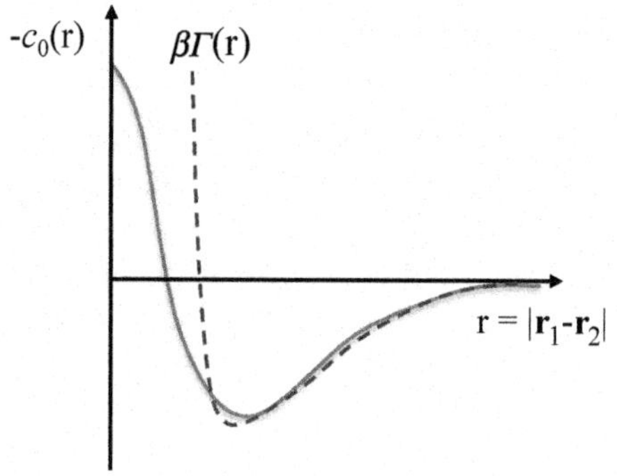

Fig. 2 A schematic representation of the direct correlation function of a simple fluid $c_0(r)$. At large distance, the direct correlation function is virtually identical to the reduced pair potential $-\beta\Gamma(r)$. At short distance, the pair potential diverges while the direct correlation function remains finite

$$c_0(r) \approx -\beta\Gamma(r) \quad \text{as } r \to \infty. \tag{33}$$

For uniform systems, Eq. (30) reduces to

$$\beta K_0(r) = \delta(r)/\rho_0 - c_0(r). \tag{34}$$

Substituting Eq. (34) into (25), we can calculate the influence parameter from the DCF of a uniform system

$$\kappa = \frac{2\pi k_B T}{3} \int_0^\infty dr\, r^4 c_0(r). \tag{35}$$

Because of the asymptotic behavior of $c_0(r)$, Eq. (35) suggests that SGA is not valid if the inter-particle potential behaves as $1/r^n,\ n < 6$ at large r.

Case study I: Interfacial tension

The interfacial tension between two coexisting phases, say α and β, is defined as the change in free energy in response to variation of the interfacial area. For two bulk phases at equilibrium, the interfacial area refers to that of an imaginary surface dividing the total mass of a particular component in the system into those corresponding to two bulk phases. The imaginary surface is called the Gibbs dividing surface.

Schematically, Fig. 3 presents a Gibbs dividing surface between two bulk phases (e.g., vapor and liquid) and the local density profile across the interface for a one-component system. Here the density profile varies only in the direction perpendicular to the interface, i.e., $\rho(\mathbf{r}) = \rho(z)$, where z represents the coordinate in perpendicular to the surface. Because the dividing surface possesses no volume, its position, here set at $z = 0$, can be determined from the one-body density profile

$$\int_{-\infty}^{0} dz[\rho(z) - \rho_\alpha] + \int_0^\infty dz[\rho(z) - \rho_\beta] = 0. \tag{36}$$

Equation (36) is also applicable to multi-component systems. In that case, the location of the dividing surface depends on a specific component selected such that its density profile satisfies Eq. (36).

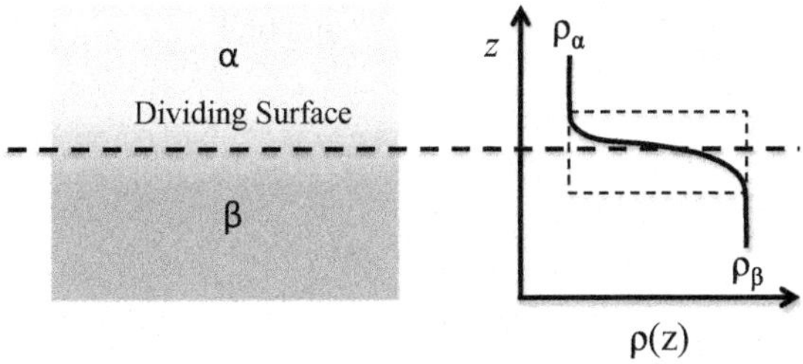

Fig. 3 The Gibbs dividing surface between two bulk phases (α and β) and a schematic of the density profile in the interfacial region

For an inhomogeneous system containing two coexisting bulk phases, SGA predicts that the grand potential per unit area is given by

$$\Omega/A = \int_{-\infty}^{\infty} dz \left\{ f_0[\rho(z)] + \frac{\kappa}{2}|\rho'(z)|^2 - \mu\rho(z) \right\} \tag{37}$$

where $\rho'(z) \equiv d\rho(z)/dz$. For the bulk systems, the grand potential reduces to $\Omega = -PV$ where pressure P is the same for the coexisting phases. The surface tension is defined as the grand potential per unit area relative to those corresponding to the bulk phases

$$\gamma \equiv \frac{\Omega - \Omega^{\alpha} - \Omega^{\beta}}{A} = \int_{-\infty}^{\infty} dz \left\{ f_0[\rho(z)] + \frac{\kappa}{2}|\rho'(z)|^2 - \mu\rho(z) - P \right\}. \tag{38}$$

To use Eq. (38), we need an equation of state for the bulk phase and the density profile. As discussed, above, the latter can be calculated by minimization the grand potential Eq. (37):

$$\mu_0[\rho(z)] - \kappa\rho''(z) - \mu = 0 \tag{39}$$

where $\mu_0 \equiv (\partial f_0/\partial\rho)_T$, $\rho''(z) \equiv d^2\rho(z)/dz^2$. Because of the inhomogeneity in local density, $\mu_0 \neq \mu$; μ_0 reduces to the bulk chemical potential only when the density is constant. Using the boundary conditions far from the interface, i.e., $\rho(-\infty) = \rho_\beta$ and $\rho(\infty) = \rho_\alpha$, one may solve the density profile from a numerical integration of Eq. (39).

To obtain an explicit expression for the surface tension, we may rewrite Eq. (39) in terms of the local grand potential density, $\omega(\rho) = f^0(\rho) - \rho\mu$,

$$\kappa\rho''(z) = \frac{\partial\omega(\rho)}{\partial\rho}. \tag{40}$$

Multiplying both sides of Eq. (40) by $\rho'(z)$ leads to

$$\frac{d}{dz}\left[\frac{\kappa}{2}\rho'(z)^2 - \omega(\rho)\right] = 0. \tag{41}$$

Using boundary conditions $\rho'(z) = 0$ and $\omega = -P$ for the bulk phases, we can integrate Eq. (41) with respect to z and find

$$\frac{\kappa}{2}\rho'(z)^2 - \omega(\rho) = -P. \tag{42}$$

Substituting Eq. (42) into (38) leads to a simplified expression for the surface tension

$$\gamma = \kappa \int_{-\infty}^{\infty} \rho'(z)^2 dz. \tag{43}$$

According to Eq. (43), the surface tension and the influence parameter have the same sign. Because the direct correlation function is negative at small separation and positive at larger distance, Eq. (35) suggests that the influence parameter may be negative under certain conditions. In that case, the surface tension is negative, favoring spontaneous formation of the interfacial area (e.g., in microemulsions as discussed below).

A nice feature of SGA for predicting surface tension is that Eq. (43) can be evaluated without knowing the density profile explicitly. According to Eq. (42), we have

$$dz = -d\rho/\sqrt{2[\omega(\rho) - P]/\kappa}. \tag{44}$$

Here a negative sign is taken with the assumption that the density declines in the z direction. We now substitute Eq. (44) into (43) and integrate by parts,

$$\gamma = \int_{\rho_\alpha}^{\rho_\beta} \sqrt{2\kappa[\omega(\rho) - P]} d\rho. \tag{45}$$

Equation (45) indicates that SGA can be used to predict the surface tension directly from the correlation functions and the equation of state for the uniform systems without computing the density profile.

SGA is able to capture the essential features of interfacial inhomogeneity and surface tension. In comparison to exact results from experiments or molecular simulations, however, its performance is mostly qualitative [16]. As shown in Fig. 4, SGA often gives a too broad one-body density profile at the interfacial region. Besides, it misses density fluctuations near the liquid side of a vapor-liquid interface.

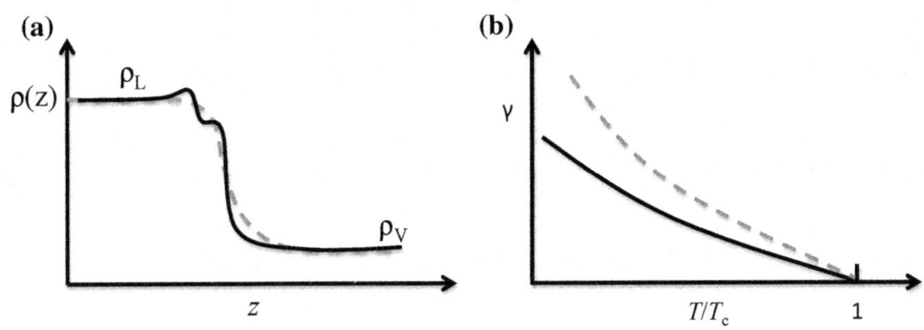

Fig. 4 A schematic comparison of the density profile at the vapor-liquid interface and the surface tension versus temperature predicted from SGA (*dashed lines*) with those from experiments (*solid lines*). Here T_c stands for the critical temperature for the vapor-liquid coexistence

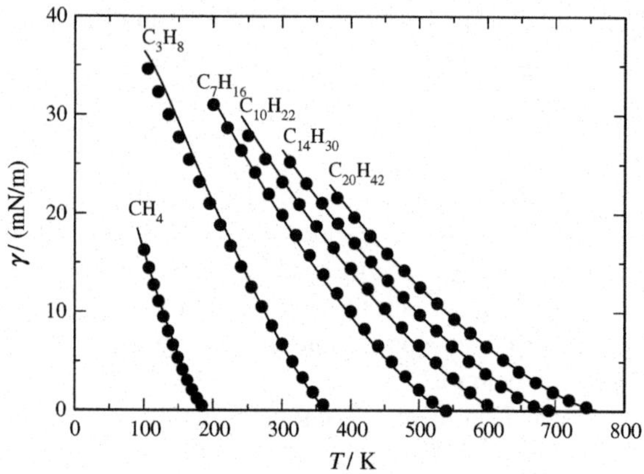

Fig. 5 The vapor-liquid interfacial tension of normal alkanes. Here the symbols are experimental data and the *solid lines* are SGA predictions. Adopted from Garrido et al. AIChE Journal (2016)

Because SGA predicts a too-broad interfacial region, it overestimates the surface tension over the entire range of the coexisting temperature. A quantitative representation of the surface tension can be accomplished by using an accurate equation of state for the bulk phases and an optimized influence parameter. For example, Fig. 5 shows that, even with the assumption that the influence parameter is independent of the local density, the SGA is able to describe vapor-liquid interfacial tensions of several normal alkanes in excellent agreement with experimental data [17].

6 Microemulsions

Microemulsions are thermodynamically stable heterogeneous mixtures of oil, water and surfactants forming microscopic structures of various sizes and shapes. In microemulsions, oil and water droplets are dispersed in "water in oil (w/o)", "oil in water (o/w)", or bicontinuous structures stabilized by pure or mixed surfactants adsorbed at the oil-water interfaces. Microemulsions have industrial applications such as polymer synthesis, drug delivery, and enhanced oil recovery in the petrochemical industry [18].

To capture the gross features of microemulsions, Teubner and Strey proposed the following phenomenological equation for the deviation of the intrinsic Helmholtz energy from that of a uniform system

$$\Delta F = \int d\mathbf{r} \left\{ a_2 \phi^2 + c_1 |\nabla \phi(\mathbf{r})|^2 + c_2 |\nabla \cdot \nabla \phi(\mathbf{r})|^2 \right\} \tag{46}$$

where $\phi(\mathbf{r})$ stands for an order parameter to account for the deviation of the local water or oil concentration from the corresponding mean value, $a_2 > 0$, $c_1 < 0$ and $c_2 > 0$ are phenomenological parameters with their signs fixed to ensure thermodynamic stability of the inhomogeneous system. It is worth noting that $c_1 < 0$ implies a negative surface tension in microemulsions, favoring spontaneous formation of surface area. In Eq. (46), the 4th-order gradient term is introduced to ensure that the system will be stabilized without unlimited growth of the interface area.

The polynomial form given by Eq. (46) is commonly known as the Landau expansion for the free energy of an inhomogeneous system. Intuitively, it may be understood as an empirical gradient expansion relative to an unstructured uniform phase without invoking any specific knowledge on the local Helmholtz energy density and correlation functions. As a result, the Landau expansion is applicable as a simple mathematical procedure to structure formation in any thermodynamic systems. The Taylor expansion is expected to be adequate when the order parameter is small as in the early stage of phase transitions or structure formation in heterogeneous systems.

A conventional experimental approach to monitor the structure of microemulsions is by small angle neutron or X-ray scattering. The scattering experiments provide information on the structures of microemulsions at nanometer or even smaller length scales. The intensity of neutron or X-ray scattering is proportional to the Fourier transform of the density-density correlation function

$$I(q) \sim \tilde{\chi}(q). \tag{47}$$

As shown in Appendix, $\tilde{\chi}(q)$ is inversely proportional to the vertex function in the Fourier space, $\tilde{K}(q)$

$$\beta\tilde{K}(q)\tilde{\chi}(q) = 1. \tag{48}$$

As the order parameter is defined linearly proportional to the local density of water or oil molecules, Eq. (46) can be used to derive the density-density correlation function. Taking a second-order functional derivative of the empirical Helmholtz energy with respect to $\phi(\mathbf{r})$ and making the Fourier transform, we find the vertex function in the Fourier space

$$\beta\tilde{K}(q) \sim (a_2 + c_1 q^2 + c_2 q^4). \tag{49}$$

Accordingly, the density-density correlation function is

$$\tilde{\chi}(q) \sim \frac{1}{a_2 + c_1 q^2 + c_2 q^4} \tag{50}$$

Equation (50) provides a theoretical basis for interpreting the physical meanings of the scattering spectra obtained from neutron or X-ray experiments. The spectra obtained from small angle scattering experiments is most relevant to the long-range limit of the density-density correlation function of the inhomogeneous system. According to Eq. (50), the asymptotic limit of the density-density correlation function may be written in the real space as

$$\chi(r) \sim \frac{de^{-r/\xi}}{2\pi r}\sin\left(\frac{2\pi r}{d}\right), \quad r/\xi >> 1 \tag{51}$$

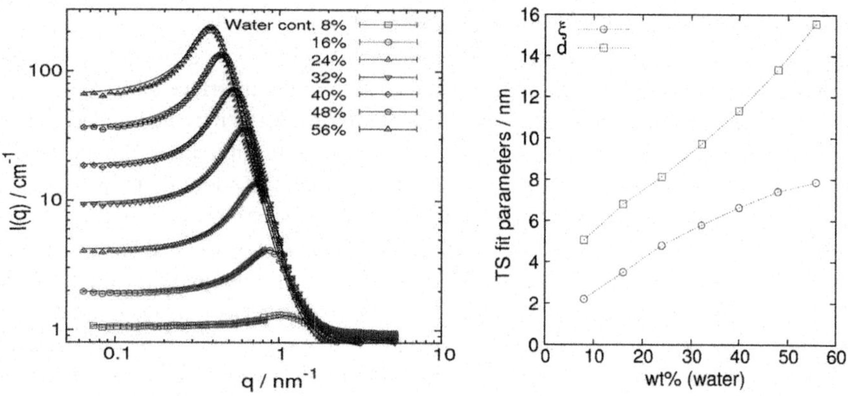

Fig. 6 Small angle neutron scatting (SNAS) spectra for nonionic microemulsions containing water, Brij 96 surfactant, ethyl oleate and hexanol [19]. The solid lines represent correlations with the Teubner-Strey (TS) model with two fit parameters ξ and d changing with the water content (right panel). (Adapted from Kaur et al. Langmuir 2012)

where

$$\xi = \left[\frac{1}{2}\left(\frac{a_2}{c_2}\right)^{1/2} + \frac{1}{4}\frac{c_1}{c_2}\right]^{-1/2} \tag{52}$$

$$d = 2\pi\left[\frac{1}{2}\left(\frac{a_2}{c_2}\right)^{1/2} - \frac{1}{4}\frac{c_1}{c_2}\right]^{-1/2}. \tag{53}$$

Equation (51) suggests that d is related to a characteristic domain size of the microemulsion, and ξ may be understood as the correlation length for the density fluctuations. For water in oil (W/O) or oil in water (O/W) microemulsions, the domain size corresponds to the diameter of spherical droplets for the dispersed phase, and the correlation length reflects density fluctuation within each spherical domain. For bicontinuous microemulsions, the sinusoidal term accounts for the alternating domains of oil and water phases with an average periodicity of d, and the exponential term is related to the short-range correlation within the water or oil domain.

The Teubner-Strey (TS) model has been routinely used to describe the scattering spectra of a wide variety of microemulsions. The domain size and correlation length obtained from the fit parameters provide insights into the microscopic structure. For example, Fig. 6 shows the small angle neutron scatting (SANS) spectra for nonionic microemulsions containing water, a surfactant (Brij 96), ethyl oleate and hexanol [19]. The system is relevant for a number of pharmaceutical and cosmetic formulations. We see that the scatting intensity curves are near perfectly reproduced by the TS model. From the fit parameters, the domain size and correlation length at different water contents can be determined. Figure 6 shows that the domain size is linearly increasing with the amount of water contained, implying that the microscopic structure swells proportionally upon the addition of water. This example illustrates how the scattering experiments shed lights for systematic formulation of microemulsions with desired microscopic structures. As the structure is not directly detected, the statistical-mechanical model becomes indispensible for interpretation of the experimental spectra.

7 Polymer Blends

Pierre-Gilles de Gennes was often credited as the first to use the square-gradient approximation (SGA) for inhomogeneous polymer blends [20]. The so-called Flory-Huggins-de Gennes (FHdG) model remains a popular choice for describing the thermodynamic properties of inhomogeneous polymeric systems and phase transactions. The FHdG theory may also be relevant to biological systems because recent studies suggest that the physics of polymer phase transitions are applicable to the formation of intracellular membrane apartments [21].

To illustrate the basic concepts behind the FHdG theory, we consider an inhomogeneous blend of two polymers A and B. According to SGA, the intrinsic Helmholtz energy is given by

$$F = \int d\mathbf{r} \left\{ f_0[\rho_A, \rho_B] + \frac{\kappa_{AA}}{2} [\nabla \rho_A(\mathbf{r})]^2 + \kappa_{AB} \nabla \rho_A(\mathbf{r}) \cdot \nabla \rho_B(\mathbf{r}) + \frac{\kappa_{BB}}{2} [\nabla \rho_B(\mathbf{r})]^2 \right\} \tag{54}$$

where $\rho_A(\mathbf{r})$ and $\rho_B(\mathbf{r})$ stand for the segment densities, κ_{ij} are influence parameters to account for the effect of local density inhomogeneity, andf_0 represents the Helmholtz energy density of a homogeneous polymer at the local segment densities.

For uniform polymer systems, the Flory-Huggins theory is commonly used to describe the Helmholtz energy of mixing. The reduced Helmholtz energy per polymer segment, relative to those of pure species, is given by

$$\beta v_0 \Delta f_0 = \frac{\phi_A}{N_A} \ln \phi_A + \frac{\phi_B}{N_B} \ln \phi_B + \chi_F \phi_A \phi_B, \tag{55}$$

where v_0 represents the volume per lattice site, N_i denotes the degree of polymerization for polymer $i = A$ or B, $\phi_i = \rho_i v_0$ stands for the polymer volume fraction, and χ_F is the Flory parameter.

As discussed above, the influence parameters can be determined from the vertex functions

$$\kappa_{ij} = \lim_{q \to 0} [K_{0,ij}(q) - K_{0,ij}(0)]/q^2. \tag{56}$$

In terms of the polymer volume fractions, the vertex function is related to the density fluctuations in the binary polymer mixture

$$K_{ij}^{-1}(\mathbf{r}_1,\mathbf{r}_2) = \beta < \delta\hat{\rho}_i(\mathbf{r}_1)\delta\hat{\rho}_j(\mathbf{r}_2) >= \frac{\beta}{v_0^2} < \delta\hat{\phi}_i(\mathbf{r}_1)\delta\hat{\phi}_j(\mathbf{r}_2) > \tag{57}$$

where $\hat{\phi}_i(\mathbf{r}) = \hat{\rho}_i(\mathbf{r}_1)v_0$, and $\delta\hat{\phi}_i(\mathbf{r}) \equiv \hat{\phi}_i(\mathbf{r}) - \phi_{i,0}$ stands for the deviation of the instantaneous local volume fraction from the mean value $\phi_{i,0}$.

The Flory-Huggins lattice model assumes that the polymer mixture is incompressible. In other words, each lattice site is occupied by one and only one polymer segment of either type A or B such that the local volume fraction is normalized

$$\hat{\phi}_{\mathrm{A}}(\mathbf{r}) + \hat{\phi}_{\mathrm{B}}(\mathbf{r}) = 1. \tag{58}$$

Using the identity

$$\nabla\phi_{\mathrm{A}}(\mathbf{r}) = -\nabla\phi_{\mathrm{B}}(\mathbf{r}), \tag{59}$$

we may simplify Eq. (54) and derive the Helmholtz energy of mixing for the inhomogeneous system

$$\Delta F = \int d\mathbf{r}\left\{\Delta f_0(\phi) + \frac{\kappa}{2}[\nabla\phi]^2\right\} \tag{60}$$

where $\phi = \phi_A$, and κ is an effective influence parameter given by

$$\kappa \equiv (\kappa_{AA} + \kappa_{BB} - 2\kappa_{AB})/v_0^2. \tag{61}$$

Because of the incompressibility hypothesis, the densities of polymer segments are inter-related and the Helmholtz energy for the binary mixture resembles that for a one-component system (viz. Eq. (24)).

To derive the influence parameter, the FHdG theory assumes further that, in a polymer melt, the local fluctuation of the polymer volume fractions behaviors as that corresponding to non-interacting polymers (viz. Gaussian chains). In other words, the density-density correlation functions are determined by the intra-chain connectivity of polymer segments. Because segments from different polymer chains are uncorrelated, we have $\chi_{0,AB}(r) = \chi_{0,BA}(r) = 0$.

As shown in the Appendix, the intra-chain density-density correlation for polymer A can be approximated by

$$\tilde{\chi}_{0,AA}(q) \approx \frac{N_{\mathrm{A}}\phi_{0,\mathrm{A}}}{v_0^2}\left(1 - \frac{q^2R_{\mathrm{A}}^2}{3}\right) \tag{62}$$

where $R_{\mathrm{A}} \equiv N_{\mathrm{A}}l_A^2/6$ is the radius of gyration for an ideal polymer chain, and l_A represents the bond length for polymer A. Substituting Eq. (62) into (56) yields the influence parameter for polymer A

$$\kappa_{AA} = \frac{1}{\beta}\lim_{q\to 0}[\chi_{0,AA}^{-1}(q) - \chi_{0,AA}^{-1}(0)]/q^2 = \frac{v_0^2R_{\mathrm{A}}^2}{3\beta N_{\mathrm{A}}\phi_{0,\mathrm{A}}} = \frac{v_0^2l_{\mathrm{A}}^2}{18\beta\phi_{0,\mathrm{A}}}. \tag{63}$$

Similarly, the influence parameter for polymer B is

$$\kappa_{BB} = \frac{v_0^2 l_B^2}{18\beta\phi_{0,B}} \tag{64}$$

where l_B represents the bond length for polymer B. Because there is no correlation between segments from different ideal polymer chains, the cross inference parameters are

$$\kappa_{AB} = \kappa_{BA} = 0. \tag{65}$$

Substituting Eqs. (64)–(65) into (61), we derive the effective influence parameter for the polymer blend

$$\kappa = \frac{1}{18\beta}\left(\frac{l_A^2}{\phi_{0,A}} + \frac{l_B^2}{\phi_{0,B}}\right). \tag{66}$$

Case study II. Kinetics of polymer phase separation

A binary polymer mixture may exist either as a single uniform phase or as two coexisting phases, depending on its composition and temperature. As shown in Fig. 7, the binodal curve separates regions of the phase diagram into single and two phases. In the two-phase region, spontaneous phase separation occurs when the system exists inside the spinodal line. The demixing process is called spinodal decomposition.

When a polymer blend undergoes spinodal decomposition, its morphology, i.e., the inhomogeneous distribution of polymer segments, is controlled by the dynamics of the phase-separation. The change of the polymer composition can be described by the phenomenological diffusion equation

$$\frac{\partial\phi(\mathbf{r},\ t)}{\partial t} = -\nabla\cdot\mathbf{J}, \tag{67}$$

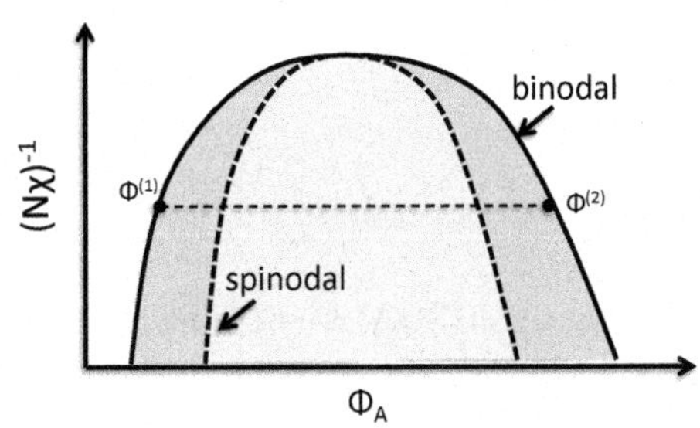

Fig. 7 A schematic phase diagram for a binary polymer blend. Here $\Phi^{(1)}$ and $\Phi^{(2)}$ represent the volume fraction of polymer A in two coexisting curves. The system exists as one single phase outside the *binodal curve* and two phases otherwise. Inside the *spinodal curve*, the mixture is spontaneously separated into two phases

where $\phi(\mathbf{r}, t)$ is the volume fraction of polymer A at position $\mathbf{r}$ and time t, $\mathbf{J}$ is the local flux of polymer A. The polymer flux may be related to the local chemical potential μ_A through the generalized Fick's law

$$\mathbf{J} = -\Lambda \nabla \beta \mu_A \tag{68}$$

where Λ represents the Onsager coefficient [9]. In general, Λ is nonlocal, depending on the polymer size and the self-diffusion coefficient.

If polymers A and B are symmetric, i.e., $N_A = N_B = N$ and $l_A = l_B = l$, the Helmholtz energy functional for the inhomogeneous polymer blend becomes

$$\beta \Delta F = \frac{1}{v_0} \int d\mathbf{r} \left\{ \frac{1}{N} \ln \phi + \frac{1-\phi}{N} \ln \phi + \chi_F \phi (1-\phi) + \frac{l^2 [\nabla \phi]^2}{36 \phi (1-\phi)} \right\} \tag{69}$$

Accordingly, the local chemical potential for polymer A is

$$\beta \mu_A(\mathbf{r}) = \frac{1}{N} \ln \frac{\phi}{1-\phi} + \chi_F (1-2\phi) - \frac{l^2}{18} \frac{\nabla^2 \phi}{\phi(1-\phi)} + \frac{l^2 (\nabla \phi)^2}{36} \left[\frac{1}{\phi^2} - \frac{1}{(1-\phi)^2} \right] \tag{70}$$

Substituting Eq. (70) into Eqs. (67) and (68), we can calculate the evolution of the segment density profile during phase transitions. In general, spinodal decomposition in a polymer blend results in a highly interconnected bicontinuous structure at the early stage of phase separation. Spherical structures are possible at relatively late stages. The fully developed morphology is mainly determined by the polymer composition, and less by other parameters such as polymer chain lengths and binary interaction parameters.

Nauman and coworkers applied the FHdG theory to investigate the morphologies of ternary polymer blends after spinodal decomposition [22]. They found that, as shown in Fig. 8, the morphologies of the polymer blends predicted by Eq. (68) closely resemble those obtained from experiments. The theoretical results may help design polymer blends with minimal experimentation.

8 Electronic Systems

The square-gradient approximation (SGA) is independent of the physical significance of the density functional or inter-particle potentials, making it universally applicable to both quantum as well as classical systems. For its application to electronic systems, a quantity of central importance is the exchange-correlation functional, $E_{XC}[\rho(\mathbf{r})]$, which is introduced in the Kohn-Sham (KS) *ansatz* for calculating the energy and the electronic structure at the ground state [23]. In essence, the KS theory asserts the existence of a non-interacting reference system with an effective external potential $v_s(\vec{r})$ such that both its ground-state energy and the density profile are the same as those corresponding to the real system.

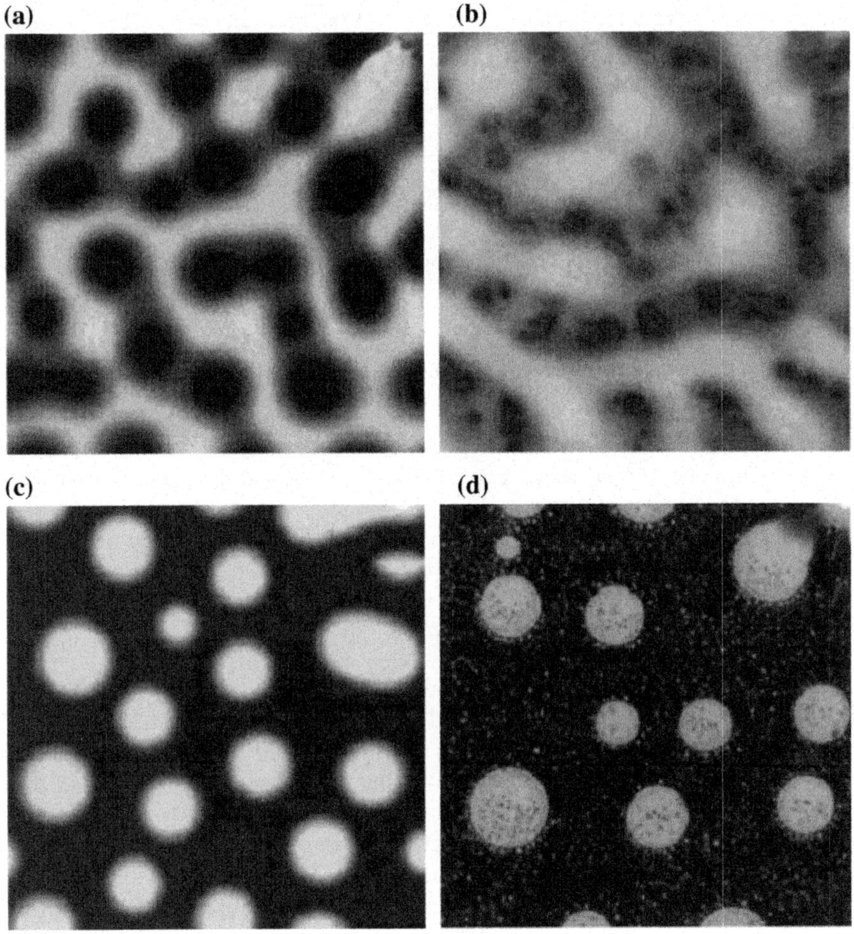

Fig. 8 Morphologies of polymer blends from theoretical predictions (**a**, **c**) and from electron microscopy (**c**, **d**). Here (**a**) and (**b**) are for a 40/40/20 blend of poly (methyl methacrylate) (PMMA), polystyrene (PS) and polybutadiene (PB); (**c**) and (**d**) are for a 34/33/33 blend of PS-PB-polyisoprene (PI). The morphologies shown here are fully developed (long time). (Adapted from Brunswick et al. Journal of Applied Polymer Science (1998))

For a non-interacting electronic system in the presence of a one-body potential $v_s(\vec{r})$, the wave function of electrons can be solved from the KS equation (viz. the single-particle Schrödinger equation)

$$\left[-\frac{\hbar^2\nabla^2}{2m} + v_s(\mathbf{r})\right]\psi_i = \varepsilon_i\psi_i \tag{71}$$

where $\hbar$ denotes the Dirac constant, m is the electron mass, $\varepsilon_i, i = 1, 2, \ldots N$, stands for the ith lowest energy of the non-interacting system, and N is the number of electrons. The one-body electron density is related to the wave function

$$\rho(\mathbf{r}) = \sum_{i=1}^{N} |\psi_i(\mathbf{r})|^2. \tag{72}$$

The exchange-correlation functional is introduced to reproduce the ground-state energy of the real system

$$E[\rho(\mathbf{r})] = T_0[\rho(\mathbf{r})] + \int d\mathbf{r}\rho(\mathbf{r})V^{\text{ext}}(\mathbf{r}) + \frac{1}{2}\int d\mathbf{r}\int d\mathbf{r}'\frac{\rho(\mathbf{r})\rho(\mathbf{r}')}{|\mathbf{r}-\mathbf{r}'|} + E_{XC}[\rho(\mathbf{r})] \tag{73}$$

where T_0 stands for the kinetic energy of non-interacting electrons, and $V^{\text{ext}}(\mathbf{r})$ represents the one-body external potential. In Eq. (73) and thereafter, all physical quantities are given in atomic units.

From the KS wave functions, we can calculate the kinetic energy of the non-interacting system

$$T_0 = -\frac{1}{2}\sum_{i=1}^{N}\int d\mathbf{r}\ \psi_i^*(\mathbf{r})\nabla^2\psi_i(\mathbf{r}). \tag{74}$$

The external energy and the classical electron-electron repulsion energy on the right side of Eq. (73) are directly related to the one-body density profile.

While there is no a priori knowledge on the exchange-correlation energy, E_{XC} $[\rho(\mathbf{r})]$ is clearly a functional of the one-body density as all other terms in Eq. (73) are. Intuitively, we may divide E_{XC} in terms of contributions from the difference between the kinetic energy of the real system and that of the non-interacting electrons, the exchange (Pauli exclusion) effects, and multi-body correlations. Because $\rho(\mathbf{r})$ minimizes the ground-state energy of both the reference and the real systems, the effective one-body potential for the non-interacting reference system, up to a constant of little relevance, can be written as

$$v_s(\mathbf{r}) = V_{ext}(\mathbf{r}) + \int d\mathbf{r}'\frac{\rho(\mathbf{r}')}{|\mathbf{r}-\mathbf{r}'|} + v_{XC}(\mathbf{r}) \tag{75}$$

where

$$v_s(\mathbf{r}) = \frac{\delta E_{XC}}{\delta\rho(\mathbf{r})}. \tag{76}$$

As discussed above (see Eq. (18)), we can derive an exact equation for the grand-state energy of the real system using the non-interacting system as a reference

$$E[\rho(\mathbf{r})] = T_0[\rho(\mathbf{r})] + \int d\mathbf{r}\rho(\mathbf{r})V^{\text{ext}}(\mathbf{r}) + \frac{1}{2}\int_0^1 d\lambda\int d\mathbf{r}\int d\mathbf{r}'\frac{\rho(\mathbf{r})\rho(\mathbf{r}')g(\mathbf{r},\mathbf{r}',\lambda)}{|\mathbf{r}-\mathbf{r}'|} \tag{77}$$

where $g(\mathbf{r}, \mathbf{r}', \lambda)$ represents the pair-correlation function between electrons with the Coulomb interaction between the electrons reduced by a factor of λ. In the electronic DFT literature, Eq. (77) is commonly known as the adiabatic connection.

A comparison of Eqs. (73) and (77) indicates that the exchange-correlation energy corresponds to the indirect energy for "charging up" the non-interacting electrons:

$$E_{XC}[\rho(\mathbf{r})] = \frac{1}{2}\int d\mathbf{r} \int d\mathbf{r}' \frac{\rho(\mathbf{r})\rho(\mathbf{r}')\bar{h}(\mathbf{r}, \mathbf{r}')}{|\mathbf{r} - \mathbf{r}'|} \tag{78}$$

where

$$\bar{h}(\mathbf{r}, \mathbf{r}') \equiv \int_0^1 d\lambda[g(\mathbf{r}, \mathbf{r}', \lambda) - 1] \tag{79}$$

stands for the average hole-correlation function (viz., the average total correlation function). Because the electrostatic energy diverges at zero separation, the hole correlation function is precisely known, $h(\mathbf{r}, \mathbf{r}, \lambda) = -1$. In addition, it must satisfy the normalization conditions because the hole correlation function is defined relative to one electron at position $\mathbf{r}$,

$$\int d\mathbf{r}' h(\mathbf{r}, \mathbf{r}', \lambda) = -1. \tag{80}$$

With an analytical expression for E_{XC} or $\bar{h}(\mathbf{r}, \mathbf{r}')$, we can determine the electron density profile and the ground-state energy from Eqs. (72) and (73), respectively. From these quantities, other ground-state properties of the electronic system can be readily calculated.

As for classical systems discussed above, the KS-DFT does not provide any systematic procedure to determine the exchange-correlation energy of the two-body correlation functions. Since the publication of the KS equation in 1965, tremendous efforts have been devoted to the development of accurate exchange-correlation functionals for inhomogeneous electrons and such efforts are still well ongoing [8]. Existing applications of the KS equation are mostly based on various forms of the generalized gradient approximations (GGA). As in SGA, the GGA functional includes an exchange-correlation energy corresponding to that of a uniform system at the local density and a gradient correction to account for the local density inhomogeneity. For a uniform electron gas, the local exchange-correlation energy as a function of the density is known from quantum Monte Carlo simulations [24]. However, the development of gradient corrections for the exchange and correlation energy is mathematically very complex. Among numerous versions of GGA, PBE [25] and B3LYP [26, 27] are two main functionals broadly used in the KS-DFT calculations for molecular and materials systems, respectively.

We may illustrate the basic procedure to use SGA for electronic systems by considering the kinetic energy of non-interacting electrons at 0 K. According to the Thomas-Fermi theory, the kinetic energy as a functional of the local density is given by

$$T_{TF}[\rho(\mathbf{r})] = \frac{3\left(3\pi^2\right)^{2/3}}{10}\int d\mathbf{r}\rho(\mathbf{r})^{5/3}. \tag{81}$$

The gradient correction to the kinetic energy can be derived again from Eq. (24) using the density-density correlation function of the corresponding uniform system as the input. For non-interacting electrons in the bulk, the density-density correlation function in the Fourier space is known as the Lindhard function [28]

$$\chi_0(q) = -\frac{k_F}{2\pi^2}\left\{1 - \frac{s}{4}\left(1 - \frac{4}{s^2}\right)\ln\left(\left|\frac{s+2}{s-2}\right|\right)\right\} \tag{82}$$

where $k_F = (6\pi^2\rho_0/g)^{1/3}$ is the Fermi momentum, $g = 2$ is the spin degeneracy for an unpolarized electronic system, and $s \equiv q/k_F$. At small q, we have

$$K_0(q) = 1/\chi_0(q) = \frac{2\pi^2}{k_F}\left(\frac{1}{2} + \frac{s^2}{24} + \cdots\right). \tag{83}$$

Therefore, the influence parameter for the non-interacting electrons is

$$\kappa = \frac{1}{3}\lim_{q\to 0}\,[K_0(q) - K_0(0)]/q^2 = \frac{\pi^2}{12k_F^3} = \frac{g}{72\rho_0}. \tag{84}$$

Together with the local kinetic energy for ideal Fermions, we have the kinetic energy functional

$$T_{TFW}[\rho(\mathbf{r})] = \int d\mathbf{r}\left\{\frac{3\left(3\pi^2\right)^{2/3}}{10}\rho(\mathbf{r})^{5/3} + \frac{|\nabla\rho(\mathbf{r})|^2}{72\rho(\mathbf{r})}\right\} \tag{85}$$

where we have replaced ρ_0 with local density $\rho(\mathbf{r})$. Equation (85) corresponds to Weiszacker's correction to the Thomas-Fermi equation for inhomogeneous electrons [29].

A similar but mathematically much more complicated procedure may be applied to derive the gradient correction for the exchange and correlation energies [30]. The mathematical complexity mainly arises from evaluation of the density-density correlation function for real electronic systems. Nevertheless, it is clear that the functional expression from the gradient expansion can be written in terms of a local density contribution and a correction for the density gradient. For example, the PBE exchange energy is given by [25]

$$E_X[\rho(\mathbf{r})] = \int d\mathbf{r}\rho(\mathbf{r})\varepsilon_X F_X(s) \tag{86}$$

where ε_X is the local exchange energy per electron for the uniform system

$$\varepsilon_X = -\frac{3}{4}\left(\frac{3}{\pi}\right)^{1/3} \rho(\mathbf{r})^{1/3} \tag{87}$$

and

$$s = |\nabla\rho(\mathbf{r})|^2/2k_F\rho(\mathbf{r}) \tag{88}$$

is a dimensionless density gradient arising from the gradient expansion. $F_X(s)$ is a semi-empirical function that was formulated to satisfy various asymptotic results for uniform electron gas

$$F_X(s) = 1.804 - \frac{0.804}{1 + 0.2730s^2}. \tag{89}$$

In derivation of Eq. (86), the local term accounts for the exchange energy with the hole correlation function calculated from that of the non-interacting reference system ($\lambda = 0$). The gradient correction is no more quadratic in the density gradient because it represents a resummation of high-order terms in the gradient expansion of the exchange-correlation energy.

9 Summary

This chapter introduces the gradient expansion method, specifically the square-gradient approximation (SGA), as a general scheme to formulate the intrinsic Helmholtz energy of inhomogeneous systems including electrons at 0 K and illustrate its applications for predicting surface tensions, microemulsion structures, and the kinetics of polymer phase transitions. Similar procedures, broadly known as the Landau expansion method, are used to describe a wide range of phase transitions including weak segregation of block copolymers, wetting and drying transitions, and phase diagrams of liquid crystals. Whereas the theoretical procedure has been well established in each individual subfield of physical sciences, little explored is the inter-connection among similarly formulated theoretical methods from different perspectives, not only in terms of mathematical concepts but also the underlying physics principles. Such a connection may be best illustrated in the context of statistical mechanics. In addition to apparent pedagogical values, this work may help to forge a common ground to comprehend fragmented developments in different subfields of statistical mechanics and promote cross-field collaborations.

Acknowledgments The author is indebted to Dr. Liu Yu for comments and suggestions. For the financial support, we are grateful to the U.S. National Science Foundation (NSF-CBET-1404046).

Appendices

Density-Gradient Expansion

Consider an inhomogeneous system with one-particle number density $\rho(\mathbf{r})$. The intrinsic Helmholtz energy can be formally expressed relative to that of a uniform system with density ρ_0 by a functional Taylor expansion with respect to the local density deviation $\Delta\rho(\mathbf{r}) = \rho(\mathbf{r}) - \rho_0$:

$$\begin{aligned} F[\rho(\mathbf{r})] = F_0 &+ \int \left.\frac{\delta F}{\delta\rho(\mathbf{r})}\right|_0 \Delta\rho(\mathbf{r})d\mathbf{r} \\ &+ \frac{1}{2}\int\int \left.\frac{\delta^2 F}{\delta\rho(\mathbf{r})\delta\rho(\mathbf{r}')}\right|_0 \Delta\rho(\mathbf{r})\Delta\rho(\mathbf{r}')d\mathbf{r}d\mathbf{r}' + \cdots \end{aligned} \tag{90}$$

where F_0 is the intrinsic Helmholtz energy of the uniform system at the same temperature, and subscript 0 denotes the uniform reference system. According to Eq. (90), the intrinsic free energy $F[\rho(\mathbf{r})]$ is fully specified by a set of functions

$$K(\mathbf{r}_1, \mathbf{r}_2, \cdots \mathbf{r}_n) = \delta^n F / \prod_{i=1}^{n} \delta\rho(\mathbf{r}_i) \tag{91}$$

where $n = 1, 2, \ldots$.

The one-body density profile satisfies the variational condition, i.e., it minimizes the grand potential Ω

$$\frac{\delta\Omega}{\delta\rho(\mathbf{r})} = 0 \tag{92}$$

The grand potential relates to the intrinsic Helmholtz energy by the Legendre transformation:

$$\Omega[\rho(\mathbf{r})] = F[\rho(\mathbf{r})] + \int \rho(\mathbf{r})u(\mathbf{r})d\mathbf{r} \tag{93}$$

where $u(\mathbf{r}) \equiv \varphi^{\text{ext}}(\mathbf{r}) - \mu$ corresponds to a one-body potential define by the chemical potential and the external potential $\varphi^{\text{ext}}(\mathbf{r})$ of the particles, μ is the chemical potential. Combing Eqs. (92) and (93) leads to

$$\frac{\delta F}{\delta\rho(\mathbf{r})} = -u(\mathbf{r}) \tag{94}$$

For a uniform system, $\varphi^{\text{ext}}(\mathbf{r}) = 0$, Eq. (94) reduces to

$$\left.\frac{\delta F}{\delta\rho(\mathbf{r})}\right|_0 = \mu. \tag{95}$$

The second-order term in the density functional expansion of the intrinsic Helmholtz energy plays a particularly important role in theoretical developments. The vertex function is defined as

$$K(\mathbf{r},\mathbf{r}') \equiv \frac{\delta^2 F}{\delta\rho(\mathbf{r})\delta\rho(\mathbf{r}')} = -\frac{\delta u(\mathbf{r})}{\delta\rho(\mathbf{r}')} \tag{96}$$

It can be shown that the density-density correlation function is related to the functional derivative of the one-body density with respect to the reduced one-body potential

$$\chi(\mathbf{r},\mathbf{r}') \equiv \langle[\hat{\rho}(\mathbf{r})-\rho(\mathbf{r})][\hat{\rho}(\mathbf{r}')-\rho(\mathbf{r}')]\rangle = -\frac{\delta\rho(\mathbf{r})}{\delta\beta u(\mathbf{r}')} \tag{97}$$

where the instantaneous density of the system

$$\hat{\rho}(\mathbf{r}) \equiv \sum_i \delta(\mathbf{r}-\mathbf{r}_i) \tag{98}$$

is expressed as a summation of Dirac delta functions, and $\langle\cdots\rangle$ represents the ensemble average.

From Eqs. (96) and (97), we see that the density-density correlation function corresponds to the inverse functional derivative of the 2nd order coefficient in the functional Taylor expansion of the intrinsic Helmholtz energy

$$\int \beta K(\mathbf{r},\mathbf{r}'')\chi(\mathbf{r}',\mathbf{r}'')d\mathbf{r}'' = \delta(\mathbf{r}-\mathbf{r}''). \tag{99}$$

For a uniform system, both $K_0(\mathbf{r},\mathbf{r}')$ and $\chi_0(\mathbf{r},\mathbf{r}')$ depend only on the distance $|\mathbf{r}-\mathbf{r}'|$. In that case, we may apply the translational and rotational symmetry for the correlation functions:

$$\begin{cases} K(\mathbf{r},\mathbf{r}') = K(0,\mathbf{r}-\mathbf{r}') \equiv K_0(|\mathbf{r}-\mathbf{r}'|) \\ \chi(\mathbf{r},\mathbf{r}') = \chi(0,\mathbf{r}-\mathbf{r}') \equiv \chi_0(|\mathbf{r}-\mathbf{r}'|) \end{cases} \tag{100}$$

Substituting Eq. (100) into (99) gives

$$\int \beta K_0(|\mathbf{r}-\mathbf{r}''|)\chi_0(|\mathbf{r}'-\mathbf{r}''|)d\mathbf{r}'' = \delta(\mathbf{r}-\mathbf{r}'') \tag{101}$$

A 3D Fourier transform of Eq. (101), reveals a simple relationship between $\tilde{K}_0(q)$ and $\tilde{\chi}_0(q)$

$$\beta\tilde{K}_0(q)\tilde{\chi}_0(q) = 1 \tag{102}$$

where

$$\tilde{K}_0(q) = \tilde{K}_0(\mathbf{q}) \equiv \int K_0(r)e^{-i\mathbf{q}\cdot\mathbf{r}}d\mathbf{r} = \frac{4\pi}{q}\int_0^\infty r\sin(qr)K_0(r)dr \tag{103}$$

$$\tilde{\chi}_0(q) = \tilde{\chi}_0(\mathbf{q}) \equiv \int \chi_0(r)e^{-i\mathbf{q}\cdot\mathbf{r}}d\mathbf{r} = \frac{4\pi}{q}\int_0^\infty r\sin(qr)\chi_0(r)dr \tag{104}$$

Applying Eq. (97) to a uniform system of average density ρ_0, we have

$$\begin{aligned}\chi_0(|\mathbf{r}-\mathbf{r}'|) &= \chi(\mathbf{r},\mathbf{r}') = \langle\hat{\rho}(\mathbf{r})\hat{\rho}(\mathbf{r}')\rangle - \rho_0^2\\ &= \langle\hat{\rho}(\mathbf{r})\hat{\rho}(\mathbf{r}')\rangle_{\mathbf{r}\neq\mathbf{r}'} + \langle\hat{\rho}(\mathbf{r})\hat{\rho}(\mathbf{r}')\rangle_{\mathbf{r}=\mathbf{r}'} - \rho_0^2\\ &= \rho^{(2)}(\mathbf{r},\mathbf{r}') + \rho_0\langle\hat{\rho}(\mathbf{r}')\rangle_{\mathbf{r}=\mathbf{r}'} - \rho_0^2\\ &= \rho_0^2 h_0(|\mathbf{r}-\mathbf{r}'|) + \rho_0\delta(\mathbf{r}-\mathbf{r}')\end{aligned} \tag{105}$$

where $\rho^{(2)}(\mathbf{r},\mathbf{r}')$ is the two-body density function, and

$$h_0(|\mathbf{r}-\mathbf{r}'|) \equiv \left.\frac{\rho^{(2)}(\mathbf{r},\mathbf{r}')}{\rho(\mathbf{r})\rho(\mathbf{r}')}\right|_0 - 1 \tag{106}$$

denotes the total correlation function. In the Fourier space, Eq. (105) becomes

$$\tilde{\chi}_0(q) = \rho_0^2\tilde{h}_0(q) + \rho_0 \tag{107}$$

With the help of Eqs. (95), (102) and (107), we can evaluate the intrinsic free energy functional $F[\rho(\mathbf{r})]$ up to the quadratic term. Apparently, the density expansion is applicable not only to the intrinsic Helmholtz energy but also to other quantities with a similar mathematic form, for example, the excess free energy and the exchange-correlation energy.

The Ornstein-Zernike (OZ) Equation

Recalling that the vertex function is inversely related to the density–density correlation function $\chi(\mathbf{r}_1,\mathbf{r}_2)$, which is also related to the total correlation function $h(\mathbf{r}_1,\mathbf{r}_2)$

$$\chi(\mathbf{r}_1,\mathbf{r}_2) = \rho(\mathbf{r}_1)\rho(\mathbf{r}_2)h(\mathbf{r}_1,\mathbf{r}_2) + \rho(\mathbf{r}_1)\delta(\mathbf{r}_1-\mathbf{r}_2). \tag{108}$$

For classical systems, the Helmholtz energy of a non-interacting system is exactly known.

$$F^{ID} = k_B T\int \rho(\mathbf{r})\{\ln[\rho(\mathbf{r})\Lambda^3] - 1\}d\mathbf{r}. \tag{109}$$

Subsequently, we may express the vertex function in terms of the direct correlation function (DCF)

$$\beta K(\mathbf{r}, \mathbf{r}') = \frac{\delta^2 \beta F}{\delta \rho(\mathbf{r}) \delta \rho(\mathbf{r}')} = \frac{\delta(\mathbf{r} - \mathbf{r}')}{\rho(\mathbf{r})} - c(\mathbf{r}, \mathbf{r}') \tag{110}$$

where $c(\mathbf{r}, \mathbf{r}')$ corresponds to the second-order functional derivatives of the excess Helmholtz energy $F^{ex} \equiv F - F^{ID}$

$$c(\mathbf{r}, \mathbf{r}') \equiv -\frac{\delta^2 \beta F^{ex}}{\delta \rho(\mathbf{r}) \delta \rho(\mathbf{r}')}. \tag{111}$$

Because

$$\int d\mathbf{r}_2 \chi(\mathbf{r}_1, \mathbf{r}_2) \beta K(\mathbf{r}_3, \mathbf{r}_2) = \delta(\mathbf{r}_1 - \mathbf{r}_3), \tag{112}$$

substituting Eqs. (110) and (108) into (111) leads to the Ornstein-Zernike (OZ) equation

$$h(\mathbf{r}_1, \mathbf{r}_2) = c(\mathbf{r}_1, \mathbf{r}_2) + \int \rho(\mathbf{r}_3) h(\mathbf{r}_1, \mathbf{r}_3) c(\mathbf{r}_2, \mathbf{r}_3) d\mathbf{r}_3 \tag{113}$$

For uniform systems, the OZ equation can be simplified as

$$h(r) = c(r) + \rho_0 \int h(|\mathbf{r}_1 - \mathbf{r}_3|) c(|\mathbf{r}_2 - \mathbf{r}_3|) d\mathbf{r}_3 \tag{114}$$

or in the Fourier space

$$[1 + \rho_0 h_0(q)][1 - \rho_0 c_0(q)] = 1. \tag{115}$$

Corrections to the Local Density Approximation (LDA)

Local density approximation (LDA) assumes that the free energy density of an inhomogeneous system is the same as that of a uniform system at the local density. According to LDA, the intrinsic Helmholtz energy functional is given by

$$F^{\mathrm{LDA}}[\rho(\mathbf{r})] = \int f_0[\rho(\mathbf{r})] d\mathbf{r} \tag{116}$$

where $f_0 = F_0/V$ corresponds to the intrinsic free energy density (per volume) of a uniform system. LDA ignores the spatial correlation effect.

Because LDA assumes f_0 as a ***function*** of $\rho(\mathbf{r})$, we may express it as a regular Taylor expansion with respect to that of a uniform system

$$f_0[\rho(\mathbf{r})] = f_0(\rho_0) + \frac{\partial f_0}{\partial \rho_0}\Delta\rho(\mathbf{r}) + \frac{1}{2}\frac{\partial^2 f_0}{\partial \rho_0^2}[\Delta\rho(\mathbf{r})]^2 + \cdots \tag{117}$$

Substituting Eq. (117) into (116), we have:

$$F^{\mathrm{LDA}}[\rho(\mathbf{r})] = F_0 + \int \frac{\partial f_0}{\partial \rho_0}\Delta\rho(\mathbf{r})d\mathbf{r} + \frac{1}{2}\int \frac{\partial^2 f_0}{\partial \rho_0^2}[\Delta\rho(\mathbf{r})]^2 d\mathbf{r} + \cdots \tag{118}$$

Comparing Eq. (118) with the functional expansion form, i.e., Eq. (90), we have:

$$\begin{aligned} F[\rho(\mathbf{r})] &= F^{\mathrm{LDA}}[\rho(\mathbf{r})] \\ &+ \frac{1}{2}\int\int [K_0(|\mathbf{r}-\mathbf{r}'|) - \left(\frac{\partial \mu}{\partial \rho_0}\right)_T \delta(\mathbf{r}-\mathbf{r}')]\Delta\rho(\mathbf{r})\Delta\rho(\mathbf{r}')d\mathbf{r}d\mathbf{r}' + \cdots \end{aligned} \tag{119}$$

In writing the above equation, we have used the thermodynamic relation

$$\mu = (\partial f_0/\partial \rho_0)_T \tag{120}$$

and the mathematic identity

$$\int \frac{\partial^2 f_0}{\partial \rho_0^2}[\Delta\rho(\mathbf{r})]^2 d\mathbf{r} = \int \left(\frac{\partial \mu}{\partial \rho_0}\right)_T \Delta\rho(\mathbf{r})\Delta\rho(\mathbf{r}')\delta(\mathbf{r}-\mathbf{r}')d\mathbf{r}d\mathbf{r}' \tag{121}$$

In Eq. (119), the terms after F^{LDA} can be regarded as spatial correlation effects neglected by LDA.

Now let F_2 represent the second term on right side of Eq. (119). Using the Fourier transform, we can express F_2 as

$$F_2 = \frac{1}{2(2\pi)^3}\int [\tilde{K}_0(q) - \left(\frac{\partial \mu}{\partial \rho_0}\right)_T][\Delta\tilde{\rho}(\mathbf{q})]^2 d\mathbf{q} \tag{122}$$

According to Eqs. (102) and (107), we have

$$\beta\tilde{K}_0(q) = \frac{1}{\rho_0^2\tilde{h}_0(q) + \rho_0} \tag{123}$$

In addition, the compressibility equation gives

$$\left(\frac{\partial \rho_0}{\partial \beta\mu}\right)_T = \rho_0 + \rho_0^2\int h_0(r)d\mathbf{r} = \rho_0 + \rho_0^2\tilde{h}_0(q=0) = \frac{1}{\beta\tilde{K}_0(0)}. \tag{124}$$

Accordingly, Eq. (122) can be rewritten in a more compact form:

$$F_2 = \frac{1}{16\pi^3}\int [\tilde{K}_0(q) - \tilde{K}_0(0)][\Delta\tilde{\rho}(\mathbf{q})]^2 d\mathbf{q} \tag{125}$$

With Eq. (125), we formulate the additional correlation term beyond LDA. Similar to the functional expansion, such procedure can be extended to other quantities.

To connect Eq. (125) with the square-gradient expansion, we recall that

$$\tilde{K}_0(q) = \tilde{K}_0(\mathbf{q}) \equiv \int K_0(r)e^{-i\mathbf{q}\cdot\mathbf{r}}\mathrm{d}\mathbf{r} = \int \frac{\sin(qr)}{qr}K_0(r)\mathrm{d}\mathbf{r} \tag{126}$$

Using the Taylor series

$$\frac{\sin(qr)}{qr} = 1 - \frac{(qr)^2}{3!} + \cdots \tag{127}$$

we have

$$\begin{aligned}\tilde{K}_0(q) &= \int \left[1 - \tfrac{(qr)^2}{3!}\right] K_0(r)\mathrm{d}\mathbf{r} \\ &= \tilde{K}_0(0) - \tfrac{q^2}{3!} \int r^2 K_0(r)\mathrm{d}\mathbf{r}\end{aligned} \tag{128}$$

Substituting Eq. (128) into (127) gives

$$F_2 = -\frac{\kappa}{16\pi^3}\int q^2[\Delta\tilde{\rho}(\mathbf{q})]^2\mathrm{d}\mathbf{q} \tag{129}$$

where κ is the influence parameter defined as

$$\kappa \equiv -\frac{1}{3!}\int r^2 K_0(r)\mathrm{d}\mathbf{r} = \frac{1}{3}\lim_{q\to 0}[K_0(q) - K_0(0)]/q^2 \tag{130}$$

Note

$$\int \nabla\rho(\mathbf{r})e^{i\mathbf{q}\cdot\mathbf{r}}d\mathbf{r} = -i\mathbf{q}\Delta\tilde{\rho}(\mathbf{q}) \tag{131}$$

and

$$\frac{1}{(2\pi)^3}\int q^2[\Delta\tilde{\rho}(\mathbf{q})]^2\mathrm{d}\mathbf{q} = \int \mathrm{d}\mathbf{r}_1 \int d\mathbf{r}_2\nabla\rho(\mathbf{r}_1)\nabla\rho(\mathbf{r}_2)\delta(\mathbf{r}_1 - \mathbf{r}_2) \tag{132}$$

we arrive the square-gradient correction to the LDA

$$F_2 = \frac{\kappa}{2}\int \mathrm{d}\mathbf{r}[\nabla\rho(\mathbf{r})]^2. \tag{133}$$

In some applications, we use the static structure factor $\tilde{S}_0(q) = \tilde{\chi}_0(q)/\rho_0$ instead of the vertex function. In that case,

$$\tilde{K}_0(q) = [\beta\rho_0\tilde{S}_0(q)]^{-1}. \tag{134}$$

thus the influence parameter is given by

$$\kappa = -\frac{1}{3!}\int r^2 K_0(r)d\mathbf{r} = \frac{1}{3\beta\rho_0}\lim_{q\to 0}[\tilde{S}_0^{-1}(q) - \tilde{S}_0^{-1}(0)]/q^2. \quad (135)$$

Intra-Chain Correlation Function of a Gaussian Chain

In a polymer blend A and B, the intra-chain correlation for polymer A as an ideal chain is given by

$$\chi_{0,AA}(|\mathbf{r}-\mathbf{r}'|) = \frac{1}{v_0^2} < \delta\hat{\phi}_A(\mathbf{r})\delta\hat{\phi}_A(\mathbf{r}') > = \frac{\phi_{0,A}}{v_0^2} P_A(|\mathbf{r}-\mathbf{r}'|) \quad (136)$$

where P_A represents the probability to find a segment at position $\mathbf{r}$ given that another segment from the same polymer chain is located at $\mathbf{r}'$. A similar expression can be written for polymer B.

For a non-interacting polymer, P_A corresponds to a Gaussian average of all segment pairs separated by distance r

$$P_A(r) = \frac{1}{N_A V}\sum_{i\neq j}^{N_A}\int d\mathbf{r}_i \int d\mathbf{r}_j p_{ij}(\mathbf{r})\delta[\mathbf{r} - (\mathbf{r}_i - \mathbf{r}_j)] \quad (137)$$

where $\mathbf{r}_i$ and $\mathbf{r}_j$ represent the position of segment i and j from the same polymer chain, respectively. In Eq. (137), the Gaussian distribution function is given by the random walk model [20]

$$p_{ij}(\mathbf{r}) = \left(\frac{3}{2\pi|i-j|l_A^2}\right)^{3/2} \exp\left(-\frac{3r^2}{2|i-j|l_A^2}\right) \quad (138)$$

where l_A stands for step length or the polymer bond length. Applying the 3-D Fourier transform to both side of Eq. (136) yields

$$\tilde{P}_A(q) = \frac{1}{N_A}\sum_{i\neq j}^{N_A}\exp\left(-\frac{q^2|i-j|l_A^2}{6}\right) \quad (139)$$

For a long polymer chain, $N_A \gg 1$, the double summations in Eq. (139) can be replaced by integrations

$$\begin{aligned}\tilde{P}_A(q) &= \tfrac{1}{N_A}\int_0^{N_A} dx \int_0^{N_A} dy \exp\left(-\tfrac{q^2|x-y|l_A^2}{6}\right) \\ &= N_A D(qR_A)\end{aligned} \quad (140)$$

where $R_A \equiv N_A l_A^2/6$ is the radius of gyration for an ideal polymer chain (Gaussian chain), and

$$D(x) = \frac{2}{x^4}\left(e^{-x^2} + x^2 - 1\right) \tag{141}$$

is known as the Debye function. For small x, $D(x) \approx 1 - x^2/3$, we can derive from Eq. (136) the intra-chain correlation in the Fourier space

$$\tilde{\chi}_{0,AA}(q) = \frac{\phi_{0,A}}{v_0^2}\tilde{P}_A(q) \approx \frac{N_A\phi_{0,A}}{v_0^2}\left(1 - \frac{q^2R_A^2}{3}\right). \tag{142}$$

References

1. D. Chandler, *Introduction to Modern Statistical Mechanics* (Oxford University Press, New York, 1987) p. xiii, 274 p
2. R. Feynman, *Statistical Mechanics: A Set of Lectures* 2 edition ed. (Westview Press, 1998)
3. D.P. Landau, K. Binder, *A guide to Monte Carlo Simulations in Statistical Physics*, 3rd ed. (Cambridge University Press, Cambridge, New York, 2009), p. xv, 471 p
4. J.S. Rowlinson, Development of theories of inhomogeneous fluids, in *Fundamentals of Inhomogeneous Fluids*, ed. by D. Henderson (Marcel Dekker, New York, 1992), pp. 1–22
5. P. Hohenberg, W. Kohn, Phys. Rev. **136**, B864–B871 (1964)
6. N.D. Mermin, Phys. Rev. **137**(5A), A1441–1443 (1965)
7. R. Evans, Adv. Phys. **28**(2), 143–200 (1979)
8. A.D. Becke, J. Chem. Phys. (USA) **140**(18) (2014)
9. J.Z. Wu, Z.D. Li, Annu. Rev. Phys. Chem. **58**, 85–112 (2007)
10. R. Evans, Density functionals in the theory of nonuniform fluids, in *Fundamentals of Inhomogeneous Fluids*, ed. by D. Henderson (Marcel Dekker, New York, 1992), pp. 85–175
11. J.Z. Wu, Density functional theory for liquid structure and thermodynamics. Struct. Bond **131**, 1–73 (2009)
12. P.M. Chaikin, T.C. Lubensky, *Principles of Condensed Matter Physics* (Cambridge University Press, Cambridge, New York, NY, USA, 1995) p. xx, 699 p
13. K. Lum, D. Chandler, J.D. Weeks, J. Phys. Chem. B **103**, 4570–4577 (1999)
14. J.M. Prausnitz, R.N. Lichtenthaler, E.G.D. Azevedo, *Molecular Thermodynamics of Fluid-Phase Equilibria*, 3rd edn. (Upper Saddle River, N.J, Prentice-Hall PTR, 1999)
15. Y. Tang, J. Chem. Phys. (USA) **127**(16) (2007)
16. J.S. Rowlinson, B. Widom, *Molecular Theory of Capillarity*, Dover ed. (Dover Publications, Mineola, N.Y., 2002) p. xi, 327 p
17. J.M. Garrido, M. Piñeiro, F. Blas, E. Muller, A. Mejia, AIChE J. (2016) In press
18. C. Solans, H. Kunieda, *Industrial Applications of Microemulsions* (M. Dekker, New York, 1997) p. ix, 404 p
19. G. Kaur, L. Chiappisi, S. Prevost, R. Schweins, M. Gradzielski, S.K. Mehta, Langmuir **28**(29), 10640–10652 (2012)
20. P.G. de Gennes, *Scaling Concepts in Polymer Physics* (Cornell University Press, Ithaca, N.Y., 1979)
21. C.P. Brangwynne, P. Tompa, R.V. Pappu, Nat. Phys. **11**(11), 899–904 (2015)
22. A. Brunswick, T.J. Cavanaugh, D. Mathur, A.P. Russo, E.B. Nauman, J. Appl. Polym. Sci. **68**(2), 339–343 (1998)
23. W. Kohn, L.J. Sham, Phys. Rev. **140**(4A), A1133–A1138 (1965)
24. D.M. Ceperley, B.J. Alder, Phys. Rev. Lett. **45**(7), 566–569 (1980)

25. J.P. Perdew, K. Burke, M. Ernzerhof, Phys. Rev. Lett. **77**(18), 3865–3868 (1996)
26. A.D. Becke, J. Chem. Phys. (USA) **98**(7), 5648–5652 (1993)
27. C.T. Lee, W.T. Yang, R.G. Parr, Phys. Rev. B **37**(2), 785–789 (1988)
28. G. Giuliani, G. Vignale, *Quantum Theory of the Electron Liquid* (Cambridge University Press, Cambridge, UK, New York, 2005) p. xix, 777 p
29. R.G. Parr, W. Yang, *Density-Functional Theory of Atoms and Molecules* (Oxford University Press, New York, 1989)
30. E. Engel, *Dreizler* (An Advanced Course, R.M., Density Functional Theory, 2011)

Classical Density Functional Theory for Molecular Systems

Jianzhong Wu

1 Molecular Models and Force Fields

A molecular system consists of a large number of atoms stoichiometrically linked with covalent bonds. While a precise of definition of chemical bonds is not an easy task from a fundamental perspective, atoms interact with each other through electrons and nuclei following quantum-mechanical (QM) descriptions. Because electrons have a negligible mass in comparison to nuclear particles,[1] they reach equilibrium distribution almost instantaneously at the time scale relevant to the dynamics of a molecular system. The discrepancy in electronic and nuclear relaxation times allows for the cascade consideration of electronic and atomic motions, as assumed by the *adiabatic* or the Born–Oppenheimer approximation. At any moment, the force acting upon each nuclear particle can be determined by the positions of other nuclear particles in the system and by the local electron density, $\rho(\mathbf{r})$,

$$\mathbf{F}_i(\mathbf{r}_i) \equiv -\nabla_{\mathbf{r}_i} V(\mathbf{r}_1, \ldots, \mathbf{r}_N) = -Z_i \int d\mathbf{r} \frac{\rho(\mathbf{r})e^2}{4\pi\varepsilon_0|\mathbf{r}-\mathbf{r}_i|} + \sum_{j \neq i} \frac{Z_i Z_j e^2}{4\pi\varepsilon_0|\mathbf{r}-\mathbf{r}_i|} \quad (1)$$

where $\nabla_{\mathbf{r}_i}$ represents the gradient with respect to the nuclear position $\mathbf{r}_i = (x_i, y_i, z_i)$ of the ith atom, $V(\mathbf{r}_1, \ldots, \mathbf{r}_N)$ stands for the total potential energy, Z_i is the nuclear valence, e is the unit charge, and ε_0 is the permittivity in free space. Intuitively we may understand atomic motion in terms of the positions of nuclear particles. According to Eq. (1), atoms are moving in an effective potential or *a force field* arising from the inhomogeneous distribution of electrons.

[1]The electron mass is 1/1836 of that for proton, the lightest nuclear particle. As a result, most of the atomic mass is concentrated in the nucleus.

J. Wu (✉)
Department of Chemical and Environmental Engineering and Department of Mathematics, University of California, Riverside, CA 92521, USA
e-mail: jwu@engr.ucr.edu

J. Wu (ed.), *Variational Methods in Molecular Modeling*,
Molecular Modeling and Simulation, DOI 10.1007/978-981-10-2502-0_3

Thanks to recent advances in high-performance computing and quantum chemistry [1], we can in principle predict the bonding energy and the physicochemical properties of many molecular systems based on QM calculations. Regrettably, QM methods are computationally very demanding and thus are inconvenient for high throughput applications, in particular to practical systems containing a large number of molecules. To predict the phase behavior and thermodynamic properties of molecular systems over a wide range of conditions, we mostly rely on semi-empirical models to describe bond connectivity and inter-atomic interactions [2, 3].

In QM/MM (quantum mechanics/molecular mechanics) simulations [4], the electron density is predicted from an electronic theory (e.g., the Kohn-Sham density functional theory [5], KS-DFT), and the dynamics of nuclear particles is determined from Newton's equations. In conventional molecular dynamics (MD) and Monte Carlo (MC) simulations [6], the force applied to each atom is calculated from a *semi-empirical* force field without invoking the explicit properties of electrons. Most MD and MC simulations are based on molecular models that represent atoms as classical particles linked together with mechanical bonds. A semi-empirical force field, rather than explicit electronic structure calculations, is often used in MD and MC simulations because the atomic forces are repeatedly calculated in molecular simulations. To further minimize the computational costs, we adopt simple mathematical functions to represent both intra- and inter- molecular potentials.

The earliest developments of force fields were reported in the 1960s and 70s [6]. Over the years, a number of successful force fields have been established [7]. For systems where nuclear quantum effects are negligible, a semi-empirical force field enables fast computation of atomic motions and molecular dynamics. For example, DREIDING [8] and UFF [9] cover essentially all elements in the periodic table and achieve a good quality of prediction for material structures and gas adsorption [10–12]. The classical models are convenient for practical applications owing to their simplicity in terms of both the physical concepts and computational efficiency [13].

In a typical Class I force field,[2] the intramolecular bonding potential, V_B^0, is formulated in terms of harmonic and trigonometric functions:

$$
\begin{aligned}
V_B^0 = \sum_{bond} K_b(b-b_o)^2 + \sum_{angle} K_\theta(\theta-\theta_o)^2 + \\
\sum_{dihedral} \frac{K_\phi}{2}[1+\cos(n\phi-\phi_o)] + \sum_{impr} K_\vartheta(\vartheta-\vartheta_o)^2
\end{aligned}
\quad . \tag{2}
$$

As shown schematically in Fig. 1, here b, θ, ϕ and ϑ stand for bond length, bond angle, dihedral (or torsional) angle, and improper angle, respectively; subscript "0" stands for equilibrium values, and K_b, K_θ, K_ϕ, and K_ϑ are energy parameters to account for the flexibility of the molecular configuration. The first two terms on the right side of Eq. (2) represent deformation energies arising from deviation of bond length

[2] Class II and III force fields contain cubic and/or quartic terms in the potential energy for bond lengths and angles.

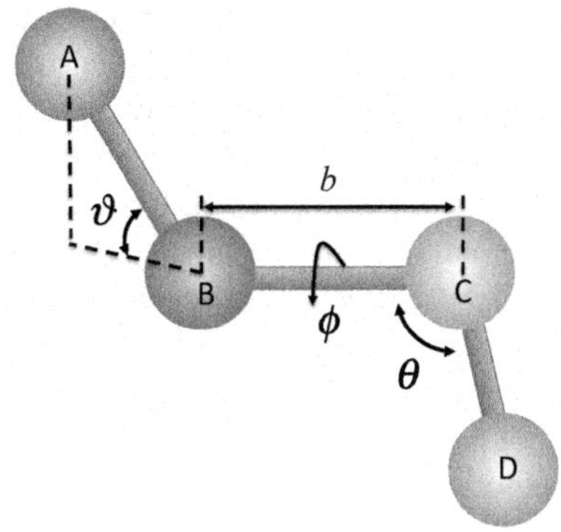

Fig. 1 The configuration of a polyatomic molecule can be described in terms of bond length b, bond angle θ, dihedral (torsional) angle ϕ, and improper angle ϑ. The improper angle is defined by the project of a bond vector (here AB) into a plane defined by 3 consecutive atoms (BCD)

b and bond angle θ from their respective equilibrium values. The cosine function reflects the potential related to the rotation of chemical bonds, with the periodicity determined by integer n. The harmonic potential related to the "improper" angle, i.e., an out-of-plane angle arising from 4 consecutive nearest-neighboring atoms, is introduced to reproduce the energetics due to the out-of-plane motions. The out-of-plane term can also be used to enforce a given molecular geometry or chirality. The simple harmonic and trigonometric forms used in Eq. (2) represent the correct molecular structure but are unable describe bond formation or breaking. To describe chemical reactions, one needs to account for the electronic degree of freedom by carrying out QM calculations or to use a reactive force field (ReaxFF) [14].

In addition to the bond potential, a molecular model specifies intra- and inter-molecular non-bonded interactions, i.e., interactions between atoms that are not directly connected by chemical bonds. With the assumption of pairwise additivity, the non-bonded interactions are commonly represented in terms of the Lennard-Jones (LJ) plus the Coulomb potentials:

$$u_{ij}(r) = 4\varepsilon_{ij}\left[\left(\frac{\sigma_{ij}}{r}\right)^{12} - \left(\frac{\sigma_{ij}}{r}\right)^{6}\right] + \frac{Z_i Z_j e^2}{4\pi\varepsilon_0 r} \tag{3}$$

where r represents the center-to-center distance between atoms i and j, $\varepsilon_{ij} = \sqrt{\varepsilon_{ii}\varepsilon_{jj}}$ and $\sigma_{ij} = (\sigma_{ii} + \sigma_{jj})/2$ are obtained from the Lorentz-Berthelot combining rules, ε_{ii} and σ_{ii} are the energy and size parameters of the LJ potential (viz., van der Waals parameters) for the ith atom, and Z_i denotes the atomic (fractional) valence. While we use Z_i to represent the valences of nuclear particles and individual atoms, it should be understood that the atomic partial charge is in general different from the nuclear charge.

The parameters affiliated with the bond length and angles are closely related to the molecular structure, which is usually determined by energy minimization for the configuration of individual molecules in vacuum. First principles methods such as various versions of KS-DFT [15] are often used to predict the optimized molecular structure and deformation energies introduced by small deviations of atomic positions from the equilibrium configuration. The partial charges of individual atoms can also be determined from QM calculations for the electronic density corresponding to the optimized molecular structure. Although conventional KS-DFT calculations are not suitable for van der Waals interactions, [16] recent advances in computational chemistry make it possible to generate accurate molecular structure and non-bonded interactions from high-level QM methods (MP2/cc-PVTZ) [17, 18]. Alternatively, the LJ parameters for each atom, ε_{ii} and σ_{ii}, can be obtained by correlation of simulation results with the thermodynamic properties of macroscopic systems (e.g., liquid density, heat of evaporation and bulk phase diagrams). With the rapid advances in first principles calculations, force field development will become less dependent on experimental input.

A conventional force field does not account for polarization effects, which arise from fluctuations of electron distributions. Such effects can be taken into account with common schemes such as fluctuating charges, Drude oscillators, or induced dipole models [19]. In general, a polarizable force field provides more accurate predictions of molecular structure and thermodynamic properties in comparison with experimental data. Besides, explicit consideration of the polarizability often leads to better transferability of model parameters, i.e., atomic parameters obtained from one set of molecules can be applied to another set without further fitting. While a polarizable model allows for a better reproduction of subtle physical phenomena, it also comes with an important disadvantage: Introduction of atomic polarizability makes the potential energy calculation significantly more time-consuming and complicated owing to iteration procedures involved in induced dipole calculations.

Although a semi-empirical force field provides a convenient starting point to investigate the thermodynamic properties of molecular systems, we should bear in mind that the molecular models entail drastic approximations. For example, it has been well documented that the pairwise additivity assumption is semi-quantitative to represent non-bonded interactions. Theoretical predictions are sensitive to the assignment of atomic partial charges based on the electronic density profiles. Besides, the correlated fluctuation of electron densities plays an important role in intermolecular interactions. From a practical perspective, systematic evaluation of the model parameters is rather challenging because they must reflect both the properties of individual molecules in vacuum and experimental data corresponding to macroscopic systems.

The force-field methods can be extended to systems containing polymers or macromolecules including biomolecular species such as proteins and nucleic acids. For macromolecular systems, the potential energy is often represented within the framework of coarse-grained or united atom models, i.e., the interaction sites are affiliated with monomeric units, functional groups, or 'united atoms'. Reducing the number of particles by use of a coarse-grained model is appealing in particular for molecular simulations. In a united atom model, we do not consider nonpolar hydrogen atoms explicitly; they are instead incorporated into the Lennard-Jones parameters of the parent atoms. In a coarse-grained model, we go one step further to represent whole chemical groups, individual monomers, or even a set of monomers as a single particle. Coarse-grained models are especially popular for studying large polymeric systems or phenomena that happen at large spatiotemporal scales.

Like atomic models, a coarse-grained or united atom model often uses the assumptions of pairwise additivity for non-bonded interactions and fixed partial charges for individual particles. While there have been rapid developments in multiscale modeling, much work is still needed to establish a seamless connection of coarse-grained models with atomic details [20]. In practical applications, coarse-grained parameters are correlated with experimental data for the thermodynamic properties of direct interest.

2 Statistical Mechanics for Polyatomic Systems

The basic concepts and key equations for the statistical mechanics of polyatomic systems are readily available from textbooks [21–23] and outstanding reviews [24, 25]. In the following, we introduce only the terminology essential for a self-contained introduction of the classical density functional theory for molecular systems. While our discussion is focused on one-component molecular systems, similar definitions can be directly extended to mixtures.

2.1 Molecular Configuration and Interaction Sites

As discussed above, in a conventional model of molecular systems, the total potential energy is defined in terms of interactions sites, i.e., atoms, coarse-grained particles, or localized partial charges. The latter represents a factitious position within a molecule where the local electronic charge does not coincide with the center of mass (COM) for a particular atom or coarse-grained particle. As electrons are not explicitly considered in a classical model, the use of the acentric partial charge helps better representation of the Coulomb interactions.

As shown in Fig. 2, the difference between molecular configuration and interaction sites may be illustrated with three representative examples. For each molecule, we define the molecular configuration in terms of the positions of individual atoms

$$\mathbf{x}_c \equiv (\mathbf{r}_1, \mathbf{r}_2, \ldots, \mathbf{r}_M) \tag{4}$$

where $\mathbf{r}_i$ represents the COM position for atom i, and M stands for the number of atoms per molecule. A similar composite vector may be used to specify the positions of interaction sites

$$\mathbf{x} \equiv (\mathbf{r}_1, \mathbf{r}_2, \ldots, \mathbf{r}_{M_s}). \tag{5}$$

In general, the number of atoms M is not necessarily the same as the number of the interaction sites M_s (e.g., $M = 3$ and $M_s = 4$ in the TIP4P model for water molecules

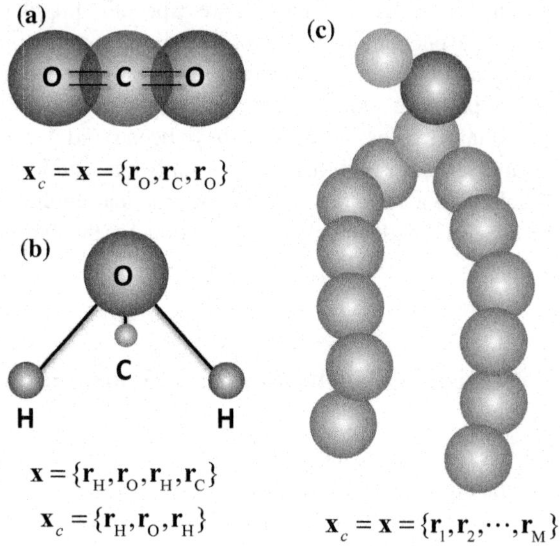

Fig. 2 Molecular configuration $\mathbf{x}_c$ and the coordinates of interacting sites $\mathbf{x}$ according to a typical molecular model. **a** carbon dioxide, **b** water, and **c** a lipid molecule. While the interaction sites coincide with the atomic/particular center of mass, a popular water model consists of four interaction sites: three of them coincide with the oxygen and hydrogen atoms, and the fourth site ('C') represents a localized charge due to the acentric distribution of electrons around the oxygen atom

[26]). Nevertheless, a conventional molecular model asserts that the positions of all interaction sites can be fully determined by the molecular configuration. In other words, the degree of freedom for the molecular configuration is the same as that for the interaction sites, even though the two vectors may have different dimensionalities. Another difference between atoms and interactions sites is that the latter may not be affiliated with any kinetic energy.

For an inhomogeneous system containing N polyatomic molecules, the total energy E is typically divided into a kinetic energy due to atomic motions, potential energies for intra- and inter- molecular interactions, and external potentials:

$$E = K + \sum_{i=1}^{N} V_B(\mathbf{x}_i) + \sum_{i=1}^{N}\sum_{j>i}^{N} \Gamma(\mathbf{x}_i, \mathbf{x}_j) + \sum_{i=1}^{N} V_{ext}(\mathbf{x}_i) \qquad (6)$$

where K stands for the total kinetic energy, $\Gamma(\mathbf{x}_i, \mathbf{x}_j)$ represents the potential for non-bonded interactions between molecules i and j, $V_B(\mathbf{x}_i)$ and $V_{ext}(\mathbf{x}_i)$ are the intramolecular and the external potentials for molecule i, respectively. For convenience, here the intramolecular potential $V_B(\mathbf{x})$ includes contributions from *both* bonding and non-bonded interactions, i.e., $V_B(\mathbf{x}) \neq V_B^0(\mathbf{x})$. Note that, in Eq. (6), the division of the total energy into kinetic and potential contributions is meaningful only in the context of a classical model. Besides, the kinetic energy affiliated with a typical molecular model is different from that with the idea-gas models introduced in standard statistical mechanics texts. Here the kinetic energy depends only on the mass and momentum of individual atoms. In other words, the total kinetic energy of a polyatomic system is approximated by that of the corresponding atomic system at the same temperature but free of both intra- and inter- molecular interactions.[3] The simplistic representation of the kinetic energy has little consequence in molecular modeling because we are mainly concerned with configurational properties, i.e., those properties related to atomic distributions.

With the pairwise-additive approximation, we can calculate the intermolecular potential $\Gamma(\mathbf{x}_i, \mathbf{x}_j)$ from the site-site LJ and Coulomb energies, as given by Eq. (3). Similarly, the external potential for each molecule can be expressed in terms of the site energies

$$V_{ext}(\mathbf{x}) = \sum_{k=1}^{M_s} \varphi_k(\mathbf{r}_k) \tag{7}$$

where $\varphi_k(\mathbf{r}_k)$ stands for the external potential for an interaction site k at position $\mathbf{r}_k$. Note that both the intramolecular and the external potentials depend on molecular configurations.

2.2 Grand Partition Function

For a one-component system consisting of polyatomic molecules at absolute temperature T, volume V and chemical potential μ, the grand partition function is given by

$$\Xi = \sum_{N=0}^{\infty} \frac{1}{N!\nu^N} \int d\Pi^N \exp\{-\beta E + \beta N\mu\} \tag{8}$$

where ν denotes the symmetry number of a polyatomic molecule, $\beta^{-1} = k_B T$, k_B is the Boltzmann constant, and $d\Pi^N$ stands for a dimensionless differential volume in the phase space, i.e., the differential volume corresponding to the positions and momenta of all atoms normalized by h^{3NM}, h is Planck's constant. In Eq. (8), symmetry number ν accounts for atomic indistinguishability within each polyatomic molecule; it is related to the number of different but equivalent arrangements of atoms that can be obtained by rotating the molecule.

A semi-empirical force field typically assumes that the degrees of freedom for the kinetic energy are independent. Therefore, we may integrate analytically the variables in Eq. (8) related to the momenta of individual atoms

$$\Xi = \sum_{N=0}^{\infty} \frac{1}{N!(\nu \prod_\alpha \Lambda_\alpha^3)^N} \int d\mathbf{x}_c^N \exp\{-\beta[\sum_{i=1}^{N} V_B(\mathbf{x}_i) + \sum_{i=1}^{N}\sum_{j>i}^{N} \Gamma(\mathbf{x}_i, \mathbf{x}_j) + \sum_{i=1}^{N} V_{ext}(\mathbf{x}_i) - N\mu]\} \tag{9}$$

[3]While quantum mechanics is used in a conventional statistical-mechanical model of polyatomic ideal gases to describe atomic motions, bond stretching and vibrations, a semi-empirical force field assumes that atoms are classical particles moving with the constraints of intramolecular potentials. As a result, a semi-empirical force field is not able to describe the ideal-gas heat capacity.

where $\Lambda_\alpha = h/\sqrt{2\pi m_\alpha k_B T}$ denotes the thermal wavelength for atom α, m_α represents the atomic mass, and $\mathbf{x}_c^N$ stands for the configurations of all molecules in the thermodynamic system. Formally, Eq. (9) resembles the grand partition function of an atomic system. The main differences are that the three-dimensional vector $\mathbf{r}$ for the position of each atom is replaced with composite vector $\mathbf{x}_c$ for molecular configuration, and that the atomic thermal wavelength is now replaced with an effective molecular thermal wavelength $\Lambda \equiv (\nu \prod_\alpha \Lambda_\alpha^3)^{1/3M}$. Whereas the numerical complexity increases quickly with the configurational dimensionality, the similarity in the mathematical forms is conceptually useful to extend formal statistical-mechanical equations from monatomic to polyatomic systems.

2.3 Molecular Density and Molecular Correlation Functions

For a one-component system, the multi-body configurational density profile is formally defined as

$$\rho(\mathbf{x}) \equiv < \sum_i \delta(\mathbf{x} - \mathbf{x}_i) > = \frac{\delta \Omega}{\delta V_{ext}(\mathbf{x})} \tag{10}$$

where $\delta(\mathbf{x})$ stands for a multi-dimensional Dirac function, $< \cdots >$ represents the ensemble average, and $\Omega \equiv -k_B T \ln \Xi$ is the grand potential. Equation (10) indicates that the configurational density corresponds to a linear response of the grand potential with respect to the one-body external potential. As $\mathbf{x}$ is a multi-dimensional vector, numerical evaluation of the configurational density is a formidable task. The computational cost rapidly grows as the number of interaction sites increases. Owing to intramolecular interactions, the configurational density profile is inhomogeneous even in a uniform system absent of the external potential.

In a uniform polyatomic system, we may specify the molecular configuration in terms of the atomic positions relative to the molecular center of mass, $\{\mathbf{r}_\alpha^{(0)}\}$,

$$\varpi \equiv [\mathbf{r}_1^{(0)}, \mathbf{r}_2^{(0)}, \ldots, \mathbf{r}_M^{(0)}]. \tag{11}$$

Note that the degree of freedom for ϖ is $M - 1$ because the molecular center of mass is already fixed. The configurational density is thus given by

$$\rho_0(\mathbf{x}) = \rho_0 \omega_0(\varpi) = < \sum_i \delta(\varpi - \varpi_i) > \tag{12}$$

where $\rho_0 = <N>/V$ is the average number density for the molecules, $\omega_0(\varpi)$ represents the intramolecular correlation function for the uniform system, and ϖ_i stands for the configuration of molecule i.

A second-order functional derivative of the grand potential with respective to the one-body potential gives the molecular density-density correlation function

$$\chi(\mathbf{x}, \mathbf{x}') \equiv -\beta^{-1} \frac{\delta\rho(\mathbf{x})}{\delta\varphi(\mathbf{x}')} =< [\sum_i \delta(\mathbf{x} - \mathbf{x}_i) - \rho(\mathbf{x})][\sum_j \delta(\mathbf{x}' - \mathbf{x}_j) - \rho(\mathbf{x}')]>. \tag{13}$$

Physically, $\chi(\mathbf{x}, \mathbf{x}')$ reflects the correlated fluctuations of configurational densities at generalized positions $\mathbf{x}$ and $\mathbf{x}'$. In dimensionless units, $\chi(\mathbf{x}, \mathbf{x}')$ is related to the total correlation function

$$h(\mathbf{x}, \mathbf{x}') = \chi(\mathbf{x}, \mathbf{x}')/\rho(\mathbf{x})\rho(\mathbf{x}') - \delta(\mathbf{x} - \mathbf{x}')/\rho(\mathbf{x}). \tag{14}$$

In Eq. (14), the first term of the right side accounts for two-body correlations, and the second term arises from self-correlation. Because $\chi(\mathbf{x}, \mathbf{x}')$ vanishes when $\mathbf{x}$ and $\mathbf{x}'$ are far apart, the density distributions are uncorrelated at large distance.

Rearranging the ensemble average on the right side of Eq. (13) yields

$$\chi(\mathbf{x}, \mathbf{x}') = \rho(\mathbf{x})\delta(\mathbf{x} - \mathbf{x}') + \rho^{(2)}(\mathbf{x}, \mathbf{x}') - \rho(\mathbf{x})\rho(\mathbf{x}'), \tag{15}$$

where the first term of the right accounts for self-correlation, and the two-body configurational density profile is defined as

$$\rho^{(2)}(\mathbf{x}, \mathbf{x}') \equiv< \sum_i \sum_{j\neq i} \delta(\mathbf{x} - \mathbf{x}_i)\delta(\mathbf{x}' - \mathbf{x}_j) > \tag{16}$$

The two-body configurational density describes the joint probability of finding a molecule with configuration $\mathbf{x}$ and *another* molecule with $\mathbf{x}'$; it is related to the response of the grand potential with respect to the intermolecular pair potential

$$\rho^{(2)}(\mathbf{x}, \mathbf{x}') = \frac{\delta\Omega}{\delta\Gamma(\mathbf{x}, \mathbf{x}')/2} \tag{17}$$

where the factor of 2 accounts for the two–body nature of intermolecular interactions. As discussed later, the correlation functions are useful to formulate non-mean-field approximations in the classical density functional theory.

In the absence of bonding potential $V_B^0(\mathbf{x})$, the grand partition function would reduce to that corresponding to a monatomic system. As a result, the properties of a polyatomic system can be considered as a "perturbation" from those corresponding to the unbounded monatomic mixture. A functional derivative of the grand potential to the exponential of the reduced bonding energy $e_B(\mathbf{x}) \equiv \exp[-\beta V_B^0(\mathbf{x})]$ yields

$$\frac{\delta\beta\Omega}{\delta e_B(\mathbf{x})} = -\rho(\mathbf{x}) \exp[\beta V_B^0(\mathbf{x})] \tag{18}$$

For a monatomic system with uniform density ρ_0, the M-body density is given by

$$\rho(\mathbf{x}) = \rho_0^M g(\mathbf{r}_1, \mathbf{r}_2, \ldots, \mathbf{r}_M) \tag{19}$$

where $g(\mathbf{r}_1, \mathbf{r}_2, \ldots, \mathbf{r}_M)$ stands for the M-body radial distribution function. Note that $g(\mathbf{x}) = \exp[-\beta V_B^0(\mathbf{x})]$ for the monatomic system in the ideal-gas limit, we have from Eqs. (18) and (19) that the functional derivative leads to the M-body cavity correlation function

$$\frac{\delta \beta \Omega}{\delta e_B(\mathbf{x})} = -\rho_0^M y(\mathbf{x}) \tag{20}$$

where $y(\mathbf{x}) \equiv g(\mathbf{x}) \exp[\beta V_B^0(\mathbf{x})]$. The cavity correlation function represents the density correlation beyond that introduced by the direct bonding potential. This function is particularly useful to describe the properties of polyatomic systems from those corresponding to monomers.

The one-body configurational density and the multi-body correlation functions provide succinct expressions to describe the complex behavior of polyatomic systems. However, these functions are inconvenient for direct computation due to the high dimensionality of molecular configuration. In general, the multi-dimensional procedure is useful only for systems of small molecules or liquid crystals where each molecule adopts only a rigid configuration. In the latter case, the one-body configurational density may be expressed in terms of the center-of-mass position $\mathbf{r}$ and orientation ω

$$\rho(\mathbf{x}) \sim \rho(\mathbf{r}, \omega) \tag{21}$$

where the proportionality constant depends on the number of atoms in each molecule and on the molecular geometry. For a linear molecule, the molecular orientation is specified by the polar angles, $\omega = (\theta, \phi)$. As a result, the orientation part of the density profile can be expanded in terms of the spherical harmonics. If the polyatomic molecule is non-linear, the molecular orientation is defined by three Euler angles, $\omega = (\theta, \phi, \gamma)$. Sophisticated numerical procedures are also available to further reduce the computational cost related to the angular degrees of freedom [27].

2.4 Site Density and Site Correlation Functions

For molecular systems where the external potential is represented by a summation of individual site energies, $\varphi_\gamma(\mathbf{r})$, the grand partition function becomes

$$\Xi = \sum_{N=0}^{\infty} \frac{e^{\beta\mu}}{N!\Lambda^{3M}} \int d\mathbf{x}_c^N \exp\{-\beta[\sum_{i=1}^{N} V_B(\mathbf{x}_i) + \sum_{i=1}^{N}\sum_{j>i}^{N} \Gamma(\mathbf{x}_i, \mathbf{x}_j) + \sum_{i=1}^{N}\sum_{\gamma_i} \varphi_{\gamma_i}(\mathbf{r}_{\gamma_i})]\} \tag{22}$$

where $\mathbf{r}_{\gamma_i}$ stands for the position of site γ from molecule i. As for the one-body configurational density $\rho(\mathbf{x})$, a functional derivative of the grand potential with respect to the external (site) potential results in the one-body site density profile

$$\rho_\alpha(\mathbf{r}) = \frac{\delta\Omega}{\delta\varphi_\alpha(\mathbf{r})}. \tag{23}$$

A second-order functional derivative of the grand potential leads to the site-site correlation function

$$\chi_{\alpha\gamma}(\mathbf{r},\mathbf{r}') \equiv -\beta^{-1}\frac{\delta\rho_\alpha(\mathbf{r})}{\delta\varphi_\gamma(\mathbf{r}')} =< [\sum_i \delta(\mathbf{r}-\mathbf{r}_{\alpha_i}) - \rho_\alpha(\mathbf{r})][\sum_j \delta(\mathbf{r}'-\mathbf{r}_{\gamma_i}) - \rho_\gamma(\mathbf{r}')] > . \tag{24}$$

Similar to the molecular density-density correlation function, $\chi_{\alpha\gamma}(\mathbf{r},\mathbf{r}')$ describes the response of the one-body density with respect to the one-body external potential; it represents the correlated fluctuations of local densities for α and γ sites at positions $\mathbf{r}_\alpha$ and $\mathbf{r}_\gamma$, respectively. The site correlation involves both intra- and inter- molecular interactions, i.e., α and γ sites can be located within the same molecule or from different molecules.

2.5 Classical Models for Polyatomic Ideal-gas Systems

In an ideal gas of polyatomic molecules, there is no intermolecular interaction. As a result, the grand partition function can be simplified to

$$\Xi^{IP} = \sum_{N=0}^{\infty} \frac{1}{N!\Lambda^{3MN}} \left[\int d\mathbf{x} \exp[-\beta W(\mathbf{x})\right]^N \tag{25}$$

where superscript "IP" denotes ideal polyatomic gas, $W(\mathbf{x}) \equiv V_B(\mathbf{x}) + V_{ext}(\mathbf{x}) - \mu$ stands for an effective one-body configurational potential. As noted above, $V_B(\mathbf{x})$ includes both bond energy $V_B^0(\mathbf{x})$ and non-bonded intra-molecular interactions.

Using the mathematical identity $e^x = \sum_{n=0}^{\infty} x^n/n!$, we can evaluate the summation in the grand potential for the ideal-gas system

$$\beta\Omega^{IP} = -\ln \Xi^{IP} = -\frac{1}{\Lambda^{3M}} \int d\mathbf{x} \exp[-\beta W(\mathbf{x})] \tag{26}$$

Recall that a functional derivative of the grand potential with respect to $V_{ext}(\mathbf{x})$ gives the one-body configurational density profile

$$\rho(\mathbf{x}) = \frac{\delta\Omega^{IP}}{\delta V_{ext}(\mathbf{x})} = \frac{1}{\Lambda^{3M}} \int d\mathbf{x}'\delta(\mathbf{x}-\mathbf{x}') \exp[-\beta W(\mathbf{x}')] = \frac{1}{\Lambda^{3M}} \exp[-\beta W(\mathbf{x})] \tag{27}$$

Equation (27) corresponds to the Boltzmann equation for the configurational distribution of ideal gas molecules; it predicts that the molecular configuration depends on the intramolecular potential as well as the external field. In terms of the one-body molecular density profile, the grand potential can be written as

$$\beta\Omega^{IP} = -\int d\mathbf{x}\rho(\mathbf{x}) = -\langle N\rangle \tag{28}$$

where $\langle N\rangle$ stands for the average number of molecules in the ideal-gas system.

For a uniform system free of the external potential, we can express the configurational density of molecules in terms of the molecular center of mass (COM)

$$\rho_0 = \frac{\langle N \rangle}{V} = \int d\mathbf{x}\delta(\mathbf{r}-\mathbf{r}_{\text{com}})\rho(\mathbf{x}) = \frac{\exp(\beta\mu)}{\Lambda^{3M}}\int d\varpi \exp[-\beta V_B(\varpi)] \tag{29}$$

where ϖ stands for the molecular configuration with the center of mass fixed at the origin. Rearranging Eq. (29) leads to

$$\beta\mu^{IP} = \ln(\rho_0\Lambda_{M'}^3) \tag{30}$$

where $\Lambda_{M'} \equiv \Lambda^M/\{\int d\varpi \exp[-\beta V_B(\varpi)]\}^{1/3}$ may be understood as an effective thermal wave length. Because the configurational integration yields a constant, Eq. (30) is virtually the same as the familiar equation for the chemical potential of a uniform monatomic ideal gas. Unlike that for a monatomic system, however, $\Lambda_{M'}$ as a function of temperature is not analytically known.

For systems where the external potential can be expressed as a summation of the site energies, we can express the one-body site-density profile in terms of the functional derivative of the grand potential. Recall that the grand partition function for an ideal-gas system is

$$\Xi^{IP} = \sum_{N=0}^{\infty}\frac{1}{N!}\left[\frac{e^{\beta\mu}}{\Lambda^{3M}}\int d\mathbf{x}\exp[-\beta V_B(\mathbf{x}) - \beta\sum_{\gamma}\varphi_\gamma(\mathbf{r}_\gamma)\right]^N \tag{31}$$

The ideal grand potential is thus given by

$$\beta\Omega^{IP} = -\ln\Xi^{IP} = -\frac{e^{\beta\mu}}{\Lambda^{3M}}\int d\mathbf{x}\exp\{-\beta V_B(\mathbf{x}) - \beta\sum_{\gamma}\varphi_\gamma(\mathbf{r}_\gamma)\} \tag{32}$$

A functional derivative of the grand potential with respective to the one-body site potential leads to

$$\rho_\alpha(\mathbf{r}) = \frac{\delta\Omega^{IP}}{\delta\varphi_\alpha(\mathbf{r})} = \frac{e^{\beta\mu}}{\Lambda^{3M}}\int d\mathbf{x}\delta(\mathbf{r}-\mathbf{r}_\alpha)\exp[-\beta V_B(\mathbf{x}) - \beta\sum_{\gamma}\varphi_\gamma(\mathbf{r}_\gamma)] \tag{33}$$

Integrating Eq. (33) on both sides yields

$$\beta\Omega^{IP} = -\int d\mathbf{r}\rho_\alpha(\mathbf{r}) \tag{34}$$

As the grand potential is immaterial to the identity of specific interaction sites, Eq. (34) implies that

$$\beta\Omega^{IP} = -\int d\mathbf{x}\rho(\mathbf{x}) = -\frac{1}{M_s}\sum_{\alpha=1}^{M_s}\int d\mathbf{r}\rho_\alpha(\mathbf{r}). \tag{35}$$

Equation (35) is valid for both bulk and inhomogeneous systems.

In a polyatomic ideal-gas system, the site-site correlation arises from intramolecular interactions (both bonded and non-bonded). It can be shown from Eq. (33) that the site-site density correlation function is related to the intramolecular correlation function

$$\chi_{\alpha\gamma}^{IP}(\mathbf{r},\mathbf{r}') = -\beta^{-1}\frac{\delta\rho_\alpha(\mathbf{r})}{\delta\varphi_\gamma(\mathbf{r}')} = \rho_\alpha(\mathbf{r}) < \delta(\mathbf{r}_\gamma - \mathbf{r}_\alpha) >_{IP_\alpha} \tag{36}$$

where $< \cdots >_{IP_\alpha}$ stands for the ensemble average in the ideal polymeric system. In terms of the intramolecular potential, the ensemble average is

$$< \delta(\mathbf{r}_\gamma - \mathbf{r}_\alpha) >_{IP_\alpha} = \frac{\int d\mathbf{x}\delta(\mathbf{r}-\mathbf{r}_\alpha)\delta(\mathbf{r}_\gamma - \mathbf{r}_\alpha)\exp[-\beta V_B(\mathbf{x}) - \beta\sum_\gamma \varphi_\gamma(\mathbf{r}_\gamma)]}{\int d\mathbf{x}\delta(\mathbf{r}-\mathbf{r}_\alpha)\exp[-\beta V_B(\mathbf{x}) - \beta\sum_\gamma \varphi_\gamma(\mathbf{r}_\gamma)]} \tag{37}$$

Equation (37) represents the probability of finding site γ at $\mathbf{r}_\gamma$ given that site α from the same molecule is located at position $\mathbf{r}_\alpha$.

The intramolecular site-site correlation function is defined as

$$\omega_{\alpha\gamma}^{IP}(\mathbf{r},\mathbf{r}') \equiv \chi_{\alpha\gamma}^{IP}(\mathbf{r},\mathbf{r}')/\rho_\alpha(\mathbf{r}). \tag{38}$$

As shown in Eq. (37), $\omega_{\alpha\gamma}^{IP}(\mathbf{r},\mathbf{r}')$ takes into account *both* bond and non-bonded interactions in an ideal-gas system. For inhomogeneous systems, we have in general $\omega_{\alpha\gamma}^{IP}(\mathbf{r},\mathbf{r}') \neq \omega_{\gamma\alpha}^{IP}(\mathbf{r},\mathbf{r}')$ because the local site densities may not be the same for different sites. Due to coupling with the intermolecular interactions, the intramolecular site-site correlation of an ideal system is different from that in a real polyatomic system.

3 Density Functional Theory

Density functional theory (DFT) was introduced into the statistical mechanics for classical systems first in the context of inhomogeneous simple fluids [28–32]. Tutorial materials are available on the basic concepts of classical DFT for atomic systems, i.e., those consist of monomeric particles [33, 34]. In this section, we focus on formalisms for the applications of classical DFT to polyatomic systems. More comprehensive discussions on recent developments of classical DFT for polymeric systems may be found in the next chapter and from earlier reviews [35–45].

3.1 Hohenberg-Kohn-Mermin Theorem

The theoretical basis underpinning all kinds of DFT methods can be established within the framework of the Hohenberg-Kohn-Mermin (HKM) theorem. The central idea is that the external potential of a multi-body system is uniquely determined by the one-body density profile, regardless of the physical nature of the inter-particle interactions [46, 47]. Although the mathematical proof was originally given for inhomogeneous electronic systems at 0 K, the HKM theorem is equally applicable to thermodynamic systems containing electrons or classical particles at finite temperature [48]. In other words, the HKM theorem is independent of the physical nature of the particles [29, 47].

While conventional methods in statistical mechanics start with the numerical evaluation of ensemble averages (e.g., *via* molecular simulations [49]) or with approximations for the partition functions (e.g., polymer field theories [50]), DFT seeks a variational solution of the one-body density profile by minimizing the grand potential functional. The variational principle facilitates direct calculations of the equilibrium density profiles and, subsequently, multi-body correlation functions and thermodynamic properties without an explicit consideration of the partition functions.

As discussed above, the grand partition function for a polyatomic system is formally identical to that of a monatomic system. With the one-body potential and density $V_{ext}(\mathbf{r})$ and $\rho(\mathbf{r})$ replaced by $V_{ext}(\mathbf{x})$ and $\rho(\mathbf{x})$, the HKM theorem for monatomic systems can be directly applicable to polyatomic systems. Similar to that for a monoatomic system, the grand potential functional for a one-component polyatomic system is defined as

$$\Omega_V[\rho(\mathbf{x})] = F[\rho(\mathbf{x})] + \int d\mathbf{x}[V_{ext}(\mathbf{x}) - \mu]\rho(\mathbf{x}). \tag{39}$$

In Eq. (39), $F[\rho(\mathbf{x})]$ stands for the *intrinsic* Helmholtz energy, which is related to the probability distribution for the microstates p_ν and the *intrinsic* energy of the multi-body system E'_ν

$$F[\rho(\mathbf{x})] \equiv k_B T \sum_\nu p_\nu \ln p_\nu + \sum_\nu p_\nu E'_\nu \tag{40}$$

where the prime symbol denoting the exclusion of the external potential. Here an intrinsic property is referred to as a quantity depending only on the identities of polyatomic molecules and the density profile. For a given molecular system, the *intrinsic* Helmholtz energy is a unique functional of the configurational (or atomic) density profiles. Note that the microstate probability p_ν is an intrinsic property because, according to the HKM theorem, the external potential in p_ν is completely determined by the configurational density profile, $\rho(\mathbf{x})$.

3.2 DFT for Ideal-Gas Systems

For an ideal gas of polyatomic molecules, the configurational density profile and the grand potential are exactly known, as given by Eqs. (27) and (28). Therefore, we can readily derive the intrinsic Helmholtz energy from Eq. (39)

$$\beta F^{IP}[\rho(\mathbf{x})] = \int d\mathbf{x}\rho(\mathbf{x})\{\ln[\rho(\mathbf{x})\Lambda^{3M}] - 1\} + \beta \int d\mathbf{x}\rho(\mathbf{x})V_B(\mathbf{x}). \tag{41}$$

Note that for an ideal-gas system, E'_ν is fully determined by the intramolecular potential, i.e.,

$$\sum_\nu p_\nu E'_\nu = \int d\mathbf{x}\rho(\mathbf{x})V_B(\mathbf{x}). \tag{42}$$

A comparison of Eqs. (41) and (40) results in the generalized Sackur–Tetrode equation for the ideal entropy

$$S^{IP}/k_B = -\sum_\nu p_\nu \ln p_\nu = -\int d\mathbf{x}\rho(\mathbf{x})\{\ln[\rho(\mathbf{x})\Lambda^{3M}] - 1\}. \tag{43}$$

Substituting Eq. (41) in Eq. (35) yields

$$\Omega_V^{IP}[\rho(\mathbf{x})] = k_B T \int d\mathbf{x}\rho(\mathbf{x})\{\ln[\rho(\mathbf{x})\Lambda^{3M}] - 1\} + \int d\mathbf{x}[V_B(\mathbf{x}) + V_{ext}(\mathbf{x}) - \mu]\rho(\mathbf{x}). \tag{44}$$

It is straightforward to show that minimizing $\Omega_V^{IP}[\rho(\mathbf{x})]$ with respect to $\rho(\mathbf{x})$ leads to

$$\rho(\mathbf{x}) = \frac{1}{\Lambda^{3M}} \exp\{-\beta[V_B(\mathbf{x}) + V_{ext}(\mathbf{x}) - \mu]\}. \tag{45}$$

As predicted by the HKM theorem, $\rho(\mathbf{x})$ is indeed identical to the equilibrium profile, i.e., the minimum Ω_V corresponds to the grand potential, $\Omega = \Omega_V[\rho(\mathbf{x})]$. While application of the DFT to an ideal-gas system does not yield much new results, it illustrates that the density functional appeared in the HKM theorem can be derived from conventional statistical-mechanic methods and that, with an expression for the intrinsic Helmholtz energy as a functional of the one-body density profile, minimization of the grand potential functional leads to the one-body density profile and from which all thermodynamic properties can be subsequently derived.

3.3 Excess Helmholtz Energy

For most systems of practical interest, we do not have *a priori* knowledge on the intrinsic Helmholtz energy. Nevertheless, reasonable approximations can be formulated by following complementary statistical-mechanical procedures. Because the HKM theorem indicates that the one-body external potential is uniquely determined by the one-body density profile, we can identify a reference system such that its one-body density profile is the same as that of the more complicated systems of practical interest but without complex intermolecular interactions. The existence of exact mapping between real and reference systems represents one of the most important ideas of DFT calculations.

Because analytical expressions are readily available for both structural and thermodynamic properties, the ideal-gas model provides a good reference to study non-ideal molecular systems. To formulate the density functional, we define the *excess* Helmholtz energy, $F^{ex}[\rho(\mathbf{x})]$, in terms of the difference between the intrinsic Helmholtz energy of a real system and that of an ideal-gas reference system with the same configurational density profile

$$F[\rho(\mathbf{x})] = k_B T \int d\mathbf{x}\rho(\mathbf{x})\{\ln[\rho(\mathbf{x})\Lambda_M^3] - 1\} + \int d\mathbf{x}\rho(\mathbf{x})V_B(\mathbf{x}) + F^{ex}[\rho(\mathbf{x})]. \tag{46}$$

According to the HKM theorem, the one-body density profile satisfies the variational principle

$$\delta\Omega_V[\rho(\mathbf{x})]/\delta\rho(\mathbf{x}) = 0. \tag{47}$$

Substituting Eqs. (39) and (46) into Eq. (47), after carrying out the functional derivative, we may derive an analytical relation between the density profile and one-body potentials:

$$\rho(\mathbf{x}) = \frac{1}{\Lambda^{3M}} \exp\{-\beta[V_B(\mathbf{x}) + V_{ext}(\mathbf{x}) - \mu + \mu^{ex}(\mathbf{x})]\} \tag{48}$$

where $\mu^{ex}(\mathbf{x}) = \delta F^{ex}/\delta\rho(\mathbf{x})$.

Equation (48) is equivalent to the familiar Boltzmann equation except that it is concerned with the configurational density of polyatomic molecules; it may be compared with the corresponding equation for the ideal-gas reference system

$$\rho(\mathbf{x}) = \frac{1}{\Lambda^{3M}} \exp\{-\beta[V_B(\mathbf{x}) + V_{ext}^{IP}(\mathbf{x}) - \mu^{IP}]\} \tag{49}$$

where $V_{ext}^{IP}(\mathbf{x})$ is the external potential of the reference system determined by $\rho(\mathbf{x})$, and μ^{IP} is the chemical potential of the ideal-gas reference system. By equaling the right sides of Eqs. (48) and (49), we have

$$\mu = \mu^{IP} + \mu^{ex}(\mathbf{x}) + V_{ext}(\mathbf{x}) - V_{ext}^{IP}(\mathbf{x}). \tag{50}$$

whereas the above procedure is formally exact, numerical solution of $\rho(\mathbf{x})$ from Eq. (48) requires simulation methods to sample the configuration of polyatomic molecules [51, 52]. The numerical procedure is not much different from that used in the polymer field theory [50], except that the molecular configuration is typically more complicated than that in a coarse-grained polymer model. For molecules with rigid configurations, the configurational density profile can be expressed in terms of position and orientational angle, $\rho(\mathbf{x}) \sim \rho(\mathbf{r}, \omega)$. In that case, efficient numerical algorithms have been developed for calculating $\rho(\mathbf{r}, \omega)$ without invoking molecular simulations [53, 54]. Equation (49) indicates that the configurational density profile of a non-ideal system can be determined from that of an ideal reference system but with a different chemical potential and the one-body external potential. Once we determine the density profile from the ideal-gas system, the intrinsic Helmholtz energy, the grand potential, and all other thermodynamic properties can be subsequently determined from conventional statistical-mechanical relations.

While the numerical procedure for calculating the properties of an ideal system is much simplified in comparison to that for the non-ideal system, it should be noted that the ideal and real systems do not have the same thermodynamic properties, even though they share the same density profile. The difference can be easily seen from the ideal grand potential, $\beta\Omega^{IP} = -\int d\mathbf{x}\rho(\mathbf{x})$, which is in general not true for non-ideal systems.

In the bulk limit, the external potential vanishes for both the ideal reference system and the real system. In that case, the chemical potential for the ideal-gas system is exactly known viz., Eq. (30), and $\mu^{ex}(\mathbf{x})$ reduces to the conventional expression for the excess chemical potential, μ^{ex},

$$\mu = k_B T \ln(\rho_0 \Lambda_{M'}^3) + \mu^{ex}. \tag{51}$$

As defined before, ρ_0 is the average number density, and $\Lambda_{M'}$ denotes an effective thermal wavelength. The latter accounts for the effect of intramolecular correlations for individual polyatomic molecules. Equation (51) provides a connection between the excess Helmholtz energy and the excess chemical potential, which can be readily determined from an equation of state for bulk systems.

3.4 Direct Correlation Function

To provide further insights into the intrinsic Helmholtz energy, we rearrange Eq. (48) to give

$$-\beta\mu^{ex}(\mathbf{x}) = \ln[\rho(\mathbf{x})\Lambda^{3M}] + \beta V_B(\mathbf{x}) + \beta[V_{ext}(\mathbf{x}) - \mu] \tag{52}$$

The functional derivative of $-\beta\mu^{ex}(\mathbf{x})$ with respective to $\rho(\mathbf{x})$ defines the two-body direct correlation function, $c^{(2)}(\mathbf{x}, \mathbf{x}')$,

$$c^{(2)}(\mathbf{x}, \mathbf{x}') \equiv -\frac{\delta \beta \mu^{ex}(\mathbf{x})}{\delta \rho(\mathbf{x}')} = -\frac{\delta^2 \beta F^{ex}}{\delta \rho(\mathbf{x}) \delta \rho(\mathbf{x}')}. \tag{53}$$

Applying the functional derivative to both sides of Eq. (52) leads to

$$c^{(2)}(\mathbf{x}, \mathbf{x}') = \frac{\delta(\mathbf{x} - \mathbf{x}')}{\rho(\mathbf{x})} - \chi^{-1}(\mathbf{x}, \mathbf{x}') \tag{54}$$

where $\chi(\mathbf{x}, \mathbf{x}') \equiv -\delta\rho(\mathbf{x})/\delta\beta V_{ext}(\mathbf{x}')$ is two-body density-density correlation function.

By applying the inverse functional derivative, we can derive from Eqs. (14) and (54) an explicit relation between the direct and total correlation functions

$$h(\mathbf{x}_1, \mathbf{x}_2) = c^{(2)}(\mathbf{x}_1, \mathbf{x}_2) + \int d\mathbf{x}_3 \rho(\mathbf{x}_3) h(\mathbf{x}_1, \mathbf{x}_3) c^{(2)}(\mathbf{x}_3, \mathbf{x}_2). \tag{55}$$

Equation (55) is known as the Ornstein-Zernike (OZ) equation for molecular systems. The integral equation provides a good starting point to derive the properties of bulk systems [23]. With an additional relation (closure), the OZ equation can be used to solve for both the total and direct correlation functions.

From the excess Helmholtz energy, we can similarly define three and multi-body direct correlation functions

$$c^{(n)}(\mathbf{x}_1, \mathbf{x}_2, \ldots \mathbf{x}_n) = -\frac{\delta \beta F^{ex}}{\delta \rho(\mathbf{x}_1) \delta \rho(\mathbf{x}_2) \cdots \delta \rho(\mathbf{x}_n)} \tag{56}$$

Like the two-body direct correlation function, the multi-body direct correlation functions can be calculated from conventional statistical-mechanical methods based on the variation of the multi-body density profiles in response to the changes in the external potential. The direct correlation functions then can be used to construct the excess Helmholtz energy. In principle, one may use the procedure to construct the excess Helmholtz energy systematically with increasing accuracy. This would address one common criticism of DFT for the lack of systematic procedure to construct or improve the intrinsic Helmholtz energy.

3.5 *Exact Functionals and Approximations*

Formal expressions of the excess Helmholtz energy can be constructed from the functional integration with respect to the direct correlation functions or the intermolecular potentials. Using the excess Helmholtz energy of a bulk system as a reference, the functional integration the direct correlation functions with respect to the density inhomogeneity leads to

$$\beta F^{ex}[\rho] = \beta F^{ex}[\rho_0] - \int d\mathbf{x}\Delta\rho(\mathbf{x})c^{(1)}(\rho_0, \mathbf{x}) - \int_0^1 d\lambda(\lambda - 1)\int\int d\mathbf{x}_1 d\mathbf{x}_2 \Delta\rho(\mathbf{x}_1)\Delta\rho(\mathbf{x}_2)c^{(2)}(\rho_\lambda, \mathbf{x}_1, \mathbf{x}_2) \tag{57}$$

where $c^{(1)}(\rho_0, \mathbf{x}) = -\beta\mu^{ex}(\rho_0)$, $\Delta\rho(\mathbf{x}) = \rho(\mathbf{x}) - \rho_0(\mathbf{x})$, $\rho_\lambda = \rho_0\omega_0(\mathbf{x}) + \lambda\Delta\rho(\mathbf{x})$, and $\rho_0(\mathbf{x})$ is the configurational density for the bulk system.

In the hypernetted chain (HNC) approximation or equivalently, the homogeneous reference fluid (HRF) approximation, [55] we replace $c^{(2)}(\rho_\lambda, \mathbf{x}_1, \mathbf{x}_2)$ in Eq. (57) with that of the uniform bulk system

$$\beta F^{HNC}[\rho] \approx \beta F^{ex}[\rho_0] - \int d\mathbf{x}\Delta\rho(\mathbf{x})c^{(1)}(\rho_0, \mathbf{x}) - \frac{1}{2}\int\int d\mathbf{x}_1 d\mathbf{x}_2 \Delta\rho(\mathbf{x}_1)\Delta\rho(\mathbf{x}_2)c^{(2)}(\rho_\lambda, \mathbf{x}_1, \mathbf{x}_2) \tag{58}$$

Equation (58) corresponds to a quadratic expansion of the excess Helmholtz energy functional in terms of $\Delta\rho(\mathbf{x})$. All high-order terms in the functional Taylor expansion is referred to as the bridge functional, i.e.,

$$F^B[\rho] \equiv F^{ex}[\rho] - F^{HNC}[\rho] \tag{59}$$

Because the grand potential corresponds to a minimum at equilibrium, the HNC approximation is often sufficient to represent the excess Helmholtz energy quantitatively. One major caveat of the quadratic approximation is that it is able to describe only one free-energy minimum, thus insufficient for representing phase transitions.

Alternatively, we can find the intrinsic Helmholtz energy by integrating the two-body density function [Eq. (17)]

$$F[\rho(\mathbf{x})] = F^{ID}[\rho(\mathbf{x})] + \int_0^1 d\lambda \int\int d\mathbf{x}_1 d\mathbf{x}_2 \rho^{(2)}(\Gamma_\lambda, \mathbf{x}_1, \mathbf{x}_2)\Gamma(\mathbf{x}_1, \mathbf{x}_2) \tag{60}$$

where $F^{ID}[\rho(\mathbf{x})]$ represents the intrinsic Helmholtz energy of the ideal-gas reference system, $\rho^{(2)}(\Gamma_\lambda, \mathbf{x}_1, \mathbf{x}_2)$ is the two-body density function when the pair potential is given by $\lambda\Gamma(\mathbf{x}_1, \mathbf{x}_2)$. In a typical mean-field approximation, a hard-sphere model is used to represent the excess Helmholtz energy due to the repulsive component of the pair potential, and that due to the attractive component ignores all correlations (viz. interacting in a "mean field"):

$$F[\rho(\mathbf{x})] = F^{ID}[\rho(\mathbf{x})] + F^{HS}[\rho(\mathbf{x})] + \int\int d\mathbf{x}_1 d\mathbf{x}_2 \rho(\mathbf{x}_1)\rho(\mathbf{x}_2)\Gamma_A(\mathbf{x}_1, \mathbf{x}_2) \tag{61}$$

Because the density profile is self-consistently calculated by minimization of the grand potential, Eq. (61) is analogous to the self-consistent mean-field theory for polymer systems.

Functional integration can also be used to connect the intrinsic Helmholtz energy of a polyatomic system with that of a monatomic system of the same atomic density profiles $\rho_\alpha(\mathbf{r}_\alpha)$. Because the external and bonding potentials are uncorrelated, we have from Eq. (20)

$$\varsigma(\mathbf{x}) \equiv \rho_0(\mathbf{x})y(\mathbf{x}) = -\frac{\delta\beta F}{\delta e_B(\mathbf{x})} \tag{62}$$

where $\rho_0(\mathbf{x}) \equiv \prod_\alpha \rho_\alpha(\mathbf{r}_\alpha)$, and $y(\mathbf{x}) = g(\mathbf{x})\exp[\beta V_B(\mathbf{x})]$ represents the multi-body cavity correlation function of the inhomogeneous system. Using the unbounded monatomic system as a reference, the functional integration of $\varsigma(\mathbf{x})$ with respect to $e_B(\mathbf{x})$ leads to

$$\beta F[\rho(\mathbf{x})] = \beta F[\rho_\alpha(\mathbf{r}_\alpha)] + \beta\Delta F^{\mathrm{com}} - \int_0^1 d\lambda \int d\mathbf{x}\varsigma_\lambda(\mathbf{x})[e_B(\mathbf{x}) - 1] \tag{63}$$

where $F[\rho_\alpha(\mathbf{r}_\alpha)]$ corresponds to the intrinsic Helmholtz energy of the monatomic system, $\beta\Delta F^{\mathrm{com}}$ represents the change in the communal entropy, i.e., changes in molecular symmetry due to bond formation, and $\varsigma_\lambda(\mathbf{x})$ corresponds to the condition when $e_{B,\lambda}(\mathbf{x}) = 1 + \lambda[e_B(\mathbf{x}) - 1]$. The communal entropy terms arises from the distinguishability of classical particles; it can be shown that

$$\beta\Delta F^{\mathrm{com}} = \int d\mathbf{x}\rho(\mathbf{x})\{\ln[\rho(\mathbf{x})\Lambda_M^3] - 1\} - \sum_\alpha \int d\mathbf{r}\rho_\alpha(\mathbf{r})\{\ln[\rho_\alpha(\mathbf{r})\Lambda_\alpha^3] - 1\}. \tag{64}$$

For an ideal polyatomic system without non-bonded interactions, $\varsigma^{IP}(\mathbf{x}) = \rho_0(\mathbf{x})\exp[\beta V_B^0(\mathbf{x})]$, it is straightforward to show that the functional integration in Eq. (63) leads to the chain formation energy as expected

$$\beta F^{chain}[\rho(\mathbf{x})] = \int d\mathbf{x}\rho_0(\mathbf{x})V_B^0(\mathbf{x}). \tag{65}$$

For a non-ideal polyatomic system, applying the bonding energy to the corresponding monoatomic system of the same site densities leads to a change in the free energy that is equivalent to the multi-body potential of mean force $-k_BT\ln g(\mathbf{x})$ in the polyatomic system. In that case, the functional integration gives

$$\beta F^{chain}[\rho(\mathbf{x})] = -k_BT\int d\mathbf{x}\rho_0(\mathbf{x})\ln g(\mathbf{x}). \tag{66}$$

Equation (66) was first derived by Chandler and Pratt for the formation of intramolecular structures by chemical bonding [56]. With $g(\mathbf{x})$ replaced by that of the monomer system and the superposition approximation for the multi-body correlation function, Eq. (66) reduces to the chain formation energy from Wertheim's thermodynamic perturbation theory [57].

4 Interaction Site Formulism

4.1 Variational Principle

As mentioned above, the HKM theorem is immaterial to the specific form of energy arising from interaction between particles, which can be either quantum or classical. In other words, the HKM theorem is equally applicable to monatomic and polyatomic systems even though the latter contains both intra- and inter- molecular interactions. In both cases, the HKM theorem indicates that a deterministic relation can be formally established between the one-body (site) potential and the one-body (site) density.

Application of the classical DFT to polyatomic fluids was introduced first by Chandler, McCoy and Singer [58]. Within the framework of the interaction site model, the grand potential functional can be formally defined as

$$\Omega_V[\rho_\alpha(\mathbf{r})] = F[\rho_\alpha(\mathbf{r})] + \sum_\alpha \int d\mathbf{r}[\varphi_\alpha(\mathbf{r}) - \mu_\alpha]\rho_\alpha(\mathbf{r}). \tag{67}$$

In writing Eq. (67), we assume all interactions sites are distinguishable, and the site chemical potential is defined as $\mu_\alpha = \mu/M_s$. As for monatomic systems, $\Omega_V[\rho_\alpha(\mathbf{r})]$ is minimized with respect to the site density $\rho_\alpha(\mathbf{r})$ without any constraints, and the minimum corresponds to the grand potential. Because the thermodynamic potential is independent of configurational or site representations of the one-body density profile, we have $F[\rho_\alpha(\mathbf{r})] = F[\rho(\mathbf{x})]$ for systems where the external potential can be expressed as a summation of the site energies.

The functional derivative of the grand potential functional with respect to the site density results in the Euler-Lagrange equation for the site densities:

$$\frac{\delta\Omega_V}{\delta\rho_\alpha(\mathbf{r})} = \frac{\delta F}{\delta\rho_\alpha(\mathbf{r})} + \varphi_\alpha(\mathbf{r}) - \mu_\alpha = 0. \tag{68}$$

With an analytical expression for the intrinsic Helmholtz energy, Eq. (68) allows us to solve for the site densities by numerical iterations.

4.2 Reference Systems in Site Formalism

As discussed above, the intrinsic Helmholtz energy of a non-ideal system is often formulated relative to those of simpler reference systems. For a polyatomic system, we may use several references. As shown schematically in Fig. 3, the thermodynamic properties for a system of practical interest may be compared with those corresponding to an ideal monatomic (IM) system with the density profiles the same as the site densities, or an ideal polyatomic (IP) system of the same site densities, or a uniform polyatomic system of the same bulk chemical potential.

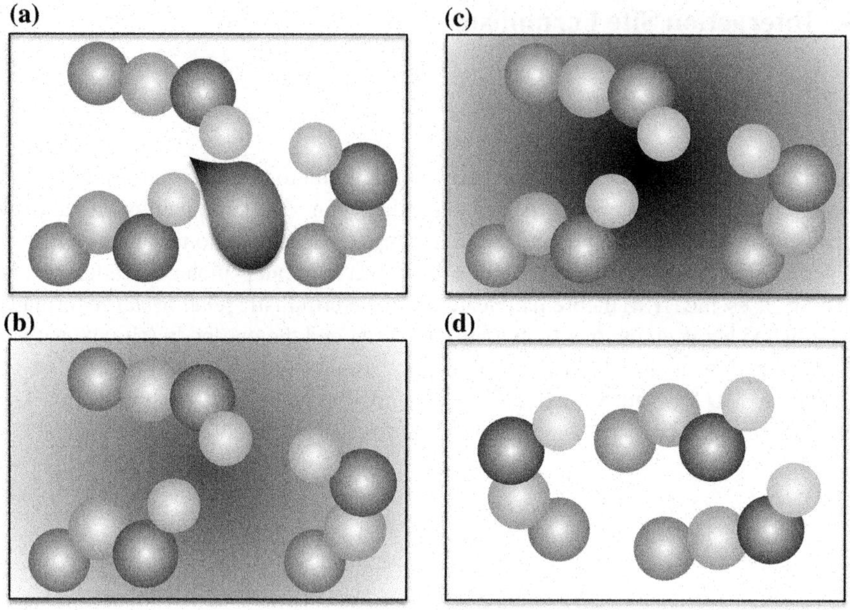

Fig. 3 Thermodynamic properties of a polyatomic system in the presence of an external field, here shown as interaction with a spherical particle (*Panel A*). The reference systems may include: (1) an ideal-gas system of monatomic particles with the density profiles the same as the site densities but with different external field (shaded background) (*Panel B*), (2) an ideal-gas system of polyatomic molecules with the same site densities and different extneral potential (*Panel C*), and (3) a uniform system of the same bulk chemical potential (*Panel D*). The background colors in *panels B* and *C* denote different external potentials

By applying Eq. (68) to the reference system of non-interacting polyatomic molecules (see Panel C in Fig. 3), we may obtain the Euler-Lagrange equation for an ideal-gas polyatomic system

$$\frac{\delta \Omega_V^{IP}}{\delta \rho_\alpha(\mathbf{r})} = \frac{\delta F^{IP}}{\delta \rho_\alpha(\mathbf{r})} + \varphi_\alpha^{IP}(\mathbf{r}) - \mu_\alpha^{IP} = 0. \tag{69}$$

where $\varphi_\alpha^{IP}(\mathbf{r})$ and μ_α^{IP} stand for, respectively, the site external potential and the site chemical potential of the ideal polyatomic system. As predicted by the HKM theorem, $\varphi_\alpha^{IP}(\mathbf{r})$ is uniquely determined by the one-body density profile $\rho_\alpha(\mathbf{r})$, and μ_α^{IP} corresponds to the chemical potential of the ideal system in the bulk limit. To match the site density profiles, the one-body site potentials of the real and ideal systems must satisfy

$$u_\alpha^{IP}(\mathbf{r}) = u_\alpha(\mathbf{r}) + \mu_\alpha^{ex}(\mathbf{r}) \tag{70}$$

where $u_\alpha(\mathbf{r}) \equiv \varphi_\alpha(\mathbf{r}) - \mu_\alpha$ represents a one-body site potential, and

$$\mu_\alpha^{ex}(\mathbf{r}) \equiv \delta(F - F^{IP})/\delta \rho_\alpha(\mathbf{r}) \tag{71}$$

is the local excess chemical potential for site α.

No analytical expression is available for the site density of the ideal polyatomic system owing to the intramolecular bond connectivity and correlations. To find an even simpler reference system, we consider an ideal gas of monatomic particles (IM, see Panel B in Fig. 3) that reproduce the site densities of the polyatomic system. The density profiles of the uncorrelated particles in the ideal monatomic system are given by the Boltzmann equation

$$\rho_\alpha(\mathbf{r}) = \rho_\alpha^{IM} \exp\{-\beta w_\alpha^{IM}(\mathbf{r})\}, \tag{72}$$

where ρ_α^{IM} and $w_\alpha^{IM}(\mathbf{r})$ stand for the bulk density and the external potential, respectively. For the ideal monatomic system, the intrinsic Helmholtz energy is related to the one-body densities

$$F^{IM}[\rho_\alpha(\mathbf{r})] = k_B T \sum_\alpha \int d\mathbf{r} \rho_\alpha(\mathbf{r}) \{\ln[\rho_\alpha(\mathbf{r})\Lambda_\alpha^3] - 1\}. \tag{73}$$

Strictly speaking, the thermal wavelength of an interaction site Λ_α may not be defined. But this should not be an issue because they can be related to the total kinetic energy of the polyatomic molecules as discussed above.

For a non-ideal inhomogeneous system of interest (Panel B in Fig. 3), we define the auxiliary intrinsic Helmholtz energy as the difference between those of the real system and an ideal monatomic system of the same site density profiles:

$$F[\rho_\alpha(\mathbf{r})] = k_B T \sum_\alpha \int d\mathbf{r} \rho_\alpha(\mathbf{r}) \{\ln[\rho_\alpha(\mathbf{r})\Lambda_\alpha^3] - 1\} + F^{au}[\rho_\alpha(\mathbf{r})] \tag{74}$$

where superscript "au" designates auxiliary. From Eqs. (67) and (74) for the grand potential functional and the intrinsic Helmholtz energy, respectively, we can derive a formal expression for the site density profiles

$$\rho_\alpha(\mathbf{r}) = \frac{1}{\Lambda_\alpha^3} \exp\{-\beta[\varphi_\alpha(\mathbf{r}) - \mu_\alpha + \mu_\alpha^{au}(\mathbf{r})]\} \tag{75}$$

where $\mu_\alpha^{au}(\mathbf{r}) \equiv \delta F^{au}/\delta\rho_\alpha(\mathbf{r})$. Comparing to the site density profile for the inhomogeneous distribution of monatomic particles in an inhomogeneous ideal-gas system with bulk site density ρ_α^{IM}, we obtain an analytical expression for the effective external potential for the reference monatomic system

$$w_\alpha^{IM}(\mathbf{r}) = \varphi_\alpha(\mathbf{r}) - \mu_\alpha + \mu_\alpha^{au}(\mathbf{r}) + k_B T \ln(\rho_\alpha^{IM}\Lambda_\alpha^3). \tag{76}$$

Clearly, the one-body potential for the ideal monatomic reference system is different from that of the ideal polyatomic system [Eq. (70)], even when the two reference systems have the same site densities.

4.3 Direct Site Correlation Functions

To find an explicit expression for $F^{au}[\rho_\alpha(\mathbf{r})]$, we follow a mathematical procedure formally the same as that used for monatomic systems. The one-body site direct correlation function can be defined in terms of the functional derivatives of the auxiliary Helmholtz energy

$$D_\alpha^{(1)}(\mathbf{r}) \equiv -\delta\beta F^{au}/\delta\rho_\alpha(\mathbf{r}) = -\beta\mu_\alpha^{au}(\mathbf{r}) = \ln[\rho_\alpha(\mathbf{r})\Lambda_\alpha^3] + \beta[\varphi_\alpha(\mathbf{r}) - \mu_\alpha] \quad (77)$$

Similarly, the two-body site direct correlation function is

$$D_{\alpha\gamma}^{(2)}(\mathbf{r}, \mathbf{r}') = -\frac{\delta\beta\mu_\alpha^{au}(\mathbf{r})}{\delta\rho_\gamma(\mathbf{r}')} \equiv \frac{\delta_{\alpha\gamma}(\mathbf{r}-\mathbf{r}')}{\rho_\alpha(\mathbf{r})} - \chi_{\alpha\gamma}^{(-1)}(\mathbf{r}, \mathbf{r}') \quad (78)$$

where $\chi_{\alpha\gamma}^{(-1)}(\mathbf{r}, \mathbf{r}') = -\delta\beta\varphi_\alpha(\mathbf{r})/\delta\rho_\gamma(\mathbf{r}')$. As for inhomogeneous monatomic systems, the auxiliary intrinsic Helmholtz energy may be expressed as a functional integration relative to that of a reference molecular system of uniform site densities ρ_α^0 (see Panel D in Fig. 3):

$$\begin{aligned}\beta F^{au}[\rho_\alpha] =& \beta F^{au}[\rho_\alpha^0] - \sum_\alpha \int d\mathbf{r}\Delta\rho_\alpha(\mathbf{r}) D_\alpha^{(1)}(\rho_\alpha^0) \\ &- \sum_{\alpha\gamma}\int_0^1 d\lambda(\lambda-1)\int\int d\mathbf{r}_1 d\mathbf{r}_2 \Delta\rho_\alpha(\mathbf{r}_1)\Delta\rho_\alpha(\mathbf{r}_2) D_{\alpha\gamma}^{(2)}(\rho_{\alpha,\lambda}, \mathbf{r}_1, \mathbf{r}_2)\end{aligned} \quad (79)$$

where $\Delta\rho_\alpha(\mathbf{r}) = \rho_\alpha(\mathbf{r}) - \rho_\alpha^0$, and $D_{\alpha\gamma}^{(2)}(\rho_{\alpha,\lambda}, \mathbf{r}_1, \mathbf{r}_2)$ is the two-body direct correlation function of an inhomogeneous system with site density $\rho_{\alpha,\lambda}(\mathbf{r}) = \rho_\alpha^0 + \lambda\Delta\rho_\alpha(\mathbf{r})$. For the uniform molecular system, $\beta\varphi_\alpha(\mathbf{r}) = 0$, and $D_\alpha^{(1)}(\mathbf{r})$ depends only on the average site density ρ_α^0.

The same procedure can be applied to an ideal polyatomic system with the same site density profiles. Relative to that of an ideal monatomic system as discussed above, the auxiliary intrinsic Helmholtz energy for the ideal polyatomic system (IP) is given by

$$\begin{aligned}\beta F_{IP}^{au}[\rho_\alpha] =& \beta F_{IP}^{au}[\rho_\alpha^0] - \sum_\alpha \int d\mathbf{r}\Delta\rho_\alpha(\mathbf{r}) D_\alpha^{(1IP)}(\rho_\alpha^0) \\ &- \sum_{\alpha\gamma}\int_0^1 d\lambda(\lambda-1)\int\int d\mathbf{r}_1 d\mathbf{r}_2 \Delta\rho_\alpha(\mathbf{r}_1)\Delta\rho_\alpha(\mathbf{r}_2) D_{\alpha\gamma}^{(2IP)}(\rho_{\alpha,\lambda}, \mathbf{r}_1, \mathbf{r}_2)\end{aligned} \quad (80)$$

where$D_\alpha^{(1IP)}(\rho_\alpha^0)$ corresponds to the one-body site direct correlation function of the ideal polyatomic system with uniform site density ρ_α^0, and $D_{\alpha\gamma}^{(2IP)}(\rho_{\alpha,\lambda}, \mathbf{r}_1, \mathbf{r}_2)$ is the two-body site direct correlation function of inhomogeneous molecular ideal system with $\rho_{\alpha,\lambda}(\mathbf{r}) = \rho_\alpha^0 + \lambda\Delta\rho_\alpha(\mathbf{r})$. A comparison of Eqs. (79) and (80) yields

$$\beta F^{ex}[\rho_\alpha] = \beta F^{ex}[\rho_\alpha^0] - \sum_\alpha \int d\mathbf{r} \Delta\rho_\alpha(\mathbf{r})[D_\alpha^{(1)}(\rho_\alpha^0) - D_\alpha^{(1IP)}(\rho_\alpha^0)]$$
$$- \sum_{\alpha\gamma} \int_0^1 d\lambda(\lambda - 1) \int\int d\mathbf{r}_1 d\mathbf{r}_2 \Delta\rho_\alpha(\mathbf{r}_1)\Delta\rho_\gamma(\mathbf{r}_2) c_{\alpha\gamma}^{(2)}(\rho_{\alpha,\lambda}, \mathbf{r}_1, \mathbf{r}_2)] \tag{81}$$

where $c_{\alpha\gamma}^{(2)}(\rho_{\alpha,\lambda}, \mathbf{r}_1, \mathbf{r}_2)$ is the site-site direct correlation function defined as

$$c_{\alpha\gamma}^{(2)}(\mathbf{r}, \mathbf{r}') \equiv \chi_{\alpha\gamma}^{(-1)IP}(\mathbf{r}, \mathbf{r}') - \chi_{\alpha\gamma}^{(-1)}(\mathbf{r}, \mathbf{r}'). \tag{82}$$

4.4 Reference Interaction Site Model

The reference interaction site model (RISM) may be understood as an extension of the Ornstein-Zernike (OZ) equation for monatomic fluids to molecular systems [59]. The RISM equations are useful to understand the structure of molecular systems from scattering measurements and to establish analytical statistical-mechanical models to represent non-ideal thermodynamic properties.

For a uniform polymeric system, the site-site correlation function may be partitioned into an ideal part and the excess:

$$\chi_{\alpha\gamma^{(r)}} = \chi_{\alpha\gamma}^{IP}(r) + \chi_{\alpha\gamma}^{ex}(r) \tag{83}$$

The ideal part is free of intermolecular interactions, i.e., it is the same as the site-site correlation function of an ideal-gas system of polyatomic molecules

$$\chi_{\alpha\gamma}^{IP}(r) = \rho_0 \omega_{\alpha\gamma}^{IP}(r) \tag{84}$$

The intramolecular site-site correlation function for the bulk ideal system is

$$\omega_{\alpha\gamma}^{IP}(r) \equiv \frac{\int d\varpi \delta(|\mathbf{r}_\alpha - \mathbf{r}_\gamma|) \exp[-\beta V_B^*(\varpi)]}{\int d\varpi \exp[-\beta V_B^*(\varpi)]} =< \delta(|\mathbf{r}_\alpha - \mathbf{r}_\gamma|) >_{IP}. \tag{85}$$

where ϖ represents molecular configuration. If the polyatomic molecule is rigid, the intramolecular correlation function is simply represented by the one-dimensional Dirac functions $\delta_{\alpha\gamma}(r - r_{\alpha\gamma})$, where $r_{\alpha\gamma}$ is the separation between sites α and γ from the same molecule.

In general, the site-site correlation function can be divided into intra- and intermolecular contributions

$$\chi_{\alpha\gamma}(r) = \rho_\alpha \omega_{\alpha\gamma}(r) + \rho_\alpha \rho_\gamma h_{\alpha\gamma}(r) \tag{86}$$

where $\omega_{\alpha\gamma}(r) \equiv < \delta(|\mathbf{r}_\alpha - \mathbf{r}_\gamma|) >$ is defined as an ensemble of the interacting system, $h_{\alpha\gamma}(r) = g_{\alpha\gamma}(r) - 1$ stands for the total correlation functions between interaction sites α and γ from *different* molecules, and $g_{\alpha\gamma}(r)$ is the standard radial distribution function.

For systems containing rigid or small polyatomic molecules, the intramolecular correlation is insensitive to the system conditions, i.e., $\omega_{\alpha\gamma}(r) = \omega_{\alpha\gamma}^{IP}(r)$. As a result, the excess site-site correlation function, $\chi_{\alpha\gamma}^{ex}(r)$, corresponds to the correlation due to intermolecular interactions

$$\chi_{\alpha\gamma}^{ex}(r) = \rho_\alpha \rho_\gamma h_{\alpha\gamma}(r) = \rho_0^2 h_{\alpha\gamma}(r). \tag{87}$$

For simplicity, here we assume that all interaction sites are distinguishable, $\rho_\alpha = \rho_\gamma = \rho_0$. Recall that the site-site direct correlation function is defined as

$$c_{\alpha\gamma}^{(2)}(r) \equiv \chi_{\alpha\gamma}^{(-1)IP}(r) - \chi_{\alpha\gamma}^{(-1)}(r) \tag{88}$$

or in matrix notation

$$\boldsymbol{\chi}^{-1} = \boldsymbol{\chi}_{IP}^{-1} - \mathbf{c} = \boldsymbol{\rho}_0^{-1}\boldsymbol{\omega}_{IP}^{-1} - \mathbf{c} \tag{89}$$

where the elements of $\mathbf{c}$ is $c_{\alpha\gamma}^{(2)}(r)$, and similar for matrices $\boldsymbol{\chi}^{-1}$, $\boldsymbol{\chi}_{IP}^{-1}$ and ω_{IP}^{-1}.

From Eq. (86), we have the matrix relation

$$\boldsymbol{\chi} = \boldsymbol{\rho}_0\boldsymbol{\omega} + \boldsymbol{\rho}_0^2\mathbf{h} \tag{90}$$

With the *assumption* of $\boldsymbol{\omega} = \boldsymbol{\omega}_{IP}$, the functional inversion of the site-site correlation functions leads to

$$(\boldsymbol{\rho}_0\hat{\boldsymbol{\omega}} + \boldsymbol{\rho}_0^2\hat{\mathbf{h}})(\boldsymbol{\rho}_0^{-1}\hat{\boldsymbol{\omega}}^{-1} - \mathbf{c}) = \mathbf{I} \tag{91}$$

where 'hat' symbol denotes 3-D Fourier transform, and $\mathbf{I}$ stands for the unit matrix. Rearrangement of Eq. (91) leads to the RISM equation

$$\hat{h}(q) = \hat{\omega}(q)\hat{c}(q)\hat{\omega}(q) + \rho_0\hat{\omega}(q)\hat{c}(q)\hat{h}(q) \tag{92}$$

Note that in the original work by Chandler and Andersen [59], the direct correlation function is defined as

$$\boldsymbol{\chi}^{-1} = \boldsymbol{\rho}_0^{-1}\boldsymbol{\omega}^{-1} - \mathbf{c} \tag{93}$$

In that case, we may say that RISM is *not* derived from the microscopic principles, but instead it follows from a set of *definitions*. If the DCFs are defined by the RISM equation, we would lose the connection between the site-site direct correlation functions and the excess Helmholtz energy functional.

4.5 Site Density Profiles

As discussed earlier in Eq. (46), $F^{ex} \equiv F - F^{IP}$ represents the difference between the intrinsic Helmholtz energies of the real and ideal polyatomic systems. Applying the functional derivative with respect to the site density on both sides of Eq. (81) gives

$$\beta\mu_{\alpha}^{ex}(\mathbf{r}) = -D_{\alpha}^{(1)}(\rho_{\alpha}^{0}) + D_{\alpha}^{(1ID)}(\rho_{\alpha}^{0}) + \delta\beta F_C/\delta\rho_{\alpha}(\mathbf{r}) \tag{94}$$

where F_C stands for the correlation energy

$$\beta F_C \equiv -\sum_{\beta\gamma}\int_0^1 d\lambda(\lambda-1)\int\int d\mathbf{r}_1 d\mathbf{r}_2 \Delta\rho_{\beta}(\mathbf{r}_1)\Delta\rho_{\gamma}(\mathbf{r}_2)c_{\beta\gamma}^{(2)}(\rho_{\beta,\lambda},\mathbf{r}_1,\mathbf{r}_2)] \tag{95}$$

Using Eqs. (33), (70) and (94), we may determine the site density profiles from that corresponding to the ideal polyatomic system

$$\begin{aligned}\rho_{\alpha}(\mathbf{r}) =& \frac{1}{\Lambda^{3M}}\int d\mathbf{x}\delta(\mathbf{r}-\mathbf{r}_{\alpha})\exp\{-\beta V_B(\mathbf{x})\\ &-\sum_{\gamma}[\beta u_{\gamma}(\mathbf{r}) - \Delta D_{\gamma}^{(1)} + \delta\beta F_C/\delta\rho_{\gamma}(\mathbf{r})]\}\end{aligned} \tag{96}$$

where $\Delta D_{\gamma}^{(1)} \equiv D_{\gamma}^{(1)}(\rho_{\gamma}^{0}) - D_{\gamma}^{(1ID)}(\rho_{\gamma}^{0})$. The unknown constants in Eq. (96) can be removed by applying the same equation to a uniform polyatomic system of the same bulk density ρ_{α}^{0}

$$\rho_{\alpha}^{0} = \frac{1}{\Lambda^{3M}}\int d\mathbf{x}\delta(\mathbf{r}-\mathbf{r}_{\alpha})\exp\{-\beta V_B(\mathbf{x}) - \beta\mu_{\alpha} + \sum_{\gamma}\Delta D_{\gamma}^{(1)}\} \tag{97}$$

A comparison of Eqs. (96) and (97) yields

$$\begin{aligned}\rho_{\alpha}(\mathbf{r}) =& \frac{\int d\mathbf{x}\delta(\mathbf{r}-\mathbf{r}_{\alpha})\exp\{-\beta V_B(\mathbf{x}) - \beta\sum_{\gamma}\lambda_{\gamma}(\mathbf{r}_{\gamma})\}}{\int d\mathbf{x}\delta(\mathbf{r}-\mathbf{r}_{\alpha})\exp\{-\beta V_B(\mathbf{x})\}}\\ =& \rho_{\alpha}^{0}e^{-\beta\lambda_{\alpha}(\mathbf{r})} < \exp[-\beta\sum_{\gamma\neq\alpha}\lambda_{\gamma}(\mathbf{r}_{\gamma})] >_{IP}\end{aligned} \tag{98}$$

where $\lambda_{\gamma}(\mathbf{r}) \equiv \varphi_{\gamma}(\mathbf{r}) + \delta F_C/\delta\rho_{\alpha}(\mathbf{r})$, and $< \cdots >_{IP}$ represents the configurational average for tagged molecule in the ideal molecular system with the site α position fixed at $\mathbf{r}$. Equation (98) represents one of the most important equations in site formulism of the molecular DFT.

4.6 Thermodynamic Potentials

To derive thermodynamic properties from the site density profiles, we may compare the grand potential for an inhomogeneous polyatomic system with that of a uniform system of the same bulk density (Panels B and D in Fig. 3, respectively)

$$\Delta\Omega \equiv \Omega[\rho_\alpha(\mathbf{r})] - \Omega[\rho_\alpha^0] = \Delta F^{IP} + \Delta F^{ex} + \sum_\alpha \int d\mathbf{r}[\varphi_\alpha(\mathbf{r})\rho_\alpha(\mathbf{r}) - \mu_\alpha\rho_\alpha(\mathbf{r}) - \mu_\alpha\rho_\alpha^0]\}. \tag{99}$$

For the ideal polymeric system (Panel C in Fig. 3), the grand potential is given by Eq. (35). The difference between the inhomogeneous and uniform intrinsic Helmholtz energy, ΔF^{IP}, is thus given by

$$\Delta F^{IP} = -\frac{k_B T}{M}\sum_\alpha \int d\mathbf{r}[\rho_\alpha(\mathbf{r}) - \rho_\alpha^0] + \sum_\alpha \int d\mathbf{r}[\varphi_\alpha^{IP}(\mathbf{r})\rho_\alpha(\mathbf{r}) - \mu_\alpha^{IP}\rho_\alpha(\mathbf{r}) - \mu_\alpha^{IP}\rho_\alpha^0]\}. \tag{100}$$

From Eq. (81), we may find the difference between the excess intrinsic Helmholtz energies of the inhomogeneous and uniform polyatomic systems:

$$\beta\Delta F^{ex}[\rho_\alpha] = -\sum_\alpha \int d\mathbf{r}\Delta\rho_\alpha(\mathbf{r})[D_\alpha^{(1)}(\rho_\alpha^0) - D_\alpha^{(1IP)}(\rho_\alpha^0)] + \beta F_C. \tag{101}$$

Because $\rho_\alpha(\mathbf{r})$ satisfies the variational principle for both the real and ideal polyatomic systems, we have from Eqs. (68) and (101)

$$\frac{\delta F^{ex}}{\delta\rho_\alpha(\mathbf{r})} = -[\varphi_\alpha(\mathbf{r}) - \mu_\alpha] + [\varphi_\alpha^{IP}(\mathbf{r}) - \mu_\alpha^{IP}] = -\beta^{-1}[D_\alpha^{(1)}(\rho_\alpha^0) - D_\alpha^{(1ID)}(\rho_\alpha^0)] + \frac{\delta F_C}{\delta\rho_\alpha(\mathbf{r})}. \tag{102}$$

Substituting Eqs. (100) and (101) into (99), with the help of Eq. (102), we derive a succinct expression for the grand potential

$$\Delta\Omega = -\frac{k_B T}{M}\sum_\alpha \int d\mathbf{r}[\rho_\alpha(\mathbf{r}) - \rho_\alpha^0] + F_C - \sum_\alpha \int d\mathbf{r}\rho_\alpha(\mathbf{r})[\delta F_C/\delta\rho_\alpha(\mathbf{r})]. \tag{103}$$

Equation (103) indicates that the grand potential of an inhomogeneous polyatomic system can be calculated from that of a bulk system and the site density profiles.

It is worth noting that the above procedure is formally exact for calculating the site density and the grand potential. For a specific polyatomic system of practical interest, however, DFT calculations require an analytical expression for the correlation energy,F_C, and the grand potential for the bulk system, $\Omega(\rho_\alpha^0)$. The former is defined in terms of the site-site direct correlation functions of inhomogeneous systems, which can be calculated from an integral-equation theory or molecular simulations. For bulk systems, the grand potential can be easily determined from an equation of state, i.e., $\Omega_b/V = -P(T, \rho_0)$, where V is the system volume, and P stands for the bulk pressure. In practical applications, approximations will be inevitable for both the correlation energy and the bulk equation of state.

5 The Bridge Functional and Universality *Ansatz*

Without loss of generality, the functional integration for the correlation energy may be written in terms of the direct correlation function of the uniform reference system

$$\beta F_C[\rho_\alpha(\mathbf{r})] = -\frac{1}{2}\sum_{\alpha,\gamma}\int d\mathbf{r}\int d\mathbf{r}'\Delta\rho_\alpha(\mathbf{r})\Delta\rho_\gamma(\mathbf{r}')c^0_{\alpha\gamma}(|\mathbf{r}-\mathbf{r}'|) + F_B[\rho_\alpha(\mathbf{r})] \tag{104}$$

where $c^0_{\alpha\gamma}(r)$ denotes the site-site direct correlation function of a uniform system with site density ρ^0_α. The first term on the right side of Eq. (104) corresponds to a quadratic expansion of the excess intrinsic Helmholtz energy with respect to that of the uniform system [see Eq. (81)], and the second term, $F_B[\rho_\alpha(\mathbf{r})]$, serves as a definition of the bridge functional.

As stated above, the bridge functional accounts for all contributions to the excess Helmholtz free energy beyond the quadratic expansion, i.e., all high-order terms in the functional Taylor expansion of the excess intrinsic Helmholtz energy. The correlation Helmholtz energy and the bridge functional can be similarly defined in terms of the quadratic expansion with respect to the configurational density profiles

$$\beta F_C[\rho(\mathbf{x})] = -\frac{1}{2}\int d\mathbf{x}\int d\mathbf{x}'\Delta\rho(\mathbf{x})\Delta\rho(\mathbf{x}')c^0(|\mathbf{x}-\mathbf{x}'|) + F_B[\rho(\mathbf{x})]. \tag{105}$$

Both Eqs. (104) and (105) are formally exact and immaterial to the additivity of the intermolecular potentials. Because the molecular and site direct correlation functions are not equivalent, the corresponding bridge functionals are in general not the same, i.e., $F_B[\rho(\mathbf{x})] \neq F_B[\rho_\alpha(\mathbf{r})]$.

As shown schematically in Fig. 4, the intermolecular potential can be generally divided into short and long-range contributions as proposed by van der Waals many years ago. Although *a priori* knowledge is not available for the bridge functional, it has been well documented in the liquid-state literature that the high-order terms in the density functional expansion are dominated by short-range interactions, insensitive to the mathematical details of the long-range intermolecular forces [23]. The insensitivity of the bridge functional to the precise form of the intermolecular potential is known as *the bridge universality*, proposed first by Rosenfeld [60]. The *universality ansatz* was supported by the fact that the quadratic approximation (i.e., without the

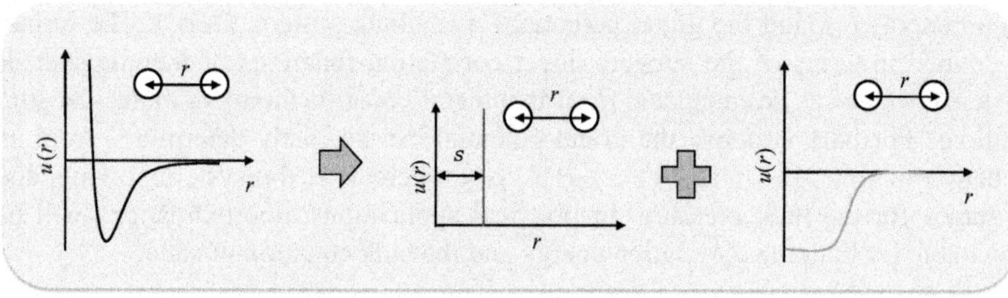

Fig. 4 Van der Waals' wisdom on thermodynamic non-ideality. The thermodynamic properties of a non-ideal system can be accounted for by a separate consideration of the molecular excluded volume and long-range attraction in the intermolecular potential

bridge functional) conforms to the exact results at both the "ideal gas" and "ideal liquid" limits for the long-range components of the intermolecular potentials.

With the universality *ansatz*, we may approximate the bridge functional with that of a reference hard-sphere (HS) system:

$$F_B[\rho_\alpha(\mathbf{r})] \approx F_{HS}^{ex}[\rho_\alpha(\mathbf{r})] - F_{HS}^{ex}(\rho_\alpha^0) - \sum_\alpha \mu_{\alpha,HS}^{ex} \int d\mathbf{r} \Delta\rho_\alpha(\mathbf{r}) + \frac{k_B T}{2} \sum_{\alpha,\beta} \int d\mathbf{r}_1 \int d\mathbf{r}_2 \Delta\rho_\alpha(\mathbf{r}_1) \Delta\rho_\beta(\mathbf{r}_2) c_{\alpha,\beta}^{HS}(|\mathbf{r}_1 - \mathbf{r}_2|) \tag{106}$$

where $F_{HS}^{ex}[\rho_\alpha(\mathbf{r})]$ represents the excess Helmholtz energy functional of the HS reference system, $\mu_{\alpha,HS}^{ex}$ and $F_{HS}^{ex}(\rho_\alpha^0)$ are, respectively, the excess chemical potential and the excess Helmholtz energy of the reference system at bulk density ρ_α^0; and $c_{\alpha\beta}^{HS}(|\mathbf{r}_1 - \mathbf{r}_2|)$ is the corresponding direct correlation functions.

Accurate expressions are available for the excess Helmholtz energy functional and correlation functions of inhomogeneous hard-sphere systems [61, 62]. As suggested by Rosenfeld [60], the effective HS diameter, σ_α, can be fixed by the Lado criterion [63, 64]

$$\sum_{\alpha,\beta} \rho_\alpha^0 \rho_\beta^0 \int d\mathbf{r}[g_{\alpha\beta}^0(r) - g_{\alpha\beta}^{HS}(r; \sigma_\alpha)] \frac{\partial b_{\alpha\beta}^{HS}(r; \sigma_\alpha)}{\partial \sigma_\alpha} = 0 \tag{107}$$

where $g_{\alpha\beta}^0(r)$ and $g_{\alpha\beta}^{HS}(r; \sigma_\alpha)$ are the radial distribution functions of the molecular and the HS reference systems at the same bulk atomic density ρ_α^0, respectively, and $b_{\alpha\beta}^{HS}(r; \sigma_\alpha)$ is the bridge function of the HS reference system.

Using the bridge functional from the fundamental measure theory (FMT) as an input, Rosenfeld demonstrated that the universality hypothesis performs well for a wide variety of monatomic systems including charged Yukawa fluids and plasma mixtures (see Ref. [65] for an overview). It has also been shown that universality hypothesis works reasonably well for ionic liquids [66] and various models for liquid water [67]. Application of the universality functional to anisotropic molecular fluids and mixtures has also been explored [54].

6 Perspectives

Conventional approaches in molecular modeling hinge on different theoretical frameworks that typically include quantum-mechanical (QM) calculations for electronic properties, molecular dynamics (MD) or Monte Carlo (MC) simulation for short-time and small-length-scale correlations, and continuous equations for larger-scale systems. Although recent developments make it possible to connect different quantum and classical approaches *via* hierarchical modeling, integration of different models for various length and time scales is far from perfect. Owing to uncertainties affiliated with calculations at each scale, a careful combination of different computational methods is critically important. An optimal multi-scale procedure should reflect not only theoretical accuracy at individual scales but, perhaps more important, a reliable linkage of different methods to attain the best overall performance.

Density functional theory (DFT) plays an important role in molecular modeling. The essential idea is to predict the properties of many-body systems using the one-body density profiles as fundamental variables. Whereas the original concepts were introduced in the context of inhomogeneous electronic systems at 0 K, the mathematical procedure is applicable to electronic systems at finite temperature [68, 69], multicomponent quantum mixtures [70], as well as to diverse thermodynamic systems ranging from those involved in materials characterization [71], electric double layers [72] to lipids [73] and liquid crystals [74]. Because DFT circumvents an explicit consideration of multi-body wave functions or microstates, its computational efficiency is far superior to conventional ab initio or simulation methods. The computational efficiency is important in particular from a practical perspective because it enables customization of molecular models for specific systems and for more systematic calibration of the theoretical results with experimental data [75].

Conceptually, DFT is equivalent to the variational principle of thermodynamics: instead of expressing thermodynamic variables as functions of a small set of macroscopic parameters such as temperature, pressure and composition, DFT defines thermodynamic potentials as functionals of density profiles (viz. structure). Given a density functional for the excess free energy, we can obtain both microscopic structure and thermodynamic properties from the variational principle of equilibrium, i.e., the density profiles minimize the free energy. To a certain degree, the shortcomings of DFT are inherent in thermodynamics. Although thermodynamics provides no *a priori* knowledge on the fundamental equation (relating for example entropy as a function of internal energy and system volume), DFT by itself does not give the density functionals. Nevertheless, like equations of state for thermodynamic systems, reliable functionals can be formulated following the basic principles of quantum and statistical mechanics, in combination with creative mathematical strategies. Importantly, quantitative predictions can often be achieved even with relatively simple approximations [75, 76].

Like conventional molecular simulations, existing molecular DFT calculations are based on the pairwise additive potentials that do not account for the polarization effects. However, the importance of non-additive interactions in molecular systems has been well recognized [1]. Inclusion of the polarization effects may drastically increase the computational costs of traditional simulation methods, typically by one order magnitude in comparison to that with non-polarizable models. From a practical point of view, there is yet no convincing procedure to properly account for the effects of polarization on solvation free energy calculations. The computational efficiency of DFT makes it an outstanding candidate for future development and application of new polarizable force fields.

References

1. A.V. Akimov, O.V. Prezhdo, Large-scale computations in chemistry: a bird's eye view of a vibrant field. Chem. Rev. **115**, 5797–5890 (2015)
2. A.S. Christensen, T. Kubar, Q. Cui, M. Elstner, Semiempirical quantum mechanical methods for noncovalent interactions for chemical and biochemical applications. Chem. Rev. **116**, 5301–5337 (2016)
3. T. Bereau, D. Andrienko, K. Kremer, Research update: computational materials discovery in soft matter. APL Mater. **4**, 053101 (2016)
4. M.W. van der Kamp, A.J. Mulholland, Combined Quantum Mechanics/Molecular Mechanics (QM/MM) methods in computational enzymology. Biochemistry **52**, 2708–2728 (2013)
5. R.O. Jones, Density functional theory: its origins, rise to prominence, and future. Rev. Mod. Phys. **87**, 897–923 (2015)
6. A.R. Leach, *Molecular Modelling: Principles and Applications*, 2nd edn. (Prentice Hall, Harlow, England, New York, 2001)
7. https://en.wikipedia.org/wiki/Force_field_(chemistry)
8. S.L. Mayo, B.D. Olafson, W.A. Goddard, Dreiding—a generic force-field for molecular simulations. J. Phys. Chem.-Us **94**, 8897–8909 (1990)
9. A.K. Rappe, C.J. Casewit, K.S. Colwell, W.A. Goddard, W.M. Skiff, Uff, a full periodic-table force-field for molecular mechanics and molecular-dynamics simulations. J. Am. Chem. Soc. **114**, 10024–10035 (1992)
10. T. Düren, L. Sarkisov, O.M. Yaghi, R.Q. Snurr, Design of new materials for methane storage. Langmuir **20**, 2683–2689 (2004)
11. K.S. Walton, R.Q. Snurr, Applicability of the BET method for determining surface areas of microporous metal-organic frameworks. J. Am. Chem. Soc. **129**, 8552–8556 (2007)
12. S. Keskin, J. Liu, R.B. Rankin, J.K. Johnson, D.S. Sholl, Progress, opportunities, and challenges for applying atomically detailed modeling to molecular adsorption and transport in metal-organic framework materials. Ind. Eng. Chem. Res. **48**, 2355–2371 (2009)
13. R. Huey, G.M. Morris, A.J. Olson, D.S. Goodsell, A semiempirical free energy force field with charge-based desolvation. J. Comput. Chem. **28**, 1145–1152 (2007)
14. T.P. Senftle, S. Hong, M.M. Islam, S.B. Kylasa, Y. Zheng, Y.K. Shin, C. Junkermeier, R. Engel-Herbert, M.J. Janik, H.M. Aktulga, T. Verstraelen, A. Grama, A.C. vanDuin, The ReaxFF reactive force-field: development, applications and future directions. Comput. Mater. **2**, Artn 15011 (2016)
15. A.D. Becke, Perspective: Fifty years of density-functional theory in chemical physics. J. Chem. Phys. **140**, Artn 18a301 (2014)
16. K. Berland, V.R. Cooper, K. Lee, E. Schroder, T. Thonhauser, P. Hyldgaard, B.I. Lundqvist, van der Waals forces in density functional theory: a review of the vdW-DF method. Rep. Prog. Phys. **78**, Artn 066501 (2015)
17. L.P. Wang, T.J. Martinez, V.S. Pande, Building force fields: an automatic, systematic and reproducible approach. J. Phys. Chem. Lett. **5**, 1885–1891 (2014)

18. S. Grimme, A General Quantum Mechanically Derived Force Field (QMDFF) for molecules and condensed phase simulations. J. Chem. Theory Comput. **10**, 4497–4514 (2014)
19. S.W. Rick, S.J. Stuart, Potentials and algorithms for incorporating polarizability in computer simulations. Rev. Comput. Chem. **18**, 89–146 (2002)
20. M.G. Saunders, G.A. Voth, Coarse-graining methods for computational biology. Annu. Rev. Biophys. **42**, 73–93 (2013)
21. C.G. Gray, K.E. Gubbins, C.G. Joslin, *Theory of Molecular Fluids: Fundamentals* (Oxford University Press, Oxford, New York, 1984)
22. C.G. Gray, K.E. Gubbins, C.G. Joslin, *Theory of Molecular Fluids II: Applications* (Oxford University Press, Oxford, New York, 2011)
23. J.P. Hansen, I.R. McDonald, *Theory of Simple Liquids*, 3rd edn. (Academic Press, London, 2006)
24. D. Chandler, Structure of molecular liquids. Annu. Rev. Phys. Chem. **29**, 441–471 (1978)
25. E.L. Ratkova, D.S. Palmer, M.V. Fedorov, Solvation thermodynamics of organic molecules by the molecular integral equation theory: approaching chemical accuracy. Chem. Rev. **115**, 6312–6356 (2015)
26. J.L.F. Abascal, C. Vega, A general purpose model for the condensed phases of water: TIP4P/2005. J. Chem. Phys. **123** (2005)
27. S.H. Chong, S. Ham, Aqueous interaction site integral-equation theory that exactly takes into account intramolecular correlations. J. Chem. Phys. **137**, Artn 154101 (2012)
28. C. Ebner, W.F. Saam, D. Stroud, Density-functional theory of simple classical fluids. 1. surfaces. Phys. Rev. A **14**, 2264–2273 (1976)
29. R. Evans, Nature of the liquid-vapor interface and other topics in the statistical-mechanics of nonuniform, classical fluids. Adv. Phys. **28**, 143–200 (1979)
30. R. Evans, in *Fundamentals of Inhomogeneous Fluids*, ed. by D. Henderson, M. Dekker (New York, 1992), pp. 85–175
31. Y. Singh, Density-functional theory of freezing and properties of the ordered phase. Phys. Rep. **207**, 351–444 (1991)
32. H.T. Davis, *Statistical Mechanics of Phases, Interfaces, and Thin Films* (VCH, New York, 1996)
33. A.P. Hughes, A.J. Archer, U. Thiele, An introduction to inhomogeneous liquids, density functional theory, and the wetting transition. Am. J. Phys. **82**, 1119–1129 (2014)
34. G. Jeanmairet, N. Levy, M. Levesque, D. Borgis, Introduction to classical density functional theory by a computational experiment. J. Chem. Educ. **91**, 2112–2115 (2014)
35. G. Jeanmairet, M. Levesque, V. Sergiievskyi, D. Borgis, in *Computational Trends in Solvation and Transport in Liquids*, vol. 28, ed. by G. Sutmann, J. Grotendorst, G. Gompper, D. Marx (Schriften des Forschungszentrums Jülich, 2015)
36. S. Sokolowski, J. Ilnytskyi, O. Pizio, Description of interfaces of fluid-tethered chains: advances in density functional theories and off-lattice computer simulations. *Condens. Matter Phys.* **17**, Artn 12601 (2014)
37. M. Schmidt, M. Burgis, W.S.B. Dwandaru, G. Leithall, P. Hopkins, Recent developments in classical density functional theory: Internal energy functional and diagrammatic structure of fundamental measure theory. *Condens. Matter Phys.* **15**, Article number 43603 (2012)
38. C.P. Emborsky, Z.Z. Feng, K.R. Cox, W.G. Chapman, Recent advances in classical density functional theory for associating and polyatomic molecules. Fluid Phase Equilibr. **306**, 15–30 (2011)
39. J. Forsman, R. Szparaga, S. Nordholm, C.E. Woodward, R. Penfold, in *Ionic Liquids—Classes and Properties*, ed. by S.T. Handy (2011), pp. 127–150. http://www.intechopen.com
40. J.F. Lutsko, Recent developments in classical density functional theory. Adv. Chem. Phys. **144**, 1–92 (2010)
41. J.Z. Wu, Z.D. Li, Density-functional theory for complex fluids. Annu. Rev. Phys. Chem. **58**, 85–112 (2007)
42. J.Z. Wu, Density functional theory for chemical engineering: from capillarity to soft materials. Aiche J. **52**, 1169–1193 (2006)

43. D.W. Oxtoby, Density functional methods in the statistical mechanics of materials. Annu. Rev. Mater. Res. **32**, 39–52 (2002)
44. L.J.D. Frink, A.G. Salinger, M.P. Sears, J.D. Weinhold, A.L. Frischknecht, Numerical challenges in the application of density functional theory to biology and nanotechnology. J. Phys.-Condens. Matter **14**, 12167–12187 (2002)
45. H. Lowen, Density functional theory of inhomogeneous classical fluids: recent developments and new perspectives. J. Phys.-Condens. Matter **14**, 11897–11905 (2002)
46. P. Hohenberg, W. Kohn, Inhomogeneous electron gas. Phys. Rev. **136**, B864–B871 (1964)
47. N.D. Mermin, Thermal properties of the inhomogeneous electron gas. Phys. Rev. **137**, A1441–1443 (1965)
48. R.G. Parr, W. Yang, *Density-functional Theory of Atoms and Molecules* (Oxford University Press, New York, 1989)
49. D. Frenkel, B. Smit, *Understanding Molecular Simulation: From Algorithms to Applications*, 2nd edn. (Academic Press, San Diego, 2002)
50. G.H. Fredrickson, *The Equilibrium Theory of Inhomogeneous Polymers* (Oxford University Press, New York, 2006)
51. D.P. Cao, T. Jiang, J. Z. Wu, A hybrid method for predicting the microstructure of polymers with complex architecture: Combination of single-chain simulation with density functional theory. J. Chem. Phys. **124**, Artn 164904 (2006)
52. X.F. Xu, D.P. Cao, X.R. Zhang, W.C. Wang, Universal version of density-functional theory for polymers with complex architecture. *Phys. Rev. E* **79**, Artn 021805 (2009)
53. S. Zhao, R. Ramirez, R. Vuilleumier, D. Borgis, Molecular density functional theory of solvation: from polar solvents to water. J. Chem. Phys. **134**, 194102 (2011)
54. G. Jeanmairet, N. Levy, M. Levesque, D. Borgis, Molecular density functional theory of water including density-polarization coupling. J. Phys.: Condens. Matter **28**, 244005 (2016)
55. R. Ramirez, M. Mareschal, D. Borgis, Direct correlation functions and the density functional theory of polar solvents. Chem. Phys. **319**, 261–272 (2005)
56. D. Chandler, L.R. Pratt, Statistical-mechanics of chemical-equilibria and intramolecular structures of nonrigid molecules in condensed phases. J. Chem. Phys. **65**, 2925–2940 (1976)
57. M.S. Wertheim, Thermodynamic perturbation-theory of polymerization. J. Chem. Phys. **87**, 7323–7331 (1987)
58. D. Chandler, J.D. McCoy, S.J. Singer, Density functional theory of nonuniform polyatomic systems. 1. General formulation. J. Chem. Phys. **85**, 5971–5976 (1986)
59. D. Chandler, H.C. Andersen, Optimized cluster expansions for classical fluids. 2. Theory of molecular liquids. J. Chem. Phys. **57**, 1930–1937 (1972)
60. Y. Rosenfeld, Free energy model for inhomogeneous fluid mixtures: Yukawa-charged hard spheres, general interactions, and plasmas. J: Chem. Phys. **98**, 8126–8148 (1993)
61. Y.X. Yu, J.Z. Wu, Structures of hard-sphere fluids from a modified fundamental-measure theory. J. Chem. Phys. **117**, 10156–10164 (2002)
62. R. Roth, R. Evans, A. Lang, G. Kahl, Fundamental measure theory for hard-sphere mixtures revisited: the White Bear version. J. Phys.: Condens. Matter **14**, 12063–12078 (2002)
63. F. Lado, S.M. Foiles, N.W. Ashcroft, Solutions of the reference-hypernetted-chain equation with minimized free energy. Phys. Rev. A **28**, 2374–2379 (1983)
64. G. Kahl, B. Bildstein, Y. Rosenfeld, Structure and thermodynamics of binary liquid mixtures: universality of the bridge functional. Phys. Rev. E **54**, 5391 (1996)
65. Y. Rosenfeld, Structure and effective interactions in multi-component hard-sphere liquids: the fundamental-measure density functional approach. J. Phys.: Condens. Matter **14**, 9141–9152 (2002)
66. J.Z. Wu, T. Jiang, D.E. Jiang, Z.H. Jin, D. Henderson, A classical density functional theory for interfacial layering of ionic liquids. Soft Matter **7**, 11222–11231 (2011)
67. Y. Liu, J. Fu, J. Wu, High-throughput prediction of the hydration free energies of small molecules from a classical density functional theory. J. Phys. Chem. Lett. **4**, 3687–3691 (2013)
68. Y. Liu, J.Z. Wu, Structure and thermodynamic properties of relativistic electron gases. Phys. Rev. E **90**, Artn 012141 (2014)

69. Y. Liu, J.Z. Wu, An improved classical mapping method for homogeneous electron gases at finite temperature. J. Chem. Phys. **141**, Artn 064115 (2014)
70. A. Chakraborty, M.V. Pak, S. Hammes-Schiffer, Development of electron-proton density functionals for multicomponent density functional theory. Phys. Rev. Lett. **101**, Artn 153001 (2008)
71. J. Landers, G.Y. Gor, A.V. Neimark, Density functional theory methods for characterization of porous materials. Colloid Surface A **437**, 3–32 (2013)
72. J. Jiang, D.P. Cao, D. Henderson, J. Z. Wu, Revisiting density functionals for the primitive model of electric double layers. J. Chem. Phys. **140**, Artn 44714 (2014)
73. L.J.D. Frink , A.L. Frischknecht, Density functional theory approach for coarse-grained lipid bilayers. Phys. Rev. E **72**, Artn 041923 (2005)
74. S. Singh, Phase transitions in liquid crystals. Phys. Rep. **324**, 108–269 (2000)
75. J. Fu, J.Z. Wu, Toward high-throughput predictions of the hydration free energies of small organic molecules from first principles. Fluid Phase Equilibr. **407**, 304–313 (2016)
76. J. Fu, Y. Liu, J. Z. Wu, Molecular density functional theory for multiscale modeling of hydration free energy. Chem. Eng. Sci. **126**, 370–382 (2015)

Classical Density Functional Theory of Polymer Fluids

Jan Forsman and Clifford E. Woodward

1 Introduction

Hohenberg and Kohn first established Density Functional Theory (DFT) in a quantum mechanical framework more than fifty years ago [1]. Their formulation relates to the ground-state energy of electrons in the presence of external fields. That work was generalized by Mermin to finite temperatures, demonstrating that the thermodynamic free energy was minimized by the equilibrium number density of the fluid [2]. This inspired the development of *classical* DFT, which now forms a powerful theoretical tool for the treatment of non-uniform fluids and solutions. While the number density for monomeric (or single atom) fluids is straightforwardly defined, the corresponding quantity for molecular or polymeric fluid models, is not unique. These, more realistic, chemical models usually consist of molecular entities containing a number of "atomic" sites, thus it is not exactly clear what "density" should be used in the density functional theory for polymeric fluids. One quantity that we could envisage using would be the density of monomeric sites (monomer site densities), which is a function of the position coordinate of the site. This approach was used by Chandler, McCoy and Singer [3] in a DFT for molecules. Another choice is the *molecular* density, which depends upon the coordinates of all sites in the molecular model [4], as was utilized by Woodward [5], in arguably the first formulation of DFT for fluids consisting of long flexible polymeric chains. Since then, most formulations of so-called *polymer* DFT (PDFT) have utilized the same theoretical framework. This is based on a separation of the free energy into ideal and excess contributions. The ideal contribution is obtained exactly in terms of the molecular density, while the excess

J. Forsman (✉)
Theoretical Chemistry, Chemical Centre, P.O.Box 124, 221 00 Lund, Sweden
e-mail: jan.forsman@teokem.lu.se

C.E. Woodward
School of Physical Environmental and Physical Sciences University College,
University of New South Wales, ADFA Canberra ACT, Kensington 2600, Australia
e-mail: c.woodward@adfa.edu.au

J. Wu (ed.), *Variational Methods in Molecular Modeling*,
Molecular Modeling and Simulation, DOI 10.1007/978-981-10-2502-0_4

term is treated as a functional of the site densities. In this way, connectivity effects, so important in polymer systems, are accurately accounted for within the ideal term, while non-bonding interactions contribute to the excess free energy.

Unlike the case of simple fluids, a closed expression for the free energy functional of ideal chains in terms of the monomer site densities has so far been elusive. On the other hand, the ideal free energy is a relatively simple functional of the multi-point molecular density. The use of this multi-point density does introduce an apparent complexity to the formalism, as it is a $3N$-dimensional function, where N is the number of monomer sites, which is generally very large. As we shall show, solution of the PDFT usually leads to integral equations and thus there is potential for $3N$-dimensional integrals to appear as part of the solution algorithm. However, for nearest-neighbor bonding constraints, such as the classic freely-jointed polymer model, the solution to the PDFT is rather straightforward. Indeed, the equilibrium monomer site densities turn out to be expressible in terms of N coupled integral equations, wherein the dimensionality reflects the symmetry of the perturbing field, not the length of the polymer. For example, the problem is trivially 1-dimensional in planar and spherical geometries and can thus be solved via straightforward numerical techniques. Having said this, the solution still requires (N times) more work than what is required to solve the DFT for single-site fluids. But this is expected and in some sense unavoidable, as in the presence of end-effects, site densities will generally differ along the length of the polymer chain and must be solved for separately, an o(N) problem. It is only for very long polymers that end-effects disappear and in this case one needs only solve a single integral equation under these conditions. Surprisingly, the same is also true for equilibrium (or living) polymers, which possess an exponential molecular weight distribution, which also lack end effects. This notwithstanding, even in the presence of end-effects these calculations are still much less demanding than direct computer simulations of the chain fluid, and a substantial number of model scenarios concerning polymer fluids and solutions under non-uniform conditions can be relatively easily tackled. Having said this, the level of difficulty in solving the DFT increases with the required specificity in the intra-molecular structure. More precisely, the dimensionality of the problem grows with the number of adjacent monomers required to describe the intra-molecular structure. For example, the PDFT for stiff polymer chains, with angularly-dependent nearest neighbor bonding, must be solved in 2-dimensions, reflecting the next nearest neighbor interactions between monomers. When the problem becomes too big for solving via direct numerical methods, single-chain simulations can be fruitfully employed.

2 Classical Density Functional for Polymers

The classical DFT for single-site monatomic liquids expresses the Helmholtz free energy, F, as a functional of the fluid density. It is useful to decompose the free energy as a sum of ideal and excess free energy terms and for a simple monatomic fluid, the functional has the following general form,

$$F[n(\mathbf{r})] = F^{id}[n(\mathbf{r})] + F^{ex}[n(\mathbf{r})] \tag{1}$$

where $n(\mathbf{r})$ is the local particle density. The ideal term is given by

$$F^{id}[n(\mathbf{r})] = k_B T \int n(\mathbf{r}) \left(\ln[n(\mathbf{r})] - 1\right) d\mathbf{r} \tag{2}$$

and the excess component, $F^{ex}[n(\mathbf{r})]$, accounts for all the contributions to the free energy due to particle interactions, which are not included in the ideal functional.

In the case of a polymeric fluid, there are multiple sites per molecule and the functional takes on a more complex form. In this exposition we will only consider simple linear chains of identical monomers (homopolymers). The generalization to more complex architectures is rather straightforward, but requires more complicated solution algorithms. A simple example is a fluid composed of chains of N tangentially bonded hard spheres (see Fig. 1). All spheres are identical with diameter σ and the bonds are freely rotating. This is the well-known freely jointed model. The free energy functional for linear chain models like this can be decomposed into ideal and excess terms, using the concept of an *equivalent* ideal fluid (EIF) [5]. The EIF is the ideal chain model (no non-bonded or intermolecular interactions), which has the same monomer site densities as the fluid of real chains. The free energy can be formally decomposed as,

$$F[N(\mathbf{R})] = F^{id}[N(\mathbf{R})] + F^{ex} \tag{3}$$

where $\mathbf{R} = \{\mathbf{r}_1, \ldots\ldots, \mathbf{r}_N\}$ denotes the coordinates of N monomers on a single polymer chain and $N(\mathbf{R})$ represents the multi-point polymeric density of the EIF. This function can be used to generate the site densities $n_i(\mathbf{r})$ using the relation,

$$n_i(\mathbf{r}) = \int \delta(\mathbf{r} - \mathbf{r_i}) N(\mathbf{R})\, d\mathbf{R} \tag{4}$$

We reiterate that $N(\mathbf{R})$ is generally not equal to the polymeric density of the *real* fluid, but that of the EIF, which has the same monomeric density. The free energy of the EIF is given by

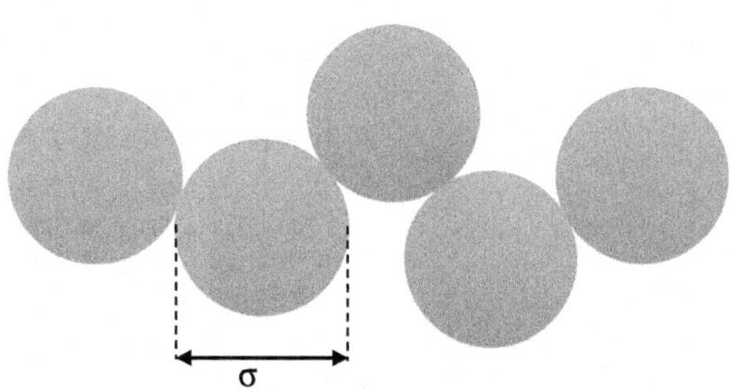

Fig. 1 A schematic diagram for the simple polymer model composed of freely and tangentially jointed spheres, each with diameter σ

$$F^{id}[N(\mathbf{R})] = k_B T \int N(\mathbf{R}) \left(\ln[N(\mathbf{R})] - 1\right) d\mathbf{R} + \int N(\mathbf{R}) \Phi^{(b)}(\mathbf{R}) d\mathbf{R} \tag{5}$$

Here we shall set $\Phi^{(b)}(\mathbf{R})$ equal to the bare bonding potential, though other choices are possible. One of the current open questions in PDFT is how to optimize $\Phi^{(b)}(\mathbf{R})$. Intuition would suggest that this would ideally be chosen to emulate the real intra-molecular distribution. For rigid molecules this would be achieved with the bare bonding potential, but that is not the case for flexible polymers. It would seem that these considerations would be usefully informed by scaling results for intra-molecular correlations of the polymer chains. We also note that different choices for $\Phi^{(b)}(\mathbf{R})$, will necessarily alter the form of the excess free energy.

For the rigid bonds of the freely jointed chain, our current choice of the bare bonding potential gives the following relation,

$$e^{-\beta \Phi^{(b)}(\mathbf{R})} = \prod_{i=1}^{N-1} \frac{\delta(|\mathbf{r}_i - \mathbf{r}_{i+1}| - \sigma)}{4\pi\sigma^2} \tag{6}$$

with $\beta = 1/k_B T$, $\delta(r)$ is the Dirac delta function and σ is also the bond length between the monomers, as the spheres are tangential. Note that we have assumed the bonding potential has an additive constant, which normalizes the Boltzmann factor. This constant can be physically interpreted as an entropic contribution due to free bond rotation.

The excess free energy, F^{ex}, accounts for the additional contributions to the free energy not included in the EIF. This will be dominated by non-bonding monomer-monomer interactions. Though, as mentioned earlier, the excess functional will also have an intrinsic intra-molecular component, which varies with the choice of $\Phi^{(b)}(\mathbf{R})$. However, we will assume this is relatively small for our current choice. The excess free energy, can always be formally expressed as a functional of the set of monomer site densities, i.e., $F^{ex}[\{n_i(\mathbf{r})\}]$ [5].

For a polymer fluid in contact with a bulk reservoir, it is appropriate to use the grand canonical ensemble. The corresponding grand free energy functional, Ω, is obtained as the Legendre transform of the Helmholtz free energy F,

$$\Omega = k_B T \int N(\mathbf{R}) \left(\ln[N(\mathbf{R})] - 1\right) d\mathbf{R} + \int N(\mathbf{R}) \Phi^{(b)}(\mathbf{R}) d\mathbf{R} + F^{ex}[\{n_i(\mathbf{r})\}] + \int [\Psi^0(\mathbf{R}) - \mu] N(\mathbf{R}) d\mathbf{R} \tag{7}$$

Here, μ is the chemical potential of polymer molecules in the bulk and we have also accounted for the presence of an external potential, $\Psi^0(\mathbf{R})$. We shall assume that this can be written as a sum of single site terms, namely,

$$\Psi^0(\mathbf{R}) = \sum_{i=1}^{N} \psi_i^0(\mathbf{r}_i) \tag{8}$$

where we have allowed for different interactions for the various sites.

The accuracy of PDFT essentially boils down to how well the excess free energy functional has been approximated, as the ideal contribution is essentially exact. For the freely jointed model considered in this example, the excess term accounts for the steric or hard sphere interactions between monomers. If longer-ranged interactions, such as dispersion and electrostatic potentials, were present the excess functional would account for those too. The steric contributions to the excess free energy describes the entropic loss due to particle repulsions at short range. Several approaches have been used to account for these, which include phenomenological approaches such as the fundamental measure theory [6–11] and hard chain equations of state [12, 13]. Other work has used functional expansions, which are more formally rigorous, but not expected to be accurate for highly structured fluids. Minimization of the grand free energy with respect to the polymeric density gives

$$\frac{\delta \Omega}{\delta N(\mathbf{R})}|_{N_{eq}(\mathbf{R})} = 0 \tag{9}$$

which yields a set of non-linear integral equations at equilibrium [5, 14],

$$N_{eq}(\mathbf{R}) = \exp[-\beta \Phi^{(b)}(\mathbf{R}) + \beta\mu - \beta\Psi^{(0)}(\mathbf{R}) - \frac{\delta \beta F^{ex}[n_i(\mathbf{r})]}{\delta N(\mathbf{R})}] \tag{10}$$

Given the relation between $N(\mathbf{R})$ and $n_i(\mathbf{r})$ in Eq. (4), the density derivative of the excess free energy is

$$\frac{\delta F^{ex}[N(\mathbf{R})]}{\delta N(\mathbf{R})} = \sum_{i=1}^{N} \frac{\delta F^{ex}[\{n_i(\mathbf{r_i})\}]}{\delta n_i(\mathbf{r}_i)} \tag{11}$$

A local excess chemical potential, $\lambda_i(\mathbf{r})$, thus acts on each monomer site i, $\{i = 1, ..., N\}$, on the polymer, given by

$$\lambda_i(\mathbf{r}) = \frac{\delta \beta F^{ex}[\{n_i(\mathbf{r})\}]}{\delta n_i(\mathbf{r})} + \beta\psi_i^{(0)}(\mathbf{r}) \tag{12}$$

One then obtains

$$N_{eq}(\mathbf{R}) = \phi_p \prod_{i=1}^{N-1} \frac{\delta(|\mathbf{r}_i - \mathbf{r}_{i+1}| - \sigma)}{4\pi\sigma^2} \prod_{i=1}^{N} \exp(\lambda_i^b - \lambda_i(\mathbf{r}_i))] \tag{13}$$

where λ_i^b is the excess chemical potential in the bulk and ϕ_p is the bulk polymer density.

2.1 End Segment Distributions

The site density of bead i on the polymer chain is obtained by the integral,

$$n_i(\mathbf{r}) = \int \delta(\mathbf{r} - \mathbf{r}_i)\phi_p \prod_{j=1}^{N-1} \frac{\delta(|\mathbf{r}_j - \mathbf{r}_{j+1}| - \sigma)}{4\pi\sigma^2} \prod_{j=1}^{N} \exp(\lambda_j^b - \lambda_j(\mathbf{r}_j))d\mathbf{R} \tag{14}$$

Due to the fact that the bonding extends only to nearest-neighbour monomers, the site densities can be re-expressed in terms of the product of end-segment distributions. Namely,

$$n_i(\mathbf{r}) = \phi_p c(N - i, \mathbf{r}) \int d\mathbf{r}' T(|\mathbf{r} - \mathbf{r}'|) c(i, \mathbf{r}') \tag{15}$$

where

$$T(|\mathbf{r} - \mathbf{r}'|) = \frac{\delta(|\mathbf{r} - \mathbf{r}'| - \sigma)}{4\pi\sigma^2} \tag{16}$$

is the nearest-neighbour bonding kernel. The end point distributions $c(i, \mathbf{r})$ correspond to the non-normalized distribution of end-monomers of chain segments of length i. They satisfy the recurrence formula,

$$c(i, \mathbf{r}) = \exp(\lambda_i^b - \lambda_i(\mathbf{r})) \int d\mathbf{r}' c(i - 1, \mathbf{r}') T(|\mathbf{r} - \mathbf{r}'|) \tag{17}$$

with the boundary condition, $c(0, \mathbf{r}) = 1$.

2.2 Estimating the Excess Free Energy: Accounting for Steric Interactions

In this example, we shall consider the chain model depicted in Fig. 1. The excess free energy $F^{ex}[\{n_i(\mathbf{r})\}]$ arises primarily from the excluded volume interaction between beads in the chains (which all have diameter σ). This excess functional is generally not known exactly and there have been many approximate forms proposed, since the earliest attempts over twenty years ago [5, 15–18]. All of these approximate choices for the excess free energy contain essentially the same physics and give rise to the same general type of mathematical expressions to solve, given that they invariably invoke the ansatz that the excess free energy is a functional of the monomer site densities. Of course different levels of accuracy may be expected, depending on the choice of functional, though this can be model dependent as well. Our general philosophy has been to utilize equation of state approaches, which are non-perturbative and therefore well-behaved at high fluid density, a quality recognized by van der Waals centuries ago.

The generalization of the van der Waals approach to non-uniform structures in simple fluids, using weighted density functionals, was pioneered more than thirty years ago by Nordholm [19–21]. Here we shall describe our adaptation of this approach to polymer fluids using the so-called generalized Flory equation of state for pure polymer fluids and mixtures. [12, 13, 22–25]. This equation of state relates the insertion probability of polymers to the excluded volume of the bulk fluid. The detailed derivation of this equation of state is not provided here, and the interested reader can consult the original article [22]. Briefly, the probability for insertion of an oligomeric chain is approximated as the product of probabilities for insertion of a dimer and then subsequent monomer beads in between, so that the original dimer forms the end beads of the linear chain. A set of shape parameters are defined relevant to monomers ($i = 1$) and dimers ($i = 2$),

$$c_i = r_i^2, \quad \omega_i = \frac{r_i s_i}{b_i}, \quad \phi_i = \frac{c_i s_i^2}{9b_i^2} \tag{18}$$

where

$$\begin{aligned} r_1 &= 0.5\sigma, \quad r_2 = 0.75\sigma \\ s_1 &= \pi\sigma^2, \quad s_2 = 2\pi\sigma^2 \\ b_1 &= \frac{\pi}{6}\sigma^3, \quad b_2 = \frac{\pi}{3}\sigma^3 \end{aligned} \tag{19}$$

These are used to define a set of coefficients

$$\begin{aligned} W_i &= -\frac{r_i^2 s_i^2}{9b_i^2}, \\ X_i &= W_i - 2\omega_i - 1 \\ Y_i &= -W_i + 2\omega_i - (3\phi_i - \omega_i - 2) \\ Z_i &= (\phi_i - 1) - \omega, \\ \Upsilon_i &= W_i + Y_i + 3Z_i \\ \Phi_i &= 0.5X_i + 2.5Y_i + 5.5Z_i + 3W_i \end{aligned} \tag{20}$$

which are used to construct the free energy densities of the monomer and dimer components, a_1^{ex} and a_2^{ex}, respectively.

$$\begin{aligned} a_i^{ex}(\eta) = & \frac{1-\eta}{\eta} ln[1-\eta] + 1 + \Upsilon_i[ln[1-\eta](\frac{1-\eta}{\eta}+1) - \frac{1}{1-\eta} - \frac{1}{2(1-\eta)^2} + \frac{5}{2}] \\ & + (\Phi_i + 2Z_i - X_i)(\frac{1}{2(1-\eta)^2} - \frac{1}{2}) + W_i(\frac{1}{2(1-\eta)^2} - \frac{2}{1-\eta} - ln[1-\eta] + \frac{3}{2}) \\ & + \Phi_i(\frac{1}{1-\eta} - \frac{1}{2(1-\eta)^2} - \frac{1}{2}) \end{aligned} \quad (21)$$

where

$$\eta = \frac{\pi\sigma^3}{6} n_m^b \quad (22)$$

is the volume fraction of monomers, with bulk density, $n_m^b = N\phi_p$. The bulk Helmholtz free energy density of the uniform hard sphere chain fluid is then approximated as,

$$\begin{aligned} f_{hs}^b = & (n_m^b - n_e^b)\frac{v_e(3) - v_e(2)}{v_e(2) - v_e(1)}[a_2^{ex}(n_m^b) - a_1^{ex}(n_m^b)] \\ & + \frac{1}{2} n_e^b a_2^{ex}(n_m^b) \end{aligned} \quad (23)$$

where $n_e^b(=2n_m^b/N)$ is the density of end monomers, $v_e(n)$ is the average excluded volume of an n-mer chain, estimated from simulation. This equation of state can be generalized to mixtures of monomers of different sizes [24], but that is not required for the model under consideration.

The conversion of the bulk equation of state to an excess density functional for non-uniform fluids follows a straightforward procedure whereby the bulk densities are replaced by their *weighted* non-uniform counterparts [12, 13, 23, 24, 26]. The device of weighted densities was originally introduced by Nordholm to reflect the non-local influence of excluded volume [19]. This phenomenological approach has been extended within so-called *fundamental measure theory* to incorporate other weight functions, which reflect geometric measures in the convex fluid [6–11]. For simplicity, we shall follow Nordholm's original formulation and use only a single *volume weighted* density. For example, the weighted total monomer density, $\bar{n}_m(\mathbf{r})$, is,

$$\bar{n}_m(\mathbf{r}) = \frac{3}{4\pi\sigma^3} \int_{|r-r'|<\sigma} n_m(\mathbf{r}')\, d\mathbf{r}' \quad (24)$$

where $n_m(\mathbf{r})$ is the sum of the local site densities.

$$n_m(\mathbf{r}) = \sum_{i=1}^{N} n_i(\mathbf{r}) \quad (25)$$

The subsequent replacement of the bulk densities appearing in Eq. (23) gives rise to a hard sphere contribution to the free energy which is a functional of the total monomer density, $n_m(\mathbf{r})$, and end site density, $n_e(\mathbf{r})$, where

$$n_e(\mathbf{r}) = n_1(\mathbf{r}) + n_N(\mathbf{r}) \tag{26}$$

Finally, we arrive at the following excess functional [13, 24],

$$F^{ex}_{hs}[n_e(\mathbf{r}), n_m(\mathbf{r})] = \int (n_m(\mathbf{r}) - n_e(\mathbf{r}))\frac{v_e(3) - v_e(2)}{v_e(2) - v_e(1)}[a^{ex}_2(\bar{n}_m(\mathbf{r})) - a^{ex}_1(\bar{n}_m(\mathbf{r}))]\, d\mathbf{r}$$
$$+\frac{1}{2}\int n_e(\mathbf{r}))a^{ex}_2(\bar{n}_m(\mathbf{r}))\, d\mathbf{r} \tag{27}$$

where the nomenclature reflects the steric origin of the excess contribution.

2.3 *The Lennard-Jones Chain Model*

Suppose now the polymer chains also attract each other via longer-ranged forces. As a generalization of the hard sphere chain model considered above, we assume that all the monomers interact instead via a pairwise additive Lennard-Jones potential,

$$\phi_{LJ}(r) = 4\varepsilon_{LJ}[(\frac{\sigma}{r})^{12} - (\frac{\sigma}{r})^{6}] \tag{28}$$

where ε_{LJ} and σ determine the attractive strength and repulsive range of the dispersion interaction respectively. We connect with the hard chain model, and assume the parameter σ is also equal to the bond-length as well as the distance of closest approach of the monomers. The excess free energy is then divided into hard sphere and attractive components. The hard sphere functional is again given by Eq. (27), and the additional attractive dispersion term is approximated by a mean-field expression,

$$F^{ex}_{disp}[n_m(\mathbf{r})] \approx \frac{1}{2}\int\int_{|\mathbf{r}-\mathbf{r}'|)\geq\sigma} n_m(\mathbf{r})n_m(\mathbf{r}')\phi_{disp}(|\mathbf{r} - \mathbf{r}'|)\, d\mathbf{r}\, d\mathbf{r}' \tag{29}$$

It is worthwhile noting that this approximation also applies for other long-ranged forces, including electrostatic interactions, in which case we arrive at a kind of polymer version of the celebrated Poisson-Boltzmann theory [27].

2.4 Solving the Density Functional Equations

A typical PDFT calculation will follow an iterative (Picard) scheme, due to the generally complex dependence of $\lambda_i(\mathbf{r})$ on the site densities. The site densities are defined on a grid of points in the spatial region to be solved, where the number of points used will be determined by the length-scale of the expected structural variation. In a simple Picard iteration scheme:
1. An initial guess is made for the $\{n_i(\mathbf{r})\}$ and the corresponding $\{\lambda_i(\mathbf{r})\}$ are obtained from Eq. (12). This is usually an integral equation of some type.
2. A new set of densities $\{n_i^{new}(\mathbf{r})\}$ are then obtained from Eqs. (13) and (4).
3. The set $\{n_i^{new}(\mathbf{r})\}$ is then mixed with the initial set (now labelled as $\{n_i^{old}(\mathbf{r})\}$) according to,

$$n_i(\mathbf{r}) = f_{mix} \cdot n_i^{new}(\mathbf{r}) + (1 - f_{mix}) \cdot n_i^{old}(\mathbf{r}) \tag{30}$$

where $f_{mix} < 1$.
4. Steps 2 and 3 are iterated until the density difference between consecutive iterations drops below a preset threshold at all spatial points for all species.

The equilibrium density can be substituted back into Eq. (7) to compute the equilibrium grand free energy, Ω_{eq}, together with other static thermodynamic properties.

3 Density Functional Theory for Polydisperse Semi-flexible Polymers

Polymer density functional calculations usually focus on monodispersed chains. This is despite the fact that laboratory samples of polymers rarely possess chains of a single molecular weight. In recent years, polydispersity has come to be viewed as a potentially useful experimental variable, that can be used to alter the effect that polymers have on surface interactions. Some examples of theoretical treatments of polymers including polydispersity have been reported, [28] including the special case of equilibrium polymers, which have an exponential molecular weight distribution [29–32]. Additionally, lattice-based self-consistent field (SCF) theory has been applied to polydisperse polymers at surfaces [33] and there have also been recent attempts to treat more general molecular weight distributions with continuum space Edwards-de Gennes SCF theories [34–36]. In the latter approaches, the molecular weight distribution is numerically integrated within the solution algorithm using adapted quadrature methods [37–39]. These methods were originally devised to consider effects of polydispersity on bulk phase diagrams. However, recent work has studied the effect of polydispersity on the depletion interaction between non-adsorbing surfaces [40].

In this section we shall outline our recent advances in introducing polydispersity into the PDFT. We shall show that, for a commonly used molecular weight distribution (namely the Schulz-Flory-Zimm distribution), the PDFT becomes very simple to solve. This is demonstrated within the framework of a more complex polymer model than that used in the previous section. Specifically, polydispersity will be introduced into a model of semi-flexible polymers.

The solution of the PDFT for monodisperse semi-flexible polymer models requires a greater computational workload when compared to calculations for flexible chains. While this workload still scales linearly with the length of the polymer, the number of numerical operations required increases significantly per additional monomer. This is because next-nearest neighbor interactions come into play, requiring storage of higher dimensional (albeit sparse) matrices. One would generally expect that introducing polydispersity into the mix would increase the computational demands even further, especially for broad molecular weight distributions. Recently, however, we showed how the PDFT could be easily generalized to include polydispersity, in the case where the molecular weight distribution is of the Schulz-Flory-Zimm (SFZ) form [41]. Quite surprisingly, we found that the algebraic structure of the PDFT was simpler than that for the monodispersed fluid and this manifested itself in a numerically more efficient solution algorithm.

We begin our derivation by writing down the expression for the ideal free energy of a fluid model consisting of semi-flexible r-mer polymer chains

$$\begin{aligned}\beta F_r^{id} = & \int d\mathbf{R} N_r(\mathbf{R}) \left(\ln[N_r(\mathbf{R})] - 1\right) + \\ & \int d\mathbf{R} N_r(\mathbf{R}) \Phi^{(b)}(\mathbf{R}) + \int d\mathbf{r} n_m(\mathbf{r}) \psi^0(\mathbf{r}) + \\ & \int d\mathbf{R} N_r(\mathbf{R}) \sum_{i=1}^{r-2} E_B(\mathbf{s}_i, \mathbf{s}_{i+1})\end{aligned} \tag{31}$$

where we have explicitly labelled the polymer density, $N_r(\mathbf{R})$ with the degree of polymerization. As before, $\Phi^{(b)}(\mathbf{R})$ describes the nearest-neighbor (non-directional bonding) i.e.,

$$\begin{aligned}\exp[-\beta\Phi^{(b)}(\mathbf{R})] &= \prod_{i=1}^{r-1} \frac{\delta(|\mathbf{r}_i - \mathbf{r}_{i+1}| - \sigma)}{4\pi\sigma^2} \\ &= \prod_{i=1}^{r-1} T(|\mathbf{r}_i - \mathbf{r}_{i+1}|)\end{aligned} \tag{32}$$

What is new in the ideal functional is the bending potential, E_B, which introduces a degree of stiffness into the polymer molecules. We shall assume the following general form for this potential,

$$\beta E_B(\mathbf{s}_i, \mathbf{s}_{i+1}) = \varepsilon\left(1 - \frac{\mathbf{s}_i \cdot \mathbf{s}_{i+1}}{\sigma^2}\right) \tag{33}$$

where, $\mathbf{s}_i$, denotes the bond vector between monomers i and $i+1$, i.e., $\mathbf{s}_i = \mathbf{r}_{i+1} - \mathbf{r}_i$, and ε is the strength of the bending potential. The incorporation of the bending potential into the freely-jointed chain model, described earlier, creates a decidedly more complex problem, due to the presence of next-nearest neighbor interactions.

To simplify the problem, we shall assume that all monomers in the chain interact in the same way with the external potential, $\psi^0(\mathbf{r})$. We consider now a grand canonical ensemble system, whereby the bulk fluid consists of a polydisperse polymer solution. The chemical potential of the r-mer in the bulk, μ_r, is given by the expression,

$$\beta\mu_r = ln[\phi_p F(r)] + \beta\mu^{ex}(r) \tag{34}$$

where ϕ_p is now the *total* bulk density of polymer molecules and $\mu^{ex}(r)$ is the excess chemical potential of the r-mer. The latter is ultimately determined by the equation of state used in the model and in our model it will be linearly dependent on r. Finally, $F(r)$, is the normalized, molecular weight distribution of the bulk fluid. As stated above, it will be described by the SFZ distribution, which has the generic form,

$$F(r) = Kr^n e^{-\kappa r} \tag{35}$$

where

$$K = \left(\frac{n+1}{<r>_b}\right)^{n+1} \frac{1}{\Gamma(n+1)} \tag{36}$$

$$\kappa = \frac{n+1}{<r>_b} \tag{37}$$

$\Gamma(x)$ is a gamma function and $<r>_b$ is the average degree of polymerization of the bulk solution. The quantity, n, determines the width of the distribution, which becomes narrower as n increases. The case $n = 0$ corresponds to equilibrium (living) polymers. As in our earlier work, we shall assume that n is an integer.

The total grand potential functional, Ω, for the *polydisperse* polymer fluid is given by

$$\Omega = \sum_r F_r^{id}[N_r(\mathbf{R})] + F^{ex}[n_m(\mathbf{r})] - \sum_r \mu_r \int d\mathbf{R} N_r(\mathbf{R}) \tag{38}$$

Note that we have assumed that the excess free energy is only a functional of the total monomer density, $n_m(\mathbf{r})$. This approximation means that the excess chemical potential of the polymer in the bulk, is proportional to the degree of polymerization. While it is possible to include the dependence on the end monomers (as was done in the previous section) that would add a complication which would detract from the clarity of our subsequent analysis.

Minimizing the grand free energy with respect to $N_r(\mathbf{R})$ gives the following integral equation,

$$N_r(\mathbf{R}) = \phi_p F(r) \prod_{i=1}^{r-1} T(|\mathbf{r}_i - \mathbf{r}_{i+1}|) \prod_{i=1}^{r-2} \Psi(\mathbf{r}_i, \mathbf{r}_{i+1}, \mathbf{r}_{i+2}) \prod_{i=1}^{r} \exp(\lambda^b - \lambda(\mathbf{r}_i)) \quad (39)$$

where most of the terms that appear here have been defined earlier. For example, we have

$$\lambda(\mathbf{r}) = \frac{\delta \beta F^{ex}[n_m(\mathbf{r})]}{\delta n_m(\mathbf{r})} + \beta \psi^{(0)}(\mathbf{r}) \quad (40)$$

which has bulk value λ^b. The introduction of bond stiffness introduces a new, next-nearest neighbour term, given by,

$$\Psi(\mathbf{r}_i, \mathbf{r}_{i+1}, \mathbf{r}_{i+2}) = e^{-\beta E_B(\mathbf{s}_i, \mathbf{s}_{i+1})} \quad (41)$$

The total average monomer density is given by the slightly more general relation,

$$n_m(\mathbf{r}) = \sum_{r=1}^{\infty} \int d\mathbf{R} \sum_{i=1}^{r} \delta(|\mathbf{r} - \mathbf{r}_i|) N_r(\mathbf{R}) \quad (42)$$

As for the case of the freely-jointed model, it is useful to define segment end distributions, but these are more complicated in the presence of the bending potential. It is necessary to introduce the end distribution for the last two monomers in a chain. We shall denote these as end pair distributions. These satisfy the following recursion formula,

$$c(i; \mathbf{r}', \mathbf{r}) = \exp[\lambda^b - \lambda(\mathbf{r}')] \int c(i-1, \mathbf{r}'', \mathbf{r}') T(|\mathbf{r}' - \mathbf{r}|) \Psi(\mathbf{r}'', \mathbf{r}', \mathbf{r}) d\mathbf{r}'' \quad (43)$$

with boundary condition $c(0; \mathbf{r}', \mathbf{r}) = 1$. We note that the arguments $\mathbf{r}'$ and $\mathbf{r}$ of $c(i; \mathbf{r}', \mathbf{r})$, are ordered to denote the coordinates of the second to last and last monomers in the segment respectively. In terms of these functions, the monomer density can be recast into the following useful form,

$$n_m(\mathbf{r}) = \phi_p \sum_{r=1}^{\infty} F(r) \sum_{i=1}^{r} \int d\mathbf{r}' c(r-i; \mathbf{r}', \mathbf{r}) T(|\mathbf{r} - \mathbf{r}'|) c(i; \mathbf{r}, \mathbf{r}') d\mathbf{r}' \quad (44)$$

Equations (40)–(44) form a closed set of equations to be solved self-consistently. The effect of polydispersity requires the determination of end pair distributions for each r-mer in the distribution. As the molecular weight distribution is continuous, one would generally need to discretize it so as to provide a finite quadrature approximation to the integral over the distribution. This approach has been used in several applications [37, 38], where it has been shown that adaptive quadrature methods can lead to efficiently determined approximate integrals. Nevertheless, this generally leads to more expensive and complex solution algorithms, when compared with a typical calculation for monodispersed polymer fluid. Below we shall demonstrate how a much simpler approach can be used for semi-flexible polymers having the SFZ molecular weight distribution.

We begin by substituting the explicit form of the SFZ distribution into Eq. (44), which gives,

$$
\begin{aligned}
n_m(\mathbf{r}) &= \phi_p K \sum_{r=1}^{\infty} \sum_{i=1}^{r} r^n e^{-\kappa r} \int d\mathbf{r}' c(r-i; \mathbf{r}', \mathbf{r}) T(|\mathbf{r}-\mathbf{r}'|) c(i; \mathbf{r}, \mathbf{r}') \qquad (45) \\
&= \phi_p K \sum_{r=1}^{\infty} \sum_{i=1}^{r} (r-i+i)^n e^{-\kappa(r-i)} \int d\mathbf{r}' c(r-i; \mathbf{r}', \mathbf{r}) \times \\
&\quad T(|\mathbf{r}-\mathbf{r}'|) e^{-\kappa i} c(i; \mathbf{r}, \mathbf{r}') \\
&= n! \phi_p K \sum_{l=0}^{n} \sum_{r=1}^{\infty} \sum_{i=1}^{r} \int d\mathbf{r}' \frac{(r-i)^l}{l!} e^{-\kappa(r-i)} c(r-i; \mathbf{r}', \mathbf{r}) \times \\
&\quad T(|\mathbf{r}-\mathbf{r}'|) \frac{i^{n-l}}{(n-l)!} e^{-\kappa i} c(i; \mathbf{r}, \mathbf{r}')
\end{aligned}
$$

The double sum over r and i can be carried out, being careful to treat the term $i = r$ as a special case (which includes the $r = 1$ term). We are led to the very simple equation,

$$
n_m(\mathbf{r})/\phi_p K n! = \sum_{l=0}^{n} \int d\mathbf{r}' \hat{c}(l; \mathbf{r}', \mathbf{r}) T(|\mathbf{r}-\mathbf{r}'|) \hat{c}(n-l; \mathbf{r}, \mathbf{r}') + \int d\mathbf{r}' T(|\mathbf{r}-\mathbf{r}'|) \hat{c}(n; \mathbf{r}, \mathbf{r}') \qquad (46)
$$

where the transformed end-point distributions are defined by,

$$
\hat{c}(k; \mathbf{r}, \mathbf{r}') = \sum_{i=1}^{\infty} e^{-\kappa i} i^k c(i; \mathbf{r}, \mathbf{r}')/k! \qquad (47)
$$

Thus, the multiple sum in Eq. (44) is replaced by a much smaller sum of terms in Eq. (46) involving only $n + 1$ transformed end distributions. A recursion formula for these can be obtained by multiplying both sides of Eq. (43) by $(i + 1)^l e^{-\kappa(i+1)}/l!$ and summing over i. This results in the following expression,

$$
\hat{c}(l; \mathbf{r}', \mathbf{r}) = e^{-\psi(\mathbf{r}')} e^{-\kappa} \{ \sum_{m=0}^{l} \frac{1}{(l-m)!} \int d\mathbf{r}'' \hat{c}(m; \mathbf{r}'', \mathbf{r}') T(|\mathbf{r}''-\mathbf{r}'|) \Psi(\mathbf{r}'', \mathbf{r}', \mathbf{r}) + 1 \} \qquad (48)
$$

which is closed by the initial condition,

$$
\hat{c}(0; \mathbf{r}', \mathbf{r}) = e^{-\psi(\mathbf{r}')} e^{-\kappa} \{ \int d\mathbf{r}'' \hat{c}(0; \mathbf{r}'', \mathbf{r}') T(|\mathbf{r}''-\mathbf{r}'|) \Psi(\mathbf{r}'', \mathbf{r}', \mathbf{r}) + 1 \} \qquad (49)
$$

3.1 Application to Semi-flexible Polymer Solution Films

We shall apply the theory to the case of a polymer solution. To simplify matters, we treat the solvent molecules implicitly. Thus the solvent influences the problem only via *effective* interactions between the monomers and also between the monomers and the surfaces. It is worth mentioning, however, that it is relatively straightforward to generalize our approach to explicit solvents, but this will not be done in this study. The implicit solvent model is reasonable if the monomer units are relatively larger than the solvent molecules. If we further assume that the attractive dispersion interactions between monomers are strongly screened by the solvent, we arrive at essentially the same model as considered earlier, which consists of a fluid of excluding chains. In order to treat the steric interactions between the monomers in the chains, we shall once again use the generalized Flory-Dimer expression. Hence we assume that the intrinsic stiffness does not significantly affect the steric contributions to the free energy [12, 42].

In this example, we shall investigate the fluid film between two large planar surfaces, immersed in the polymer solution, at some separation h. The (solvent mediated) interaction between the surfaces and monomers is given by $\psi^{(0)}(z,h) = w(z) + w(h-z)$, where $w(z)$ is modelled as a truncated and shifted Lennard-Jones potential,

$$\begin{aligned} w(z) &= w_{LJ}(z) - w_{LJ}(z_c) \; z < z_c \\ &= 0 \quad otherwise \end{aligned} \tag{50}$$

where,

$$\beta w_{L-J}(z) = 2\pi[\frac{2}{45}(\frac{\sigma}{z})^9 - \frac{a_w}{3}(\frac{\sigma}{z})^3] \tag{51}$$

and we have set $z_c = 4\sigma$. An *adsorbing* surface is characterized by $a_w = 1$, while $a_w = 0$ creates a *non-adsorbing* surface. In order to avoid the effect of surface-surface interactions, we have only investigated separations at which there is no overlap between the potentials, i.e. $h > 8\sigma$.

Firstly, we note that the planar symmetry of the system allows a simplification of the bending contribution to the polymer distribution as follows. Using $\Delta z_i \equiv z_{i+1} - z_i$, we can write the vector product between adjacent bond vectors as,

$$\mathbf{s}_i \cdot \mathbf{s}_{i+1} = \Delta z_i \Delta z_{i+1} + (\sigma^2 - \Delta z_i^2)^{\frac{1}{2}}(\sigma^2 - \Delta z_{i+1}^2)^{\frac{1}{2}} cos\phi_{i,i+1}. \tag{52}$$

where $\phi_{i,i+1}$ is the angle between $\mathbf{s}_i$ and $\mathbf{s}_{i+1}$, as projected onto the plane of the surfaces. Averaging the corresponding Boltzmann factor over this angle we obtain,

$$\Psi(\Delta z_i, \Delta z_{i+1}) = e^{-\varepsilon(1-\frac{\Delta z_i \Delta z_{i+1}}{\sigma^2})} I_0 \left[\varepsilon(1-(\frac{\Delta z_i}{\sigma})^2)^{\frac{1}{2}}(1-(\frac{\Delta z_{i+1}}{\sigma})^2)^{\frac{1}{2}}\right] \tag{53}$$

where $I_0(x) = \frac{1}{2\pi}\int_0^{2\pi} exp\,[-xcos\phi]\,d\phi$ is a modified Bessel function, which can be evaluated from a polynomial expansion.

3.2 Interaction Free Energy Between Non-adsorbing Surfaces

We now consider the polymer solution in contact with non-adsorbing surfaces. The solution-mediated interaction between the surfaces is obtained by considering the change in the fluid free energy as a function of the surface separation. That is, the equilibrium polymer density is obtained by solving the functional in Eq. (38) at different surface separations. These densities are reinserted back into the free energy functional in order to obtain, $\Omega_{eq}(h)$, which is the total equilibrium grand potential at a given separation. The corresponding *total* free energy per unit area, $g(h)$, is obtained by adding the contribution due to volume changes in the bulk reservoir and then scaling with the area of the surface, S. This gives,

$$g(h) = \Omega_{eq}(h)/S + P_b h \tag{54}$$

where P_b is the osmotic pressure in the bulk. Figure 2a, b shows the net interaction between surfaces for a fully flexible polymer ($\varepsilon = 0$) and stiff polymers ($\varepsilon = 6$), as a function of the degree of polydispersity. The net interaction free energy, which is plotted in the figures is obtained by removing the bulk surface terms that survive at infinite surface separation, i.e., $\Delta g(h) = g(h) - g(\infty)$. In the figures we also show for comparison a typical hydrocarbon-water-hydrocarbon dispersion interaction (Hamaker constant 5x10^{-21} joules).

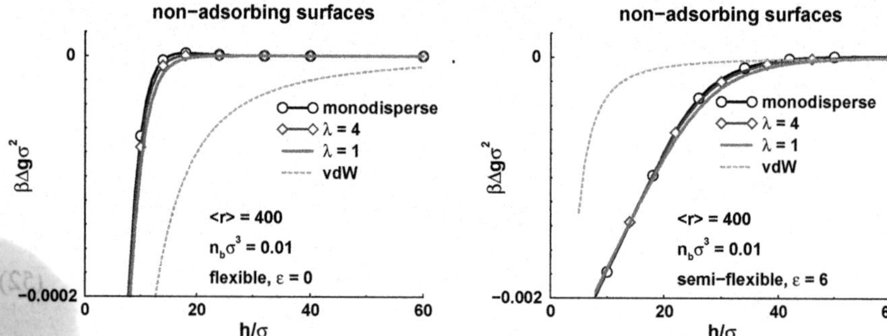

Fig. 2 Interaction free energies per unit area (Δg), between non-adsorbing surfaces, immersed in polymer solutions containing chains with an average length, $< r >_b = 400$. The parameter $\lambda = n + 1$. The bulk monomer concentration is $n_b\sigma^3 = 0.01$. The dashed line shows a typical Hamaker attraction for hydrocarbon surfaces across water. **a** Flexible polymers, with varying degree of polydispersity. **b** Semi-flexible polymers, $\varepsilon = 6$, with varying degree of polydispersity. Reproduced, with permission, from *Macromolecules* 42, 7563 (2009). Copyright 2009, American Chemical Society

These interaction curves display the depletion attraction, typical of polymer solutions in contact with excluding surfaces. Reduction of the configurational entropy of the polymers at the excluding walls decreases the internal density leading to an osmotic attraction between the surfaces. The surface interactions are much stronger and have a longer range for stiffer polymers. The increase in the strength of attraction with stiffness is due to the stronger intra-molecular correlations, which leads to relatively larger configurational entropy losses per molecule due to the surface confinement. That is, while a stiff polymer has less configurational entropy than a flexible polymer, the effect of the surfaces is more drastic on stiff polymers, leading to a greater degree of depletion. The depletion interaction begins to act at separations on the length-scale of the polymer radius of gyration, R_g. This means that the interaction free energy has a larger range with stiffer polymers, which of course have a larger R_g. In addition, we also see a small but discernable increase in the range of the attraction between the surfaces as the degree of polydispersity increases. This is because the depletion of polymer molecules longer than the average molecular weight (present in polydisperse samples) leads to an osmotic attraction at larger separations, compared with the monodispersed case. This effect is also reflected in the average polymer length $< r >$ in the slit as a function of the surface separation.

Figure 3 shows that $< r >$ is smaller between the surfaces than its bulk value, even at rather large separations. Not surprisingly, this effect is greater for the stiffer polymers, given their larger radius of gyration. We also note that the relative increase in the number of short polymers between the surfaces, as the separation decreases, is greater for samples with higher polydispersity. For example, the average polymer length has diminished to approximately 15 for stiff polymers with $n = 0$ at a separation of $h = 10\sigma$. It is four times larger for $n = 3$, at the same separation. Nevertheless, as noted above, the consequent changes to the interaction free energies are small. That is, at large separations, the osmotic pressure for polydisperse systems is only slightly more attractive than for the monodispersed case, with a cross-over occurring at smaller separations. The cross-over occurs because monodispersed polymers are now significantly depleted from between the surfaces, whilst in the polydisperse

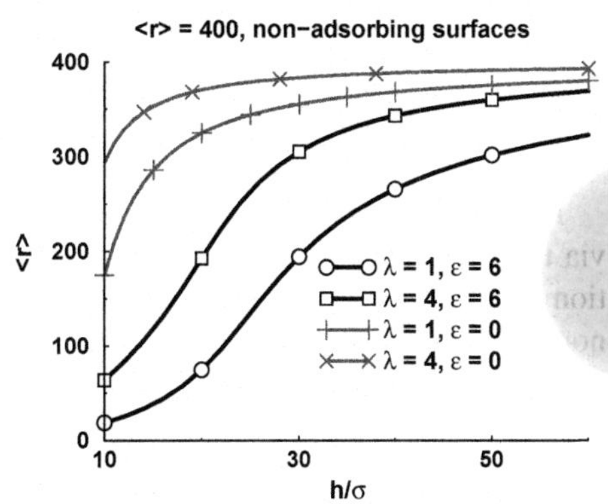

Fig. 3 The average polymer length, $< r >$, of chains in the inter-surface (slit) region, $\lambda = n + 1$. Polymer solutions as defined in Fig. 2. Reproduced, with permission, from *Macromolecules* 42, 7563 (2009). Copyright 2009, American Chemical Society

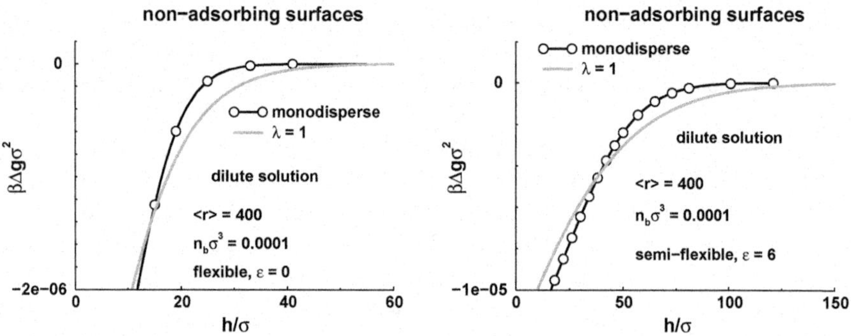

Fig. 4 Interaction free energies per unit area, between nonadsorbing surfaces, immersed in polymer solutions containing chains with an average length in the bulk, $< r >_b = 400$. The bulk monomer concentration is $n_b\sigma^3 = 0.0001$, i.e. considerably lower than in Fig. 2. **a** Flexible polymers, with varying degree of polydispersity, $\lambda = n + 1$. **b** Semi-flexible polymers, $\varepsilon = 6$, with varying degree of polydispersity. Reproduced, with permission, from *Macromolecules* 42, 7563 (2009). Copyright 2009, American Chemical Society

case the number of shorter polymers between the surfaces creates a relatively more repulsive osmotic pressure. This leads to a relatively larger free energy as the configurational entropy of these shorter polymers is reduced by the surfaces. Though polydispersity effects appears to be small under these conditions, it is well-known that if the monomer concentration is high enough, excluded volume effects on the correlation length can be significant. That is, monomer correlations become screened and the surface interaction can become relatively insensitive to chain length. At considerably lower concentrations, the range of the depletion attraction is essentially governed by the radius of gyration and the effects of polydispersity are expected to be more pronounced. This is verified by the results shown in Fig. 4 wherein the polymer concentration is reduced 100-fold. In this case, the range of the depletion attraction increases considerably with the width of the molecular weight distribution.

3.3 Interaction Free Energy Between Adsorbing Surfaces

Figure 5a, b, shows the interaction between two surfaces that adsorb monomers via the truncated L-J potential described above. The general shape of the interaction free energy curves is qualitatively different to the depletion interaction between non-adsorbing walls. We again see that the introduction of stiffness into the model increases both the strength and the range of the interaction. In addition to this, the presence of an attraction between monomers and the surfaces has caused the formation of a barrier into the interaction free energy. The nature of this barrier is somewhat well-known [25, 43–46]. It is due to the steric interaction experienced between the

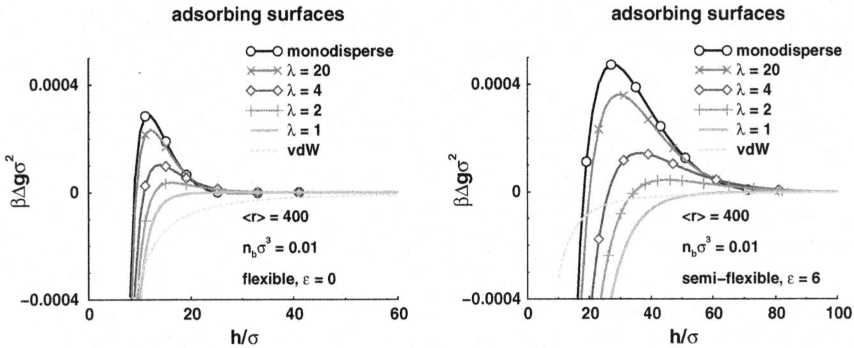

Fig. 5 Interaction free energies per unit area, between adsorbing surfaces, immersed in polymer solutions containing chains with an average length in the bulk given by, $< r >_b = 400$. The bulk monomer concentration is $n_b\sigma^3 = 0.01$. The dashed line shows a typical Hamaker attraction for hydrocarbon surfaces across water. **a** Flexible polymers, with varying degree of polydispersity, $\lambda = n + 1$. **b** Semi-flexible polymers, $\varepsilon = 6$, with varying degree of polydispersity. Reproduced, with permission, from *Macromolecules* 42, 7563 (2009). Copyright 2009, American Chemical Society

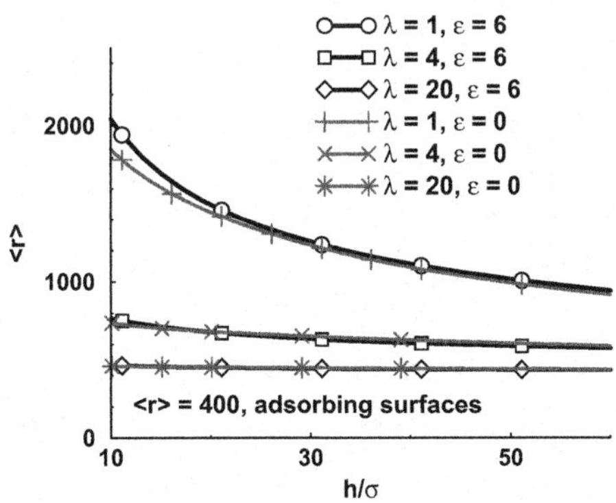

Fig. 6 The average polymer length of chains in the inter-surface (slit) region. Polymer solutions as defined in Fig. 5, $\lambda = n + 1$. Reproduced, with permission, from *Macromolecules* 42, 7563 (2009). Copyright 2009, American Chemical Society

tails of the adsorbed polymer layers on each surface. For the stiffer polymers investigated here ($\varepsilon = 6$), the tails of adsorbed polymers generally project further than for flexible polymers. Thus the barrier occurs at larger separation for the stiffer polymers. As the surfaces approach even closer, penetration of tails to the other surface occurs and attractive bridging interactions begin to dominate, leading to a strong attraction at shorter separation.

Surfaces which are attractive to monomers have a different effect on polydisperse polymers. In this case, longer polymers will tend to adsorb more strongly between attractive surfaces due to cooperative adsorption between monomers. This co-operativity is greater for stiffer polymers, which lose less configurational entropy when they adsorb onto surfaces, than flexible polymers. Thus, for polydisperse systems, the average molecular weight of polymers between the surfaces is larger than that of the bulk. This occurs at all separations, as shown in Fig. 6. Not surprisingly, the more widely polydisperse samples allow adsorption of longer polymers. For example, equilibrium polymers ($n = 0$) between the surfaces have an average length some five times higher (at separation $h = 10\,\sigma$) than in the bulk. Generally, the average polymer length between the surfaces increases as the surfaces approach. As the average length of the adsorbed chains becomes larger, the bridging attraction also increases. It is the ability of longer polymers to bridge between surfaces, at relatively small cost to their configurational entropy, which leads to their preferential accumulation in the region at and between the surfaces. The increased bridging strength leads to a reduction of the free energy barrier, as clearly seen in Fig. 5a, b. Again, this effect is stronger for stiffer chains. One interesting result is that for equilibrium polymers, the free energy barrier appears to have disappeared completely. Compared with the case of depletion attraction between non-adsorbing surfaces, the effect of both stiffness *and polydispersity* on the interaction free energies are relatively much larger.

In order to determine if the magnitude of the polymer mediated forces are relevant to phenomena such as colloidal stability, we have employed the Derjaguin approximation [47], which allows us to estimate the interaction between large particles. The expression for the interaction between two spherical colloidal particles, with radius R, is given by,

$$W(D) \approx -\pi R \int_D^\infty dx g(x) \tag{55}$$

where $g(x)$ is the free energy per unit area between the surfaces, and D is the surface to surface separation between the spheres. Recent work by us has shown that the Derjaguin approximation can be quite accurate, even for surprisingly small colloidal radii (relative to that of the polymer radius of gyration) [48]. We chose a sphere radius

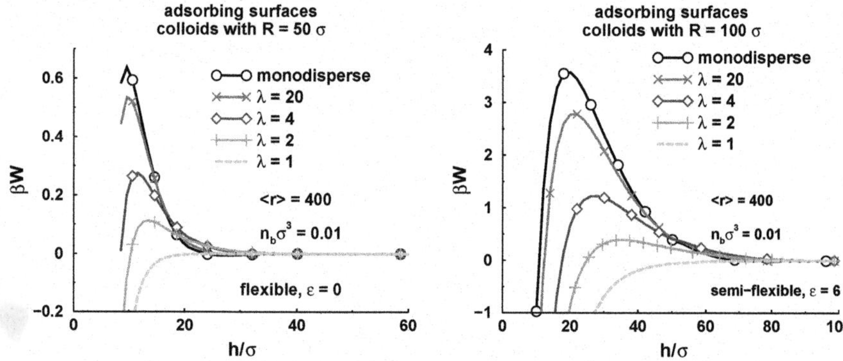

Fig. 7 Interaction free energies (W), between adsorbing spherical colloids, as obtained via the Derjaguin approximation, Eq. (55), assuming a colloid radius of 100 σ. The bulk monomer density is $n_b\sigma^3 = 0.01$. **a** Flexible polymers, with varying degree of polydispersity, $\lambda = n + 1$. **b** Semi-flexible polymers, $\varepsilon = 6$, with varying degree of polydispersity. Reproduced, with permission, from *Macromolecules* 42, 7563 (2009). Copyright 2009, American Chemical Society

of 100 σ. The resulting interactions, under adsorbing conditions, are presented in Fig. 7. The free energy barrier is about $0.6k_BT$ for monodispersed flexible polymers, and $3.5k_BT$ for monodispersed stiff polymers. Note also that the barrier occurs at larger separation in the latter case, which makes the barrier more relevant. This is because the van der Waals attraction rapidly becomes weak as the separation increases.

The most important effect of polydispersity is the progressive diminishing of this free energy barrier, as the samples are made more polydisperse, to the point of monotonic attraction, in the presence of equilibrium polymers. These barriers seem to contradict a well-known theorem by de Gennes, according to which surface interactions at full equilibrium supposedly give a monotonic attraction [36]. However, this theorem relies on the Edwards-de Gennes theory, in the limit of infinitely long chains. As demonstrated in previous work [44] a sizable free energy barrier will persist even for very long polymers. This has also been shown by van der Gucht et al. [49].

3.4 Approaching the Rod-Like Limit

In the approach to the rod-limit, the persistence length, $l_p(\approx \varepsilon)$, approaches the polymer contour length, $L = (r - 1)\sigma$. One may expect that polydispersity effects for semi-flexible polymers approaching the rod limit may be relatively larger than for flexible chains. Between non-adsorbing walls, for example, one expects to find a population of shorter (more rod-like) polymers between the surfaces, compared with the bulk. On the other hand, for adsorbing walls, longer (less rod-like) polymers will be found. Figure 8 shows the interaction free energy for adsorbing surfaces immersed in a fluid of short, monodispersed polymers. As one approaches more rod-

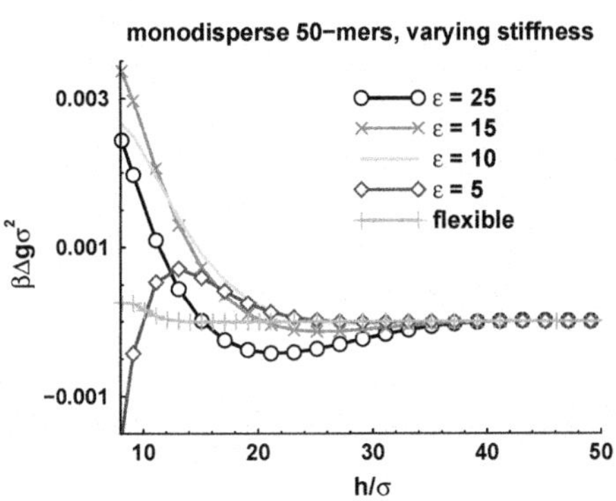

Fig. 8 Interaction free energies per unit area, between adsorbing surfaces, immersed in monodisperse polymer solutions, where $r = 50$. The bulk monomer concentration is $n_b\sigma^3 = 0.01$. Interactions are given for various values of the bend stiffness parameter, ε. Reproduced, with permission, from *Macromolecules* 42, 7563 (2009). Copyright 2009, American Chemical Society

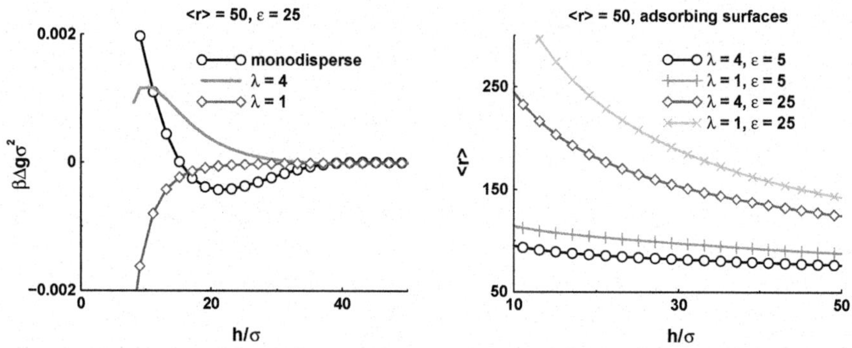

Fig. 9 Adsorbing surfaces, immersed in polydisperse polymer solutions, with average length $< r >_b = 50$ in the bulk. The bend stiffness parameter is high, $\varepsilon = 25$, which renders the chains rod-like. The bulk monomer concentration is $n_b\sigma^3 = 0.01$. **a** Interaction free energies per unit area. **b** The average chain length, in the inter-surface (slit) region. Reproduced, with permission, from *Macromolecules* 42, 7563 (2009). Copyright 2009, American Chemical Society

like behaviour (by increasing ε), we see a non-monotonic behaviour in the free energy. In the presence of flexible polymers we see the onset of a weak free energy barrier, due to the overlap of adsorbed monomers, as described earlier. Of course, this occurs at smaller separation now, as the polymers are much shorter ($r = 50$). As ε increases, the position of the barrier, and its magnitude, initially shifts out to larger separation, as expected, but then starts to move inward, as the adsorbed layers become flatter. The flattening out of the adsorbed layer is due to the increasing stiffness, which promotes the cooperative adsorption of monomers on the same chain. One also begins to see the onset of a depletion attraction regime at intermediate separations ($h \approx 20\sigma$), which is quite pronounced for the stiffest polymer investigated ($\varepsilon = 25$). This depletion comes from exclusion of non-adsorbed monomers between the surfaces at separations around the persistence length of the polymer ($l_p \approx \varepsilon$).

As shown in Fig. 9a, if the sample containing the stiffest chains is polydisperse, we find dramatically different interaction free energies. The typical rod-like characteristics of the free energy is suppressed. Instead, we observe a behaviour which is reminiscent of longer polymers between adsorbing surfaces. In particular, we see that at intermediate polydispersity ($n = 3$), the depletion attraction has disappeared and a free energy barrier occurs at a separation that indicates an interaction between two thicker adsorbed layers. For the equilibrium polymer case ($n = 0$), we once again see a purely attractive interaction that is due essentially entirely to bridging. The reason for this behaviour is clearly shown in Fig. 9b, which indicates a dramatic growth in the average length of polymer population between the surfaces. Thus the interaction free energy is similar to those found with longer semi-flexible polymers. Interactions between non-adsorbing surfaces, immersed in solutions containing rod-like polymers, are presented in Fig. 10a. The forces are much stronger in this case, as compared to those seen in Fig. 2a, b. This is due to the fact that maximal depletion forces in general are stronger with shorter polymers, at a fixed bulk monomer concen-

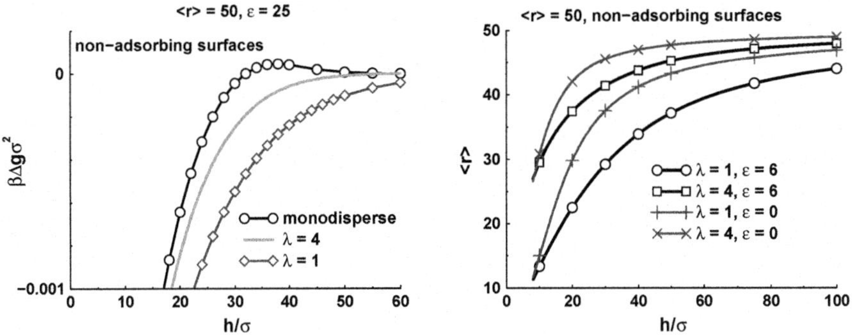

Fig. 10 Non-adsorbing surfaces, immersed in polydisperse polymer solutions, with average length $< r >_b = 50$ in the bulk. The bend stiffness parameter is high, $\varepsilon = 25$, which renders the chains rod-like. The bulk monomer concentration is $n_b\sigma^3 = 0.01$. **a** Interaction free energies per unit area. **b** The average chain length, in the inter-surface (slit) region. Reproduced, with permission, from *Macromolecules* 42, 7563 (2009). Copyright 2009, American Chemical Society

tration, since the bulk osmotic pressure is higher. For the monodispersed case, we see the occurrence of a slight free energy barrier, at a separation somewhat larger than the persistence length. Here, polymer molecules begin to lose significant configurational free energy due to exclusion by both surfaces simultaneously, but depletion is not yet large enough to create an osmotic attraction. Instead the entropic losses, including those due to an increased excluded volume, cause the free energy to increase slightly, with a resulting barrier. Once the separation is small enough to cause significant depletion, the free energy is monotonically attractive, down to contact. In a very polydisperse solution, the free energy barrier is absent. This is because long chains are replaced (to some degree) by shorter ones, as the surfaces approach. The subsequent osmotic effect leads to an attraction. This is a similar mechanism to that described above for longer polymers. In this case, however, there is an additional effect from short polymers being more rod-like.

The overall response is illustrated in Fig. 10b, which shows the average molecular weight of polymers in the region between the surfaces. If we compare with results of flexible polymers, we note that long, flexible polymers are less depleted at large separations, than are stiff molecules. At short separations, though, the greater loss of configurational entropy experienced by the flexible chains, causes them to be depleted more rapidly, and a crossover occurs at $h \approx 10\sigma$. Interestingly, this occurs at approximately the same separation for both degrees of polydispersity, $n = 0, 3$, though we cannot see a fundamental reason why this should be the case and expect this result is coincidental.

4 Other Polymeric Architectures

One of the simplifying aspects of the PDFT as implemented here is that the excess term in the functional is assumed to be largely independent of the polymer molecular architecture. In this case, the influence of architecture is manifested only in the ideal part of the functional via the bonding potential, $\Phi^{(b)}(\mathbf{R})$. One recent example of this type of formulation is in the application of PDFT to room temperature ionic liquids [50, 51]

4.1 Using the PDFT for Room Temperature Ionic Liquids

Room-Temperature ionic liquids (RTILs) are salts that remain in the liquid state at temperatures lower than 100 °C. Unlike traditional molten salts, RTILs generally possess a complex molecular structure, which precludes them easily packing into a crystal even at room temperatures [52, 53]. The often robust chemical structure of RTILs also imparts them with a high electrochemical stability. The combination of chemical stability and low melting point makes RTILs potentially good replacements for traditional molten salts in many applications. Some specific uses of RTILs include: solvents for heterogeneous and homogeneous catalysis; electrolytes in various electrochemical processes, and dispersive agents for stabilization of nanoparticles [54–56]. Additionally, RTILs have so-called "green" applications, e.g., as a solvent for cellulose, an important bio-resource [57] and as selective solvents for the removal of heavy metal contaminants in waste-water treatments [58]. Polymer density functional theory provides us with a theoretical framework with which to describe this fascinating class of fluids. In the brief exposition below, we will focus on the class of imidazolium-based RTILs, but it is clear that the theory can be rather straightforwardly applied to other RTIL types.

The *1-alkyl-3-methyl imidazolium* cation will be generally denoted as, $[C_{nMIM}^{+}]$, where n denotes the length of the alkyl chain. Using a united-atom approach we shall model the CH_2 and CH_3 groups of the alkyl chain as neutral spheres (Fig. 11). The imidazolium ring is assumed to be made up of a 4-branched star of tangential spheres, with each sphere carrying a partial charge of +0.2, to mimick charge delocalization. The *tetrafluoroborate*, $[BF_4^-]$, anion is represented by a similar star, with each spherical bead carrying a −0.2 partial charge. To simplify the treatment, the bond distance for all connected sites has the same value of $\sigma = 2.4$ Å. This reflects the approximate size of the molecules concerned [59]. We also assume that the molecules are freely-jointed, which means that there is no inherent stiffness in the oligomeric chains and the spheres are allowed to freely rotate over each others' surfaces. Clearly, both approximations of uniform bond-length and lack of stiffness can be removed in a more elaborate version of the PDFT.

The partial charges are fixed at the geometric center of each hard sphere site. In reality, however, the electron density in RTIL molecules will be polarized by fields

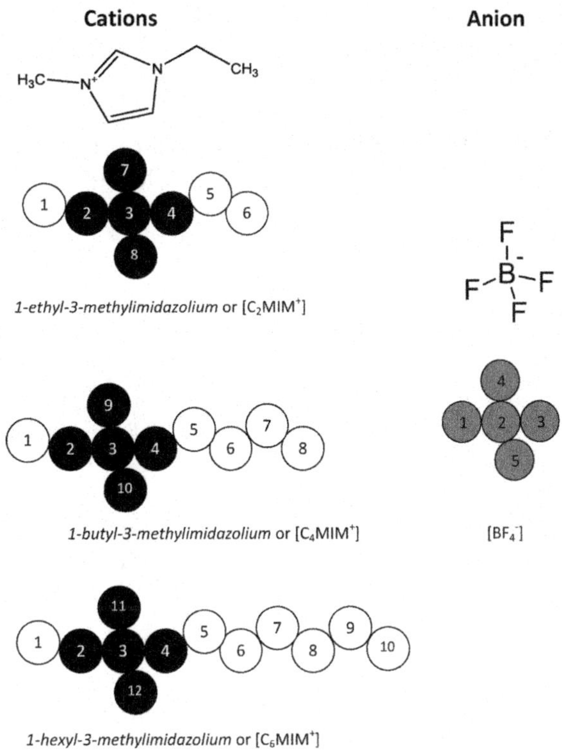

Fig. 11 Coarse-grained model of RTILs cations $[C_2MIM^+]$, $[C_4MIM^+]$ and $[C_6MIM^+]$; anion $[BF_4^-]$. *Coloured spheres* are charged and the rest are neutral (Color figure online)

induced by the local (highly charged) environment. To account for this polarizability effect, the model fluid is assumed to be immersed in a dielectric continuum with a relative dielectric constant, ε_r. Although this approximation compromises the dynamic behaviour, compared to polarizable models [60], it doesn't significantly affect many equilibrium properties, such as fluid density profiles and the differential capacitance [61–63]. We chose, $\varepsilon_r = 2.3$, which is typical of the electronic polarizability of hydrocarbon groups. Hence the electrostatic interaction between partial charges is given by,

$$\phi_{el}^{\alpha\beta}(r) = \frac{z_\alpha z_\beta e^2}{4\pi\varepsilon_0\varepsilon_r r} \tag{56}$$

where z_α and z_β are the valencies of the interacting pair of charged sites α and β separated by distance r, e is the elementary charge and ε_0 is the vacuum permittivity. In addition, *all* beads are assumed to interact via dispersion forces, modelled as a pairwise additive Lennard-Jones potential,

$$\phi_{LJ}(r) = 4\varepsilon_{LJ}[(\frac{\sigma}{r})^{12} - (\frac{\sigma}{r})^6] \tag{57}$$

where ε_{LJ} and σ determine the attractive strength and repulsive range of the dispersion interaction respectively. The parameter σ is set equal to the bond-length (=2.4 Å) and ε_{LJ} is assumed to be the same for all pairs of sites. Its value was chosen to reproduce the experimental density (for all n) in the simulated fluid model for $[C_nMIM^+][BF_4^-]$ at 1 atm and a temperature of 294 K. Remarkably, this could be achieved with essentially a *single* value, $\varepsilon_{LJ} = 100k_BK$, where k_B is Boltzmann constant.

5 A Classical Density Functional Theory for RTILs

To construct a DFT for the ionic liquid model, it is necessary to adapt the polymer theory described above. As with the PDFT, the free energy is written as a functional of the oligomeric densities describing the distribution of the cation and anion sites. The general form of the grand potential functional, Ω, can be written as,

$$\Omega = F^{id}[N_c(\mathbf{R}^c), N_a(\mathbf{R}^a)] + F^{ex}_{hs}[n_s(\mathbf{r})] + F^{ex}_{disp}[n_s(\mathbf{r}))] + F^{ex}_{corr}[n_c(\mathbf{r}), n_a(\mathbf{r})] + \sum_{\alpha=c,a} \int \Phi_\alpha(\mathbf{R}) N_\alpha(\mathbf{R})\, d\mathbf{R} - \sum_{\alpha=c,a} (\mu_\alpha + z_\alpha \Psi_D) \int N_\alpha(\mathbf{R})\, d\mathbf{R} \tag{58}$$

Here $\{\mathbf{R}^\alpha; \alpha = c, a\}$ are collective coordinates of the constituent sites on the cation (c) and anion (a) oligomers. Thus $\mathbf{R}^\alpha = (\mathbf{r}_1^\alpha, ...)$, where $\mathbf{r}_i^\alpha$ is the ith site on species α. The set $\{N_\alpha(\mathbf{R}); \alpha = c, a\}$ are the oligomeric densities for cations and anions. The function $\Phi_\alpha(\mathbf{R})$ is an externally applied potential, μ_α is the chemical potential of species α in the bulk with total charge z_α and Ψ_D is the Donnan potential, whose value ensures that the system remains electroneutral.

The ideal contribution to the free energy is exactly given by,

$$\beta F^{id}[N_c(\mathbf{R}^c), N_a(\mathbf{R}^a)] = \sum_{\alpha=c,a} \int N_\alpha(\mathbf{R}) \left(\ln[N_\alpha(\mathbf{R})] - 1\right) d\mathbf{R} + \sum_{\alpha=c,a} \int N_\alpha(\mathbf{R}) \beta \Phi_\alpha^{(b)}(\mathbf{R})\, d\mathbf{R}^\alpha \tag{59}$$

where $\Phi_\alpha^{(b)}(\mathbf{R})$ is the bond potential for molecular species α. As bonded spheres are able to roll over each others' surfaces without angular constraints and all bonds have the same length, σ, we have

$$e^{-\beta \Phi_\alpha^{(b)}(\mathbf{R}^\alpha)} = \prod_{\langle i,j \rangle_\alpha} \frac{\delta(|\mathbf{r}_i - \mathbf{r}_j| - \sigma)}{4\pi\sigma^2} \tag{60}$$

where $\langle i, j \rangle_\alpha$ runs over all pairs of bonded sites i and j on species α.

The *site* density, $n_i^{\alpha}(\mathbf{r})$, is the contribution to the monomer density of the site i on species α. It is obtained as an integral over the oligomeric density $N_{\alpha}(\mathbf{R}^{\alpha})$,

$$n_{\alpha}^{(i)}(\mathbf{r}) = \int \delta(\mathbf{r} - \mathbf{r}_i^{\alpha}) N_{\alpha}(\mathbf{R}) d\mathbf{R} \tag{61}$$

The excess functionals are assumed to depend upon certain *collective* site densities, which are sums over the individual site densities. For example, $\{n_{\beta}(\mathbf{r});\ \beta = c, a, n\}$, refer to densities of *positive*, *negative* and *neutral* sites where, e.g., $n_c(\mathbf{r})$ is the sum of density contributions over all the charged sites on the cation. The overall *sum* of site densities is, $n_s(\mathbf{r}) = n_c(\mathbf{r}) + n_a(\mathbf{r}) + n_n(\mathbf{r})$. The excess hard sphere term, $F_{hs}^{ex}[n_s(\mathbf{r})]$, is a functional of the total site density and was given earlier in Eq. (27). As was done above, we will neglect the dependence on terminal monomers. The dispersion term, $F_{disp}^{ex}[n_s(\mathbf{r})]$, has the same form as Eq. (29).

5.1 Dealing with Electrostatic Correlations

The electrostatic correlation functional is assumed to be second-order with respect to the relevant charge densities, and can be divided into like-charge and unlike-charge terms,

$$F_{corr}^{ex} \approx \frac{1}{2} \int \int \sum_{\alpha,\beta} n_{\alpha}(\mathbf{r}) n_{\beta}(\mathbf{r}') K_{corr}^{\alpha\beta}(|\mathbf{r} - \mathbf{r}'|)\, d\mathbf{r}\, d\mathbf{r}' = F_{el}^{like} + F_{el}^{unlike} \tag{62}$$

5.1.1 Correlations Between Like Charges

The mean-field expression for the electrostatic interaction between like-charged ions is given by,

$$F_{el}^{MF} = \frac{1}{2} \int \int \sum_{\alpha} n_{\alpha}(\mathbf{r}) n_{\alpha}(\mathbf{r}') \Phi_{el}^{\alpha\alpha}(|\mathbf{r} - \mathbf{r}'|)\, d\mathbf{r}\, d\mathbf{r}' \tag{63}$$

where $\Phi_{el}^{\alpha\alpha}(|\mathbf{r} - \mathbf{r}'|)$ is given by Eq. (56). This mean-field expression does not adequately account for electrostatic correlations. In dilute electrolytes, the correlation term is expected to be small and delocalized. In dense RTILs, however, it is more localized and contributes a significant repulsive interaction which should be removed. This can be viewed as a "Coulomb hole" in the like-charge density around a given particle. This is a *soft* hole as particles carrying charges of the same sign naturally tend to avoid each other due to electrostatic repulsions. To describe this effect, we shall use a so-called "hole" function, $H(r)$, with the following properties: $H(0) = 0$ and $H(r) \to 1$ as $r \to \infty$. In terms of this function, the electrostatic interactions between like-charges is given by,

$$F_{el}^{like} = \frac{1}{2}\int\int\sum_{\alpha} n_\alpha(\mathbf{r})n_\alpha(\mathbf{r}')H_\alpha(|\mathbf{r}-\mathbf{r}'|)\Phi_{el}^{\alpha\alpha}(|\mathbf{r}-\mathbf{r}'|)\,d\mathbf{r}\,d\mathbf{r}' \quad (64)$$

As a simple approximation, we choose an exponential form for the hole function, i.e.,

$$H_\alpha(r) = (1 - e^{-\lambda_\alpha r}) \quad (65)$$

The parameter λ_α is determined by a bulk condition that the hole function describes the exclusion of a single charge, i.e.,

$$n_b^{(\alpha)}\int\{H_\alpha(|\mathbf{r}|) - 1\}\,d\mathbf{r} = -1 \quad (66)$$

where $n_b^{(\alpha)}$ is the bulk density of the charged species α. Thus we obtain,

$$\lambda_\alpha = \frac{\sqrt{2}}{s_\alpha} \quad (67)$$

with

$$s_\alpha = \left[\frac{3}{4\pi n_b^{(\alpha)}}\right]^{1/3} \quad (68)$$

5.1.2 Correlations Between Opposite Charges

The correlation between oppositely charged ions is generally overestimated by the mean-field electrostatic interaction. This is due to the fact that the mean-field expression allows complete penetration by opposite charges, which in reality is excluded by the steric repulsion between particles. A simple way to correct the mean-field theory is to build in an effective hard sphere like exclusion hole around each oppositely charged pair. The unlike-charge electrostatic interaction is then given by

$$F_{el}^{unlike} = \frac{1}{2}\int\int\sum_{\alpha\neq\beta} n_\alpha(\mathbf{r})n_\beta(\mathbf{r}')\Theta(|\mathbf{r}-\mathbf{r}'|-d)\Phi_{el}^{\alpha\beta}(|\mathbf{r}-\mathbf{r}'|)d\mathbf{r}d\mathbf{r}' \quad (69)$$

where the Heaviside function is defined as: $\Theta(x) = 1$ for $x > 0$ and $\Theta(x) = 0$ for $x \leq 0$. The parameter d is a "distance of closest approach" between unlike charged species. As such it should be numerically equal to the sum of the radii of the species, σ. However, additional correlations (beyond the hard sphere diameter σ) between oppositely charged species can be mimicked to some extent by setting $d = \chi\sigma$, with χ an adjustable parameter. Given this ansatz, the question then arises as to how χ should be chosen. In the same way that ε_{LJ} was chosen so as to obtain agreement between experimental and *simulated* bulk RTIL densities, we chose χ so that the correct bulk densities were obtained from density functional theory. For

$[C_nMIM^+][BF_4^-]$ this procedure gives rise to the values $\chi = 0.64, 0.71$ and 0.80 for $n = 2, 4, 6$ respectively. We not that χ is generally less than unity, which reflects the positive adsorption between unlike ions.

5.2 Numerical Solution via End Segment Distribution

Minimizing the grand potential functional Ω in Eq. (58) with respect to the oligomeric densities $N_\alpha(\mathbf{R}^\alpha)$, leads to the now familiar looking general formula for the configurational densities,

$$N_\alpha(\mathbf{R}^\alpha) = n_b \exp[-\beta\Phi_\alpha^{(b)}(\mathbf{R}^\alpha) + \sum_j (\lambda_{\alpha j}^b - \lambda_{\alpha j}(\mathbf{r}_j))] \tag{70}$$

where n_b is the density of cations (and anions) in the bulk. Each site j on the molecular species α carries an excess chemical potential $\lambda_{\alpha j}^b - \lambda_{\alpha j}(\mathbf{r})$ where,

$$\lambda_{\alpha i}(\mathbf{r}) = \frac{\delta\beta F^{ex}[N_c(\mathbf{R}^c), N_a(\mathbf{R}^a)]}{\delta n_\alpha^{(i)}(\mathbf{r})} + \beta\phi_\alpha^{(i)}(\mathbf{r}) \tag{71}$$

Here $F^{ex}[N_c(\mathbf{R}^c), N_a(\mathbf{R}^a)]$ is the sum of all the non-ideal terms in Eq. (58), and we have included the external potential, $\phi_\alpha^{(i)}(\mathbf{r})$, which is assumed to act on individual beads. The form of this potential is determined by the system being studied, but we will usually have only three possible types of excess site terms, $\lambda_\beta(\mathbf{r})$, with $\beta = c, a, n$, corresponding to positive, negative and neutral sites. In the bulk we have $\lambda_\beta(\mathbf{r}) = \lambda_\beta^b$. Equation (70) needs to be solved self-consistently, as the $\lambda_\beta(\mathbf{r})$ are usually functionals of the site densities.

It is useful to represent the solutions to the DFT diagrammatically. To do this, we define a set of vertices and bonds, as shown in Fig. 12. The open symbols in the shape of circle, square and diamond respectively, correspond to the neutral, positive and

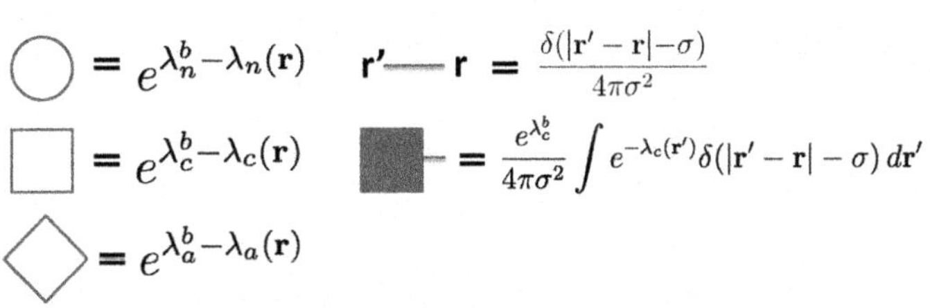

Fig. 12 Diagrams of bonds and vertices with circle for neutral, square for positive and diamond for negative beads in the model. Vertices stand for local excess chemical potential factors. The *solid line* represents the rigid bond between neighbouring beads. Colouring in a vertex means integrating the bead over spherical surface of its neighbouring bead. An example is given of integrating a positive bead over the bond terminating at **r**

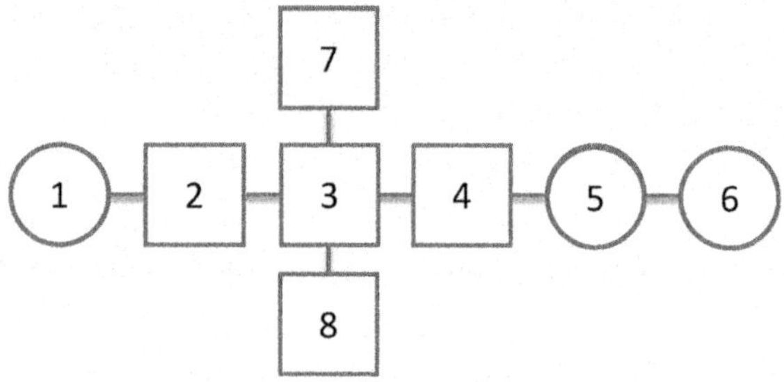

Fig. 13 Diagram of normalized density solution of $N_c(\mathbf{R}^c)/n_b$ for cation $[C_2MIM^+]$

negative factors containing the corresponding excess chemical potential terms (as indicated). The solid line represents the normalized Dirac delta function, describing the freely rotating rigid bond between sites. Graphs consisting of vertices connected by bonds represent the product of the associated functions. Colouring in a vertex indicates integrating that site over all space, as shown in Fig. 12. With these rules it is easy to see that the normalized equilibrium configurational density for the cation (from Eq. (70)), $N_c(\mathbf{R}^c)/n_b$, can be represented diagrammatically as in Fig. 13. The site density of bead i on molecule α is obtained by the integral,

$$\begin{aligned} n_\alpha^{(i)}(\mathbf{r}) &= \int \delta(\mathbf{r}-\mathbf{r}_i^\alpha) N(\mathbf{R}) d\mathbf{R}^\alpha \\ &= \int \delta(\mathbf{r}-\mathbf{r}_i^\alpha) n_b \exp[-\beta \Phi_\alpha^{(b)}(\mathbf{R}^\alpha) + \sum_j (\lambda_{\alpha j}^b - \lambda_{\alpha j}(\mathbf{r}_j^\alpha))] d\mathbf{R} \end{aligned} \quad (72)$$

in which j runs over all sites on molecule α. These can also be represented in terms of the product of simpler diagrams, as we shall now illustrate.

As before, we will define end segment distribution functions, but these will be slightly different to those described in earlier sections. One type of end distribution will be called *branch* end segment distributions, denoted as $c(\alpha i, \mathbf{r})$. The function $c(\alpha i, \mathbf{r})$ contains coloured vertices connected by bonds with one terminal site (a site with only a single adjacent bond), but no branching sites of the molecule and a dangling bond attached at site αi. A branching site is one with more than two adjacent bonds, e.g., site $c3$ on the cation in Fig. 13. Two examples of *branch* end segment distributions at $c1$ and $c2$ are given in Fig. 14. The other type of segment distribution we will call *complementary* end segment distributions. These are connected graphs of coloured vertices and a dangling bond attached at the end site. All the molecular vertices not included in the complementary graph can be connected and coloured to make a branch end segment distribution. In linear molecules the discrepancy between branch and complementary end segment distributions doesn't exist. An example of a complimentary end segment distribution, denoted here as $c'(c2, \mathbf{r})$ is also given in Fig. 14. The branch end segment distributions are obtained recursively along each chain branch. For example, we have

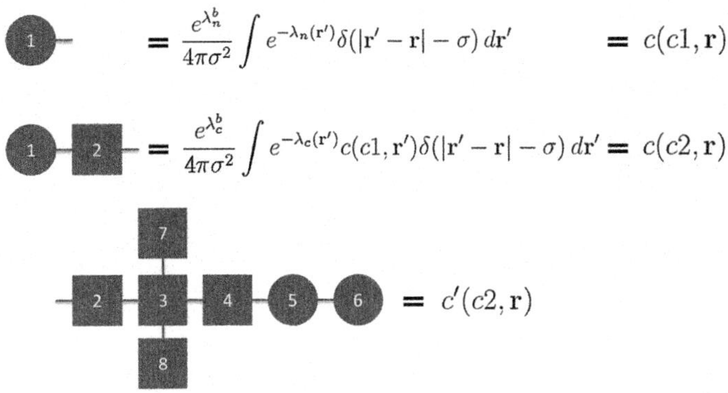

Fig. 14 Diagrams for the recursive process of generating branch end segment distribution, $c(c2, \mathbf{r})$ from $c(c1, \mathbf{r})$. Example of the complementary end segment distribution $c'(c2, \mathbf{r})$ for $[C_2MIM^+]$

$$c(c1,\mathbf{r}) = \frac{e^{\lambda_n^b}}{4\pi\sigma^2}\int e^{-\lambda_n(\mathbf{r}')}\delta(|\mathbf{r}'-\mathbf{r}|-\sigma)\,d\mathbf{r}' \tag{73}$$

The function $c(c2, \mathbf{r})$ is then generated from $c(c1, \mathbf{r})$, as illustrated in Fig. 14, according to the recursion formula.

$$c(c2,\mathbf{r}) = \frac{e^{\lambda_c^b}}{4\pi\sigma^2}\int c(c1,\mathbf{r}')e^{-\lambda_c(\mathbf{r}')}\delta(|\mathbf{r}'-\mathbf{r}|-\sigma)\,d\mathbf{r}' \tag{74}$$

On the other hand, *complementary* segment distributions must be constructed using products of branch end segment distributions. For example, the complementary graph $c'(c2, \mathbf{r})$ is given by,

$$c'(c2,\mathbf{r}) = \frac{e^{\lambda_c^b}}{4\pi\sigma^2}\int e^{-\lambda_c(\mathbf{r}')}c'(c3,\mathbf{r}')\delta(|\mathbf{r}'-\mathbf{r}|-\sigma)\,d\mathbf{r}' \tag{75}$$

with

$$c'(c3,\mathbf{r}) = \frac{e^{\lambda_c^b}}{4\pi\sigma^2}\int e^{-\lambda_c(\mathbf{r}')}c(c4,\mathbf{r}')c(c7,\mathbf{r}')c(c8,\mathbf{r}')\delta(|\mathbf{r}'-\mathbf{r}|-\sigma)\,d\mathbf{r}' \tag{76}$$

While it is not clear from the simple nomenclature used here, we note that there are three different complimentary distributions at the branching site $c3$ (as there are three branches emerging from that site). Two of these will be equivalent due to symmetry. Any site density can always be written as a product of the site vertex and branch and complementary end segment distributions. For example, the graph in Fig. 15 representing the normalized site density at $c2$, contains a product of end segment distributions, given by

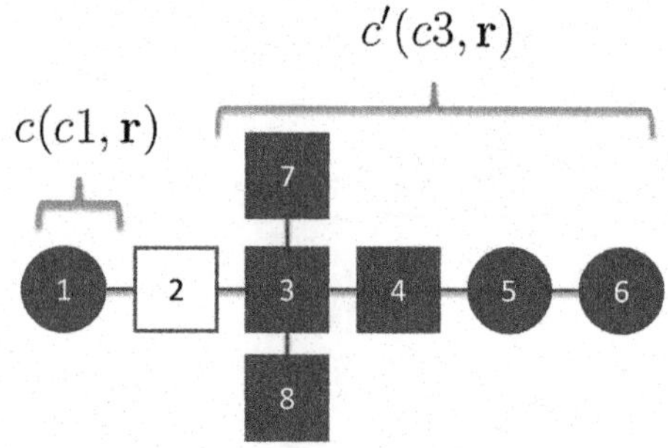

Fig. 15 Diagram for normalized site density of bead $c2$ on cation $[C_2MIM^+]$. End segment distribution $c(c1, \mathbf{r})$ and $c'(c3, \mathbf{r})$ are combined

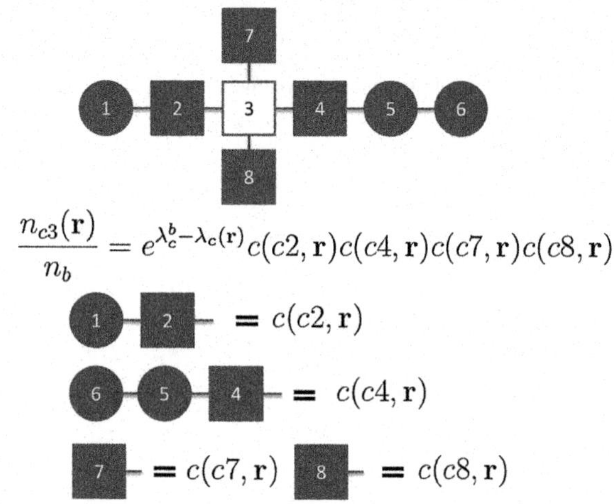

Fig. 16 Diagram for normalized site density of central bead $c3$ on cation $[C_2MIM^+]$. It is obtained as the product of 4 branch end segment distributions (as shown) and the site vertex

$$n_c^{(2)}(\mathbf{r}) = n_b e^{\lambda_c^b - \lambda_c(\mathbf{r})} c(c1, \mathbf{r}) c'(c3, \mathbf{r}) \tag{77}$$

The density contribution due to all other sites can be calculated in a similar fashion.

The monomer density at branching sites can be written in terms of the product of a vertex and end segment distribution functions. For example the graphical expression for the density at site $c3$ is depicted in Fig. 16 and is given by,

$$n_c^{(3)}(\mathbf{r}) = n_b e^{\lambda_c^b - \lambda_c(\mathbf{r})} c(c2, \mathbf{r}) c(c4, \mathbf{r}) c(c7, \mathbf{r}) c(c8, \mathbf{r}) \tag{78}$$

5.3 *RTILs at Planar Electrodes: Comparison of the DFT with Simulations*

The performance of the DFT for the RTIL is evaluated by considering the screening of charged electrodes [59] immersed in the fluid. We compare the predictions of the DFT to Monte Carlo simulations for the coarse-grained model described above. The electrodes are assumed to be planar with an inverse surface charge density given by the parameter $a(s)$. Furthermore, all sites of the RTIL fluid experience a dispersion interaction with the electrode, obtained by integrating the Lennard-Jones potential over the half-space constituting the surfaces:

$$\beta w_{LJ}(z) = 2\pi\varepsilon_{LJ}\left[\frac{2}{45}(\frac{\sigma}{z})^9 - \frac{1}{3}(\frac{\sigma}{z})^3\right] \tag{79}$$

where z is the distance from the plane of the surface, and we assume a particle density for the half-space of $1/\sigma^3$. Figure 17 compares the DFT predictions for the density profiles of positive, negative and the sum of all sites with simulation data for $[C_4MIM^+][BF_4^-]$ adjacent to electrodes with varying surface charge densities. The DFT performs remarkably well, seemingly able to capture all the important structural

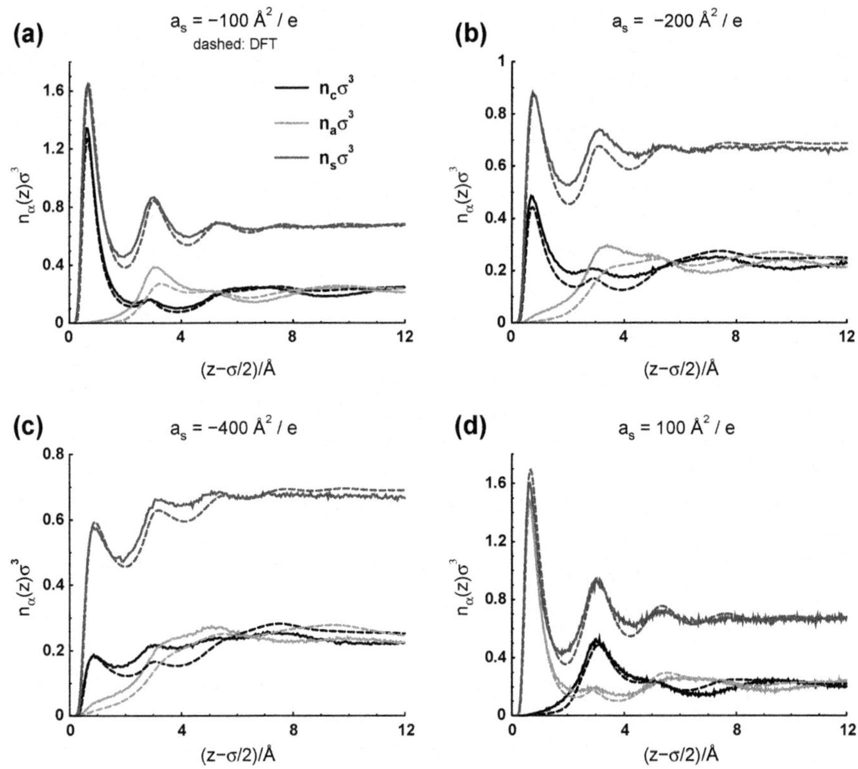

Fig. 17 Comparing simulated (*solid lines*) and calculated (*dashed lines*—by DFT) density distributions, of positive, negative and sum of all beads for $[C_4MIM^+][BF_4^-]$ adjacent to charged surfaces (at $z = 0$). At various surface charge densities: **a** $a_s = -100\,\text{Å}^2/e$; **b** $a_s = -200\,\text{Å}^2/e$ **c** $a_s = -400\,\text{Å}^2/e$; **d** $a_s = 100\text{Å}^2/e$

features of the density profiles with only minor variations from the (exact) simulation results.

6 Conclusion

Classical density functional theory for fluids has developed rapidly over the last few decades. The application to polymeric and other complex fluids, followed from its demonstrated early successes in simple fluid models. We have shown how the polymer density functional theory can be successfully applied to both monodisperse and polydisperse polymers, where the latter displays a Schulz-Flory-Zimm molecular weight distribution. In that case, the theory is significantly simplified, with relatively modest numerical requirements for its solution. This was specifically illustrated for semi-flexible polydisperse polymers in solution. Finally, we showed how more complex molecular architectures could also be handled merely by introducing the connectivity constraints in the ideal free energy contribution. This was illustrated using a coarse-grained model for room temperature ionic liquids. For these highly charged, dense fluids a relatively simple treatment of electrostatic correlations, coupled with established steric contributions to the free energy, led to a very accurate theory, as compared to simulations.

The computational advantages of classical polymer density functional theory facilitates the study of systems with boundary conditions that better mimic experimental scenarios. One example of this is dealing with the grand canonical ensemble, which ensures that equilibrium with a bulk reservoir is maintained. This condition is often very difficult to achieve in simulations. Furthermore, one is able to consider systems with a greater degree of polymerization.

One of the drawbacks of the polymer density functional theory is that it employs mean-field approximations that generally fail to capture the effect of correlations on chain configurations. This could be addressed by using a more accurate treatment of intra-molecular constraints, or else using experimental measurements to enforce say a given radius of gyration. These issues notwithstanding, density functional theory has now established itself as a valuable tool for the study of non-uniform complex fluids, for both theoreticians and experimentalist alike.

References

1. P. Hohenberg, W. Kohn, Phys. Rev. **136**, B864 (1964)
2. N.D. Mermin, Phys. Rev. **127**, A1441 (1965)
3. D. Chandler, J.D. McCoy, S.J. Singer, J. Chem. Phys. **85**, 5971 (1986)
4. L. Pratt, D. Chandler, J. Chem. Phys. **66**, 147 (1977)
5. C.E. Woodward, J. Chem. Phys. **94**, 3183 (1991)
6. Y. Rosenfeld, Phys. Rev. Lett. **63**, 980 (1989)
7. Y.X. Yu, J. Wu, J. Chem. Phys. **117**, 2368 (2002)

8. Z. Li, D. Cao, J. Wu, J. Chem. Phys. **122**(17), 174708 (2005)
9. R. Roth, J. Phys.: Condens. Matter **22**(6), 063102 (2010)
10. X. Xu, D. Cao, J. Wu, Soft Matter **6**(19), 4631 (2010)
11. S. Lamperski, M. Kaja, L.B. Bhuiyan, J. Wu, D. Henderson, J. Chem. Phys. **139**(5), 054703 (2013)
12. C.E. Woodward, A. Yethiraj, J. Chem. Phys. **100**, 3181 (1994)
13. J. Forsman, C.E. Woodward, B.C. Freasier, J. Chem. Phys. **117**, 1915 (2002)
14. J. Forsman, C.E. Woodward, Phys. Rev. Lett. **94**, 118301 (2005)
15. E. Kierlik, M.L. Rosinberg, J. Chem. Phys. **97**, 9222 (1992)
16. G.J. Fleer, M.A.C. Stuart, J.M.H.M. Scheutjens, T. Cosgrove, B. Vincent, *Polymers at Interfaces* (Chapman and Hall, London, 1993)
17. J.Z. Wu, Aiche J **52**(3), 1169 (2006)
18. J.Z. Wu, Z. Li, Annu. Rev. Phys. Chem. **58**, 85 (2007)
19. S. Nordholm, M. Johnson, B.C. Freasier, Aust. J. Chem. **33**, 2139 (1980)
20. S. Nordholm, Aust. J. Chem. **37**, 1 (1984)
21. S. Nordholm, Chem. Phys. Lett. **105**, 302 (1984)
22. J.M. Wichert, H.S. Gulati, C.K. Hall, J. Chem. Phys. **105**, 7669 (1996)
23. A. Yethiraj, C.E. Woodward, J. Chem. Phys. **102**, 5499 (1995)
24. J. Forsman, C.E. Woodward, J. Chem. Phys. **120**, 506 (2004)
25. J. Forsman, C.E. Woodward, Macromol. **39**, 1261 (2006)
26. B.C. Freasier, C.E. Woodward, S. Nordholm, J. Chem. Phys. **90**, 5657 (1989)
27. T. Åkesson, C.E. Woodward, B. Jönsson, J. Chem. Phys. **91**(4), 2461 (1989)
28. R. Tuinier, A. Petukhov, Macromol. Theory Simul. **11**, 975 (2002)
29. J. van der Gucht, N.A.M. Besseling, Phys. Rev. E **65**, 051801 (2002)
30. J. van der Gucht, N.A.M. Besseling, G.J. Fleer, J. Chem. Phys. **119**, 8175 (2003)
31. J. van der Gucht, N.A.M. Besseling, J. Phys.: Condens. Matter. **15**, 6627 (2003)
32. J. van der Gucht, N.A.M. Besseling, G.J. Fleer, Macromol. **37**, 3026 (2004)
33. S.P.F.M. Roefs, J.M.H.M. Scheutjens, G.J. Fleer, Macromol. **27**, 4180 (1994)
34. S.F. Edwards, Proc. Phys. Soc. **85**, 613 (1965)
35. P.G. deGennes, Rep. Prog. Phys. **32**, 187 (1969)
36. P.G. deGennes, Macromol. **15**, 492 (1982)
37. S.W. Sides, G.H. Fredrickson, J. Chem. Phys. **121**, 4974 (2004)
38. M.W. Matsen, Eur. Phys. J. E **21**, 199 (2006)
39. D.M. Cooke, A.C. Shi, Macromol. **39**, 6661 (2006)
40. S. Yang, H. Tan, D. Yan, E. Nies, A.C. Shi, Phys. Rev. E **75**, 061803 (2007)
41. C.E. Woodward, J. Forsman, Phys. Rev. Lett. **100**, 098301 (2008)
42. K.G. Honnell, C.K. Hall, J. Chem. Phys. **95**, 4481 (1991)
43. J. Forsman, C.E. Woodward, Macromol. **39**, 1269 (2006)
44. C.E. Woodward, J. Forsman, Phys. Rev. E **74**, 010801 (2006)
45. M. Turesson, J. Forsman, T. Åkesson, Phys. Rev. E **76**, 021801 (2007)
46. M. Turesson, C.E. Woodward, T. Åkesson, J. Forsman, J. Phys. Chem. B **112**, 9802 (2008)
47. B.V. Derjaguin, Kolloid Zeits. **69**, 155 (1934)
48. J. Forsman, C.E. Woodward, J. Chem. Phys. **131**, 044903 (2009)
49. J. van der Gucht, N.A.M. Besseling, J. van Male, M.A. Cohen-Stuart, J. Chem. Phys. **113**, 2886 (2000)
50. J. Forsman, C.E. Woodward, M. Trulsson, J. Phys. Chem. B **115**, 4606 (2011)
51. J.Z. Wu, T. Jiang, D.E. Jiang, Z.H. Jin, D. Henderson, Soft Matter **7**(23), 11222 (2011)
52. I. Krossing, J.M. Slattery, C. Daguenet, P.J. Dyson, A. Oleinikova, H. Weingartner, J. Am. Chem. Soc. **128**(41), 13427 (2006)
53. M.V. Fedorov, A.A. Kornyshev, Chem. Rev. **114**(5), 2978 (2014)
54. N.V. Plechkova, K.R. Seddon, Chem. Soc. Rev. **37**(1), 123 (2008)
55. R.D. Rogers, K.R. Seddon, Science **302**(5646), 792 (2003)
56. D. Zhao, M. Wu, Y. Kou, E. Min, Catal Today **74**(1), 157 (2002)
57. R.P. Swatloski, S.K. Spear, J.D. Holbrey, R.D. Rogers, J. Am. Chem. Soc. **124**(18), 4974 (2002)

58. G. Wei, Z. Yang, C. Chen, Anal. Chim. Acta **488**(2), 183 (2003)
59. M. Turesson, R. Szparaga, K. Ma, C.E. Woodward, J. Forsman, Soft Matter **10**(18), 3229 (2014)
60. D. Bedrov, O. Borodin, Z. Li, G.D. Smith, J. Phys. Chem. B **114**(15), 4984 (2010)
61. G. Feng, P.T. Cummings, J. Phys. Chem. Lett. **2**(22), 2859 (2011)
62. G. Feng, D.E. Jiang, P.T. Cummings, J. Chem. Theory Comput. **8**(3), 1058 (2012)
63. G. Feng, S. Li, J.S. Atchison, V. Presser, P.T. Cummings, J. Phys. Chem. C **117**(18), 9178 (2013)

Variational Perturbation Theory for Electrolyte Solutions

Leo Lue

1 Introduction

In typical treatments of the statistical mechanics of electrolyte solutions, the focus is on the particles (ions) in the system. The state of the system is labeled by the particle positions and orientations, and the partition function of the system is given by an integral over all these degrees of freedom. Another approach, which we pursue and present here, is to focus on the interaction potential generated by the particles, rather than on the particles themselves. We illustrate this approach by examining systems of particles interacting through electrostatic forces; however, these techniques can be used for a broad range of interaction potentials.

The remainder of the work is organized as follows. First we describe the representation of the partition function of an electrolyte system in terms of a functional integral over shapes of an interaction potential, rather than the positions and orientations of the ions in the system. As part of this, we will review electrostatics, and in particular, we will discuss the Green's function, which describes the propagation of the electric potential from a charge and plays a central role in the theory. The use and limitations of the mean-field approximation to evaluate the functional integral are discussed. In the following section, we present the variational perturbation approximation, which is a method that overcomes some of the limitations of the mean-field approximation. Some general aspects of this approach are discussed. The resulting theory is then illustrated in Sect. 4 by applying it to the point charge model. The first order variational perturbation theory is compared to the Poisson-Boltzmann theory, and various features of the model, which are not captured within the classic Poisson-Boltzmann theory, are discussed, including the influence of image-charge

L. Lue (✉)
Department of Chemical and Process Engineering, University of Strathclyde, James Weir Building, 75 Montrose Street, Glasgow G1 1XJ, UK
e-mail: leo.lue@strath.ac.uk

J. Wu (ed.), *Variational Methods in Molecular Modeling*,
Molecular Modeling and Simulation, DOI 10.1007/978-981-10-2502-0_5

or polarization interactions due to dielectric interfaces. Finally, the main points of this work are summarized in Sect. 5, along with other applications and extensions of the theory.

2 Development of the Field Theory

We consider a multicomponent mixture of charged particles at fixed temperature T, volume V, and species chemical potentials μ_α. Associated with each particle of type α is a rigid charge density, given by $Q_\alpha(\mathbf{r}, \boldsymbol{\Omega})$, where $\boldsymbol{\Omega}$ represents its orientation. These particles are embedded in a medium with a spatially varying dielectric constant $\varepsilon(\mathbf{r})$. In addition, there could be a fixed charge distribution $\Sigma(\mathbf{r})$ through the system, for example, due to a charged boundary.

2.1 Electrostatics

The total charge density $\mathcal{Q}(\mathbf{r})$ in the system is composed of a contribution from the mobile particles and a contribution from a fixed charge density $\Sigma(\mathbf{r})$ and can be written as

$$\mathcal{Q}(\mathbf{r}) = \sum_{k\alpha} Q_\alpha(\mathbf{r} - \mathbf{R}_{k\alpha}, \boldsymbol{\Omega}_{k\alpha}) + \Sigma(\mathbf{r}) \tag{1}$$

where $\mathbf{R}_{k\alpha}$ is the position of the kth particle of type α, and $\boldsymbol{\Omega}_{k\alpha}$ denotes its orientation.

Charges are sources for the electrostatic potential $\phi(\mathbf{r})$. These two quantities are related by the Poisson equation [1]:

$$-\frac{1}{4\pi}\nabla \cdot [\varepsilon(\mathbf{r})\nabla\phi(\mathbf{r})] = \mathcal{Q}(\mathbf{r}). \tag{2}$$

Key to the description of electrostatic interactions is the Green's function G_0 of the Poisson equation, which is defined through [1]

$$-\frac{1}{4\pi}\nabla \cdot [\varepsilon(\mathbf{r})\nabla G_0(\mathbf{r}, \mathbf{r}')] = \delta^d(\mathbf{r} - \mathbf{r}'), \tag{3}$$

The Green's function $G_0(\mathbf{r}, \mathbf{r}')$ gives the electrostatic potential $\phi(\mathbf{r})$ generated at position $\mathbf{r}$ due to a point charge located at $\mathbf{r}'$. In the specific case where the dielectric constant has a uniform value ε through the entire system, the explicit form of the Green's function, denoted by G_{free}, is

$$G_{\text{free}}(\mathbf{r}, \mathbf{r}') = \frac{1}{\varepsilon|\mathbf{r} - \mathbf{r}'|}. \tag{4}$$

The form of the Green's function is affected by the variations in the dielectric constant $\varepsilon(\mathbf{r})$ through the system, which reflects the polarizability of the background.

The manner in which the potential emanates from a general charge distribution and propagates through the system is characterized by the Green's function:

$$\phi(\mathbf{r}) = \int d\mathbf{r}' G_0(\mathbf{r}, \mathbf{r}')\mathscr{Q}(\mathbf{r}') \tag{5}$$

which is simply another manner to write the Poisson equation.

The total electrostatic energy E^{elec} of the system can be written as the product of the charge density in the system and the electrostatic potential [1]:

$$E^{\text{elec}} = \frac{1}{2}\int d\mathbf{r}\mathscr{Q}(\mathbf{r})\phi(\mathbf{r}). \tag{6}$$

Substituting the expression for the electrostatic potential in terms of the charge density, as given in Eq. (5), the energy of the electrostatic field can be written in a more familiar form:

$$E^{\text{elec}} = \frac{1}{2}\int d\mathbf{r}d\mathbf{r}'\mathscr{Q}(\mathbf{r})G_0(\mathbf{r}, \mathbf{r}')\mathscr{Q}(\mathbf{r}') \tag{7}$$

This expression for the energy generalizes Coulomb's law (where G_0 is replaced with G_{free}) for the situation where the dielectric constant is not spatially uniform.

The expression for the electrostatic energy given in Eq. (7) contains contributions from the interaction of the charge of a particle with the electrostatic potential that it generates—the self energy of the particle. The electrostatic self energy of a particle of type α is given by

$$e_\alpha^{\text{se}}(\mathbf{R}, \boldsymbol{\Omega}) = \frac{1}{2}\int d\mathbf{r}d\mathbf{r}' Q_\alpha(\mathbf{r}-\mathbf{R}, \boldsymbol{\Omega})G_{\text{free}}(\mathbf{r}, \mathbf{r}')Q_\alpha(\mathbf{r}'-\mathbf{R}, \boldsymbol{\Omega}) \tag{8}$$

The self energy is the interaction of a particle's charge distribution with the electrostatic potential it generates in the absence of spatial variations of the background dielectric constant. The introduction of inhomogeneities in the dielectric constant, such as polarizable bodies or dielectric interfaces will lead to a shift of the self energy.

Removing this self energy contribution (which is potentially infinite, such as for a point charge), the electrostatic interaction energy between particles is given by

$$E^{\text{elec}} = \frac{1}{2}\int d\mathbf{r}d\mathbf{r}'\mathscr{Q}(\mathbf{r})G_0(\mathbf{r}, \mathbf{r}')\mathscr{Q}(\mathbf{r}') - \sum_{k\alpha} e_\alpha^{\text{se}}(\mathbf{R}_{k\alpha}, \boldsymbol{\Omega}_{k\alpha}). \tag{9}$$

2.2 Partition Function

All the equilibrium static properties (e.g., structural and thermodynamic) of an open system of fixed volume V, temperature T, and species chemical potential μ_α can be determined from knowledge of the grand partition function Z_G. The grand partition function of this system is given by [2]

$$Z_G[\gamma, \Sigma] = \sum_{N_1=0}^{\infty} \cdots \sum_{N_M=0}^{\infty} \prod_{\nu} \frac{1}{N_\nu! \Lambda_\nu^{dN_\nu}} \times \int \prod_{t\tau} d\mathbf{R}_{t\tau} d\boldsymbol{\Omega}_{t\tau} e^{-\beta E^{\text{elec}} - \beta E^{\text{ref}} + \sum_{k\alpha} \gamma_\alpha(\mathbf{R}_{k\alpha}, \boldsymbol{\Omega}_{k\alpha})} \tag{10}$$

where $\beta = 1/(k_B T)$, k_B is the Boltzmann constant, Λ_α is the de Broglie wavelength of particles of type α, $\gamma_\alpha(\mathbf{R}_{k\alpha}, \boldsymbol{\Omega}_{k\alpha}) = \beta\mu_\alpha - \beta u_\alpha^{(1)}(\mathbf{R}_{k\alpha}, \boldsymbol{\Omega}_{k\alpha})$ is a dimensionless chemical potential which includes the effect of an applied one-body potential $u_\alpha^{(1)}$, and E^{ref} is the interaction energy between particles that is not due to electrostatics (e.g., excluded volume).

In order to transform the partition function from a sum over particle positions and orientations to a sum over shapes of a fluctuating field, we introduce the Hubbard-Stratonovich transformation [3, 4]. In terms of the electrolyte systems, this approach has been used by many researchers [5–8]. This transformation allows us to represent the electrostatic interaction energy E^{elec} of the charge distribution in the system in terms of the following average

$$\exp\left(-\frac{1}{2\beta} \int d\mathbf{r} d\mathbf{r}' \mathcal{Q}(\mathbf{r}) G_0(\mathbf{r}, \mathbf{r}') \mathcal{Q}(\mathbf{r}')\right) = \left\langle e^{-\int d\mathbf{r} \mathcal{Q}(\mathbf{r}) i\psi(\mathbf{r})} \right\rangle_0 \tag{11}$$

where

$$\langle(\cdots)\rangle_0 = \frac{1}{\mathcal{N}_0} \int \mathcal{D}\psi(\cdot)(\cdots) e^{-H_0[\psi]}, \tag{12}$$

the Hamiltonian H_0 is given by

$$H_0[\psi] = \frac{1}{2\beta} \int d\mathbf{r} d\mathbf{r}' \psi(\mathbf{r}) G_0^{-1}(\mathbf{r}, \mathbf{r}') \psi(\mathbf{r}'), \tag{13}$$

and $\mathcal{N}_0$ is a normalization constant which is given by

$$\mathcal{N}_0 = \int \mathcal{D}\psi(\cdot) e^{-H_0[\psi]} \propto (\det G_0^{-1})^{1/2} \tag{14}$$

Consequently, we find that the grand partition function can be written as

$$Z_G[\gamma, \Sigma] = \left\langle Z_G^{\mathrm{ref}}[\gamma - Qi\psi + \beta e^{\mathrm{se}}] \exp\left[-\int d\mathbf{r}\Sigma(\mathbf{r})i\psi(\mathbf{r})\right]\right\rangle_0 \quad (15)$$

where $Z_G^{\mathrm{ref}}[\gamma]$ is the grand partition function of a system where there are no electrostatic interactions between the particles.

The main physical interpretation of this mathematical transformation of the partition function is that the grand partition function of a system with electrostatic interactions is identical to a system without electrostatic interactions, but with each particle interacting with a "randomly" fluctuating external field. The strength of the coupling of each particle to this field is proportional to its charge:

$$\gamma_\alpha(\mathbf{R}, \boldsymbol{\Omega}) \to \gamma_\alpha(\mathbf{R}, \boldsymbol{\Omega}) - \int d\mathbf{r}\, Q_\alpha(\mathbf{r} - \mathbf{R}, \boldsymbol{\Omega})i\psi(\mathbf{r}) + \beta e_\alpha^{\mathrm{se}}(\mathbf{R}, \boldsymbol{\Omega}).$$

The field ψ is not entirely randomly distributed but, rather, is spatially correlated to itself. It is actually distributed according to a Gaussian probability distribution function with a zero mean. The correlation of a fluctuation of ψ at a position $\mathbf{r}$ and a fluctuation at a position $\mathbf{r}'$ is given by the Green's function of the Poisson equation (i.e. $\langle\psi(\mathbf{r})\psi(\mathbf{r}')\rangle = \beta G_0(\mathbf{r}, \mathbf{r}')$). The Hubbard-Stratonovich transformation is not limited to electrostatic interactions, but it can also be performed on any general pairwise additive interaction. The two-point correlation of the fluctuating field is proportional to the pairwise interaction potential of the system.

2.3 Dispersion Interactions

Before we discuss methods to evaluate the grand partition function, there are a couple of nice features that should be mentioned about the functional integral formulation. First, the "fluctuations" of the electric potential in the system were assumed to be due to the thermal motion of the ions in the system. However, we can consider the partition function for a system without any charged particles. The partition function essentially becomes the determinant of the Green's function of the Poisson equation. Changes in the arrangement of dielectric bodies in the system will alter the function $\varepsilon(\mathbf{r})$, and, consequently, will affect the Green's function $G_0(\mathbf{r}, \mathbf{r})$ of the Poisson equation. This will lead to differences in the normalization constant $\mathcal{N}$ of the functional integral, which are essentially related to the determinant of the Green's function

$$\beta F = -\ln \frac{\mathcal{N}_0}{\mathcal{N}_{\mathrm{free}}} = \frac{1}{2}\mathrm{Tr}\ln \frac{G_0}{G_{\mathrm{free}}}. \quad (16)$$

where Tr stands for the operator trace. Interestingly, in this case, the functional integral formulation reduces to the Lifshitz theory of the dispersion interaction [9]. The theory is able to nicely couple dispersion (aka van der Waals) forces and electrostatic interactions within a single framework [10].

The difference in this determinant in the situation where there are dielectric bodies versus the absence of dielectric bodies (the free state), leads to an effective interaction between the bodies due to a shift in the normal modes available to the electric potential. So one nice thing about this transformation is that it allows us to naturally couple the thermal motion of the charged particles in the system to the zero frequency dispersion interactions. For example, we naturally get screening of the dispersion interaction [7, 11–13].

2.4 Mean-Field Approximation

The functional integral formulation of the partition function is exact. Unfortunately, it is not possible to analytically evaluate the functional integral due to the nonlinear dependence of the reference partition function $\ln Z^{\text{ref}}$ on the one-body potentials $\gamma_\alpha(\mathbf{R}, \boldsymbol{\Omega})$. As with all exact truths, we really can not do anything useful until we make some sort of approximation.

One of the simplest approximations is to neglect the fluctuations in the field altogether and simply choose the value of the field which has the "largest" contribution to the free energy—the largest value of the integrand. That is, we approximate the average of a quantity of a functional $A[\psi]$ as

$$\left\langle e^{A[\psi]}\right\rangle_0 \approx e^{A[\bar{\psi}]} \tag{17}$$

where $\bar{\psi}$ is the form of the function ψ which makes the functional $e^{A[\psi]}$ stationary:

$$\frac{\delta A[\bar{\psi}]}{\delta\psi(\mathbf{r})} = 0 \tag{18}$$

This is known as the mean-field approximation.

Applying this to the evaluation of the grand partition function, we find

$$\begin{aligned}\ln Z_G[\gamma, \Sigma] \approx \frac{1}{2\beta}\int d\mathbf{r}d\mathbf{r}' i\bar{\psi}(\mathbf{r})G_0^{-1}(\mathbf{r}, \mathbf{r}')i\bar{\psi}(\mathbf{r}) - \int d\mathbf{r}\Sigma(\mathbf{r})i\bar{\psi}(\mathbf{r})\\ + \ln Z_G^{\text{ref}}[\gamma - Qi\bar{\psi}]\end{aligned} \tag{19}$$

The final term is the grand potential of the reference system, where the chemical potential (or one-body potential) of the particles has been shifted by a coupling of the particle by its charge to the mean field $\bar{\psi}$:

$$\gamma_\alpha(\mathbf{R}, \boldsymbol{\Omega}) \to \gamma_\alpha(\mathbf{R}, \boldsymbol{\Omega}) - \int d\mathbf{r}\, Q_\alpha(\mathbf{r} - \mathbf{R}, \boldsymbol{\Omega})i\bar{\psi}(\mathbf{r}). \tag{20}$$

In the mean field approximation, contributions of configurations of the system that deviate from the representative value $\bar{\psi}$ are entirely neglected.

The value of $\bar{\psi}$ is determined by the stationary condition

$$\frac{\delta \ln Z_G[\gamma, \Sigma]}{\delta \bar{\psi}(\mathbf{r})} = 0. \tag{21}$$

This leads to the Poisson equation:

$$-\frac{1}{4\pi\beta}\nabla \cdot \varepsilon(\mathbf{r})\nabla i\bar{\psi}(\mathbf{r}) = \sum_{\alpha} \int d\mathbf{R}d\boldsymbol{\Omega} Q_{\alpha}(\mathbf{r}-\mathbf{R}, \boldsymbol{\Omega})\rho_{\alpha}(\mathbf{R}, \boldsymbol{\Omega}) + \Sigma(\mathbf{r}) \tag{22}$$

where $\rho_{\alpha}(\mathbf{R}, \boldsymbol{\Omega})$ density of particles of type α in the system. From this, we find that the field $i\bar{\psi}$ can be directly related to the average electrostatic potential in the system as $i\bar{\psi}(\mathbf{r}) = \beta\phi(\mathbf{r})$.

The particle density is given by

$$\rho_{\alpha}(\mathbf{R}, \boldsymbol{\Omega}) = \rho_{\alpha}^{\mathrm{ref}}(\mathbf{R}, \boldsymbol{\Omega}; \gamma - Qi\bar{\psi}). \tag{23}$$

Within the mean-field approximation, we see that the particle density is the same as for the reference system (which does not include electrostatic interactions) but with an additional one-body interaction given by the coupling of the particle charge to the average electrostatic potential [14].

This is a good approximation when the charge of the particles and the magnitude of any fixed charges are relatively small. However, if either of these become significant (e.g., multivalent ion or surfaces with a high charge density), then fluctuations make an important contribution to the properties of the system. There are also situations where even though the fluctuations in a system may be small, they can lead to qualitative differences in the predictions of the theory. One such example is the depletion of ions in aqueous solution from an air-water interface, thereby increasing the interfacial tension.

One manner to incorporate fluctuations in the evaluation of the functional integral is to expand the Hamiltonian in terms of fluctuations in the field around the mean field value. This is known as the loop approximation. When truncated at quadratic order, this is the one-loop approximation. This type of approach has been performed analytically [7, 11, 15] and numerically (e.g., lattice field theory methods [6, 16, 17]). One difficulty with the loop expansion is that it leads to divergences, which must be dealt with by careful renormalization of the theory. In the following section, we will present another approach, known as the variational perturbation approximation, that is free of this issue and has many other nice features.

3 Variational Perturbation Approximation

In the previous section, we mentioned mean field theory, where the functional integral is approximated by only one representative distribution of the field $\bar{\psi}$. So long as the contributions of other shapes of the field are much smaller, then the mean field approximation should be good. In this section, we discuss the variational perturbation approximation [18], a method to systematically include fluctuation contributions to the partition function to increasing orders of accuracy.

In this approach, the integrand in the partition function (see Eq. (15)) is approximated with a Gaussian function:

$$H_{\mathscr{K}}[\psi] = \frac{1}{2\beta} \int d\mathbf{r} d\mathbf{r}' [\psi(\mathbf{r}) - \bar{\psi}(\mathbf{r})] G_{\mathscr{K}}^{-1}(\mathbf{r}, \mathbf{r}') [\psi(\mathbf{r}') - \bar{\psi}(\mathbf{r}')] \tag{24}$$

where $G_{\mathscr{K}}$ is a renormalized Green's function, defined as

$$G_{\mathscr{K}}^{-1}(\mathbf{r}, \mathbf{r}') = G_0^{-1}(\mathbf{r}, \mathbf{r}') + \mathscr{K}(\mathbf{r}, \mathbf{r}'). \tag{25}$$

where $\bar{\psi}$ and $\mathscr{K}$ are arbitrary functions which will be used as variational parameters.

The grand partition function is then rewritten in terms of averages with respect to the variational Hamiltonian $H_{\mathscr{K}}$ as

$$\ln Z_G[\gamma, \Sigma] = \left\langle e^{\ln Z_G^{\mathrm{ref}}[\gamma - iQ\psi + \beta e^{\mathrm{se}}] - (H[\psi] - H_{\mathscr{K}}[\psi])} \right\rangle_{\mathscr{K}} \frac{\mathscr{N}_{\mathscr{K}}}{\mathscr{N}_0} \tag{26}$$

where $\langle(\cdots)\rangle_{\mathscr{K}}$ denotes an average with respect to the Hamiltonian $H_{\mathscr{K}}$, and $\mathscr{N}_{\mathscr{K}}$ is the associated normalization constant. The ratio which follows the average is the change of the normalization constant due to the change of the average from using H_0 to using $H_{\mathscr{K}}$.

The average of the exponential term in Eq. (26) can be evaluated using a cumulant expansion:

$$\ln \left\langle e^{A[\psi]} \right\rangle \approx \langle A[\psi] \rangle^{(c)} + \frac{1}{2!} \langle A^2[\psi] \rangle^{(c)} + \frac{1}{3!} \langle A^3[\psi] \rangle^{(c)} + \cdots$$

where $\langle A^n[\psi] \rangle^{(c)}$ denotes the nth cumulant of the distribution. Truncating this expansion to the first order, we find the following inequality

$$\begin{aligned} \ln Z_G[\gamma, \Sigma] \geq & \int d\mathbf{r} d\mathbf{r}' i\bar{\psi}(\mathbf{r}) G_0^{-1}(\mathbf{r}, \mathbf{r}') i\bar{\psi}(\mathbf{r}') - \int d\mathbf{r} \Sigma(\mathbf{r}) i\bar{\psi}(\mathbf{r}) \\ & + \left\langle \ln Z_G^{\mathrm{ref}}[\gamma - iQ\psi + \beta e^{\mathrm{se}}] \right\rangle_{\mathscr{K}} + \frac{1}{2} \mathrm{Tr} \mathscr{K} G_{\mathscr{K}} + \ln \frac{\mathscr{N}_{\mathscr{K}}}{\mathscr{N}_0} \end{aligned} \tag{27}$$

The true value of the grand partition function will be larger than the estimate of the first order cumulant approximation for any choice of $\bar{\psi}$ and $\mathscr{K}$.

In the exact theory, the predictions for the properties of the system are independent of the choice of $\bar{\psi}$ and $\mathscr{K}$; however, by making approximations, the resulting theory will have a dependence on these quantities. In order to minimize this effect, we select these values such that the free energy is stationary with respect to variations in these quantities. This leads to the following variational conditions:

$$\frac{\delta F}{\delta \bar{\psi}(\mathbf{r})} = 0 \tag{28}$$

$$\frac{\delta F}{\delta \mathscr{K}(\mathbf{r}, \mathbf{r}')} = 0. \tag{29}$$

These conditions can be used to determine the "optimal" values of the quantities $\bar{\psi}$ and $\mathscr{K}$.

Equation (28) leads to the Poisson equation

$$-\frac{1}{4\pi\beta}\nabla \cdot \varepsilon(\mathbf{r})\nabla i\bar{\psi}(\mathbf{r}) = \sum_{\alpha} \int d\mathbf{R}d\boldsymbol{\Omega}\, Q_{\alpha}(\mathbf{r}-\mathbf{R}, \boldsymbol{\Omega})\rho_{\alpha}(\mathbf{R}, \boldsymbol{\Omega}) + \Sigma(\mathbf{r}) \tag{30}$$

where $\rho_{\alpha}(\mathbf{R}, \boldsymbol{\Omega})$ is the density of particles of type α and is given by

$$\rho_{\alpha}(\mathbf{R}, \boldsymbol{\Omega}) = \left\langle \rho_{\alpha}^{\text{ref}}(\mathbf{R}, \boldsymbol{\Omega}; \gamma - Qi\psi + \beta e^{\text{se}})\right\rangle_{\mathscr{K}}. \tag{31}$$

From this relation, we see that $\bar{\psi}(\mathbf{r}) = \beta\phi(\mathbf{r})$ can be identified with the average of the electrostatic potential, as in the mean-field approximation.

The variational relation with respect to the screening function $\mathscr{K}$ (see Eq. (29)) leads to

$$\begin{aligned}\mathscr{K}(\mathbf{r}, \mathbf{r}') = &\left\langle \psi(\mathbf{r})\psi(\mathbf{r}')\ln Z_G^{\text{ref}}[\gamma - Qi\psi + \beta e^{\text{se}}]\right\rangle_{\mathscr{K}} \\ &- G_{\mathscr{K}}(\mathbf{r}, \mathbf{r}')\left\langle \ln Z_G^{\text{ref}}[\gamma - Qi\psi + \beta e^{\text{se}}]\right\rangle_{\mathscr{K}}.\end{aligned} \tag{32}$$

One of the key challenges to applying this theory is the ability to solve Eq. (32). In particular, it is typically quite difficult to determine a closed, analytical form for $G_{\mathscr{K}}(\mathbf{r}, \mathbf{r}')$. One approximation strategy is to find some simple variational form for $\mathscr{K}(\mathbf{r}, \mathbf{r}')$ where the Green's function is known, and then vary the parameters of this function to minimize the free energy.

The Helmholtz free energy functional can be determined by performing a Legendre transform:

$$F[\rho, \Sigma] = \sum_{\alpha} \int d\mathbf{R}d\boldsymbol{\Omega}\, \rho_{\alpha}(\mathbf{R}, \boldsymbol{\Omega})\gamma_{\alpha}(\mathbf{R}, \boldsymbol{\Omega}) - \ln Z_G[\gamma, \Sigma]. \tag{33}$$

The resulting expression for the free energy is:

$$
\begin{aligned}
F[\rho, \Sigma] = &\sum_{\alpha} \int d\mathbf{R} d\boldsymbol{\Omega}\, \rho_\alpha(\mathbf{R}, \boldsymbol{\Omega})[\ln \rho_\alpha(\mathbf{R}, \boldsymbol{\Omega})\Lambda_\alpha^d - 1] \\
&- \frac{1}{2\beta} \int d\mathbf{r} d\mathbf{r}' i\bar{\psi}(\mathbf{r}) G_0^{-1}(\mathbf{r}, \mathbf{r}') i\bar{\psi}(\mathbf{r}') \\
&+ \int d\mathbf{r} \left[\int d\mathbf{R} d\boldsymbol{\Omega}\, \rho_\alpha(\mathbf{R}, \boldsymbol{\Omega}) Q_\alpha(\mathbf{r} - \mathbf{R}, \boldsymbol{\Omega}) + \Sigma(\mathbf{r}) \right] i\bar{\psi}(\mathbf{r}) \\
&- \ln \frac{\mathcal{N}_{\mathcal{K}}}{\mathcal{N}_0} - \frac{1}{2} \mathrm{Tr} \mathcal{K} G_0.
\end{aligned}
\tag{34}
$$

From the free energy functional, we can derive all equilibrium properties of the system.

The use of the variational approximation assumes that the fluctuations of the field are weak and can be represented by a Gaussian distribution. While this is a good approximation for fluctuations over large length scales, at short wavelengths it breaks down.

4 Point Charge Model

To concretely illustrate the application of this theoretical framework, we consider a system of point charges. This model is commonly used to describe electrolyte systems. The charge density for a point charge is given by

$$Q_\alpha(\mathbf{r}, \boldsymbol{\Omega}) = q_\alpha \delta^d(\mathbf{r}) \tag{35}$$

Half of the particles are cations (denoted by $+$) with charge q, and the other half are anions (denoted by $-$) with charge $-q$.

There are no other interactions in the system apart from electrostatic interactions. The reference system is, therefore, an ideal gas, which has a grand partition function given by

$$Z_G[\gamma] = \sum_\alpha \Lambda_\alpha \int d\mathbf{R} d\boldsymbol{\Omega}\, e^{\gamma_\alpha(\mathbf{R}, \boldsymbol{\Omega})}. \tag{36}$$

A key length scale for this system is the Bjerrum length $l_B = \beta q^2/\varepsilon$, which is the distance at which the electrostatic interaction energy between two charges equals their thermal energy $k_B T$.

4.1 Mean-Field Approximation

We first examine the mean-field approximation for the point charge system. The grand partition function for the point charge model in the mean-field approximation is given by

$$\ln Z_G[\gamma, \Sigma] \approx \frac{1}{2\beta}\int d\mathbf{r}d\mathbf{r}' i\bar{\psi}(\mathbf{r})G_0^{-1}(\mathbf{r},\mathbf{r}')i\bar{\psi}(\mathbf{r}') - \int d\mathbf{r}\Sigma(\mathbf{r})i\bar{\psi}(\mathbf{r}) + \sum_\alpha \Lambda_\alpha^d \int d\mathbf{R} e^{\gamma_\alpha(\mathbf{R}) - q_\alpha i\bar{\psi}(\mathbf{R})}. \quad (37)$$

The value of the mean field is determined from the stationary condition, given in

$$-\frac{1}{4\pi\beta}\nabla\cdot\varepsilon(\mathbf{r})\nabla i\bar{\psi}(\mathbf{r}) = \sum_\alpha q_\alpha\rho_\alpha(\mathbf{r}) + \Sigma(\mathbf{r}) \quad (38)$$

where $\rho_\alpha(\mathbf{r})$ is the density of particles of type α.

From the grand partition function, we can derive all static equilibrium properties of the system. In particular, the density of particles of type α is give by:

$$\rho_\alpha(\mathbf{R}) = \Lambda_\alpha^d e^{\gamma_\alpha(\mathbf{R}) - q_\alpha i\bar{\psi}(\mathbf{R})}. \quad (39)$$

Therefore, we see that the mean-field approximation leads to the Poisson-Boltzmann theory.

The corresponding expression for the Helmholz free energy is

$$F[\rho, \Sigma] \approx -\frac{1}{2\beta}\int d\mathbf{r}d\mathbf{r}' i\bar{\psi}(\mathbf{r})G_0^{-1}(\mathbf{r},\mathbf{r}')i\bar{\psi}(\mathbf{r}) + \int d\mathbf{r}\left[\sum_\alpha q_\alpha\rho_\alpha(\mathbf{r}) + \Sigma(\mathbf{r})\right] i\bar{\psi}(\mathbf{r}) + \sum_\alpha \int d\mathbf{R}\rho_\alpha(\mathbf{R})[\ln\rho_\alpha(\mathbf{R})\Lambda_\alpha^d - 1]. \quad (40)$$

4.2 Debye-Hückel Theory

Now we apply the variational perturbation theory at first order to the point charge model. First we consider a uniform system. In a uniform system, $\bar{\psi}(\mathbf{r}) = 0$ due to symmetry. Using this fact, combined with the expression for the renormalized Green's function, the grand partition function is determined to be

$$\frac{1}{V}\ln Z_G[\gamma] \geq \sum_\alpha z_\alpha e^{\frac{\kappa l_B}{2}} - \frac{\kappa^3}{24\pi} \quad (41)$$

where κ is inverse Debye screening length

$$\kappa^2 = \frac{4\pi\beta}{\varepsilon}\sum_\alpha q_\alpha^2 \rho_\alpha = 8\pi\rho_\pm \tag{42}$$

and $z_\alpha = \Lambda_\alpha^d e^{\gamma_\alpha}$ is the fugacity of ions of type α. Due to electroneutrality, the fugacity of the cations and anions must be equal; we denote this common value by $z_\pm$.

From the variational condition (see Eq. (29)), the renormalized Green's function is

$$G_{\mathscr{K}}(\mathbf{r}, \mathbf{r}') = \frac{e^{-\kappa|\mathbf{r}-\mathbf{r}'|}}{\varepsilon|\mathbf{r} - \mathbf{r}'|}. \tag{43}$$

and $\rho_\pm$ is the electrolyte concentration. Accounting for the fluctuations of the electrostatic potential due to the thermal motion of the point charges leads to an exponential decay of the Green's function. Physically, this corresponds to screening of charge.

The Helmholtz free energy is

$$F[\rho, \Sigma] \approx +V\sum_\alpha \rho_\alpha(\ln \rho_\alpha \Lambda_\alpha^d - 1) - V\frac{\kappa^3}{24\pi}. \tag{44}$$

The chemical potential of species α is given by

$$\beta\mu_\alpha = \ln \rho_\alpha \Lambda_\alpha^d - \frac{\beta q_\alpha^2 \kappa}{2\varepsilon}. \tag{45}$$

The first-order variational perturbation approximation for point charges leads to the Debye-Hückel theory of electrolyte solutions.

For a nonuniform systems, the Helmholtz free energy is:

$$\begin{aligned} F[\rho, \Sigma] \approx &-\frac{1}{2\beta}\int d\mathbf{r}d\mathbf{r}' i\bar{\psi}(\mathbf{r}) G_0^{-1}(\mathbf{r}, \mathbf{r}') i\bar{\psi}(\mathbf{r}') + \int d\mathbf{r}\left[\sum_\alpha q_\alpha \rho_\alpha(\mathbf{r}) + \Sigma(\mathbf{r})\right] i\bar{\psi}(\mathbf{r}) \\ &+\sum_\alpha \int d\mathbf{R}\rho_\alpha(\mathbf{R})[\ln \rho_\alpha(\mathbf{R})\Lambda_\alpha^d - 1] - \ln\frac{\mathscr{N}_{\mathscr{K}}}{\mathscr{N}_0} - \frac{1}{2}\mathrm{Tr}\mathscr{K} G_0. \end{aligned} \tag{46}$$

In this case the screening function becomes local

$$\mathscr{K}(\mathbf{r}, \mathbf{r}') = \delta^d(\mathbf{r} - \mathbf{r}')\sum_\alpha q_\alpha^2 \rho_\alpha(\mathbf{r}) \tag{47}$$

We can identify $\mathscr{K}$ with the inverse Debye screening length κ, which varies with the local density

$$\kappa^2(\mathbf{r}) = \frac{4\pi\beta}{\varepsilon}\sum_\alpha q_\alpha^2 \rho_\alpha(\mathbf{r}) \tag{48}$$

For a uniform system, this becomes a constant.

Before, we mentioned that there is a "self-energy" of the particles, due to the interaction of each charge with the electrical potential that it generates itself. This seemingly irrelevant term, which is infinite for point charges, was subtracted out. However, interestingly, it reappears again in the theory. Due to the screening of charge by the response of the other charged particles in the system, the electric potential that a point charge generates is altered. This shifts the self-energy of the point charge. What we find is that the difference in the self-energy of the point charge in the presence and in the absence of other mobile charges actually is the excess chemical potential of the particle. We can see this term in the expression for the particle density

$$\rho_\alpha(\mathbf{R}) = \Lambda_\alpha^d e^{\gamma_\alpha(\mathbf{R}) - q_\alpha i\bar{\psi}(\mathbf{R}) - \frac{\beta q^2}{2}\Delta G_{\mathrm{K}}(\mathbf{R},\mathbf{R})}. \tag{49}$$

where $\Delta G_{\mathrm{K}} = G_{\mathrm{K}} - \Delta G_{\mathrm{free}}$ is the shift in the Green's function due to the presence of fluctuations. The self energy term changes propensity of a particle to remain at a location due to fluctuations in the local environment caused by thermal motion of charge.

4.3 Stability of the Point Charge Model

In the previous section, we demonstrated that the first order variational perturbation approximation leads to the Debye-Hückel theory for electrolyte solutions. In the variational perturbation approximation, the value of the inverse screening length is determined by maximizing the grand partition function, and this leads to the standard relationship between the inverse screening length and the ionic strength of the system. The astute reader, however, may have noticed that the expression for the grand partition function (see Eq. (41)) actually diverges as $\kappa \to \infty$. From a physical perspective, this is a consequence of the fact that the point charge model is inherently unstable. The positive and negative point charges will collapse with each other, resulting in a situation where there are an infinite number of ions in the system. So then what is the relevance of the Debye-Hückel expression for κ?

To investigate this point, we show the variation of the grand partition function with κ, given by the solid lines in Fig. 1, for several different values of the electrolyte fugacity $z_\pm$. For all fugacities, $\ln Z_G$ is maximized when $\kappa \to \infty$, indicating that the point charge system wants to collapse. At high fugacities, $\ln Z_G$ increases monotonically with increasing κ. However, for sufficiently low values of $z_\pm$, which corresponds to low ion densities, the grand partition function has a local maximum. This corresponds to the Debye-Hückel theory for the point charge model.

For real electrolytes, excluded volume interactions prevent the collapse of the system. These interactions suppress the fluctuations at a short length scales. Properly accounting for their effect on the grand partition function is quite difficult, however, we can get a qualitative understanding of their influence by introduction

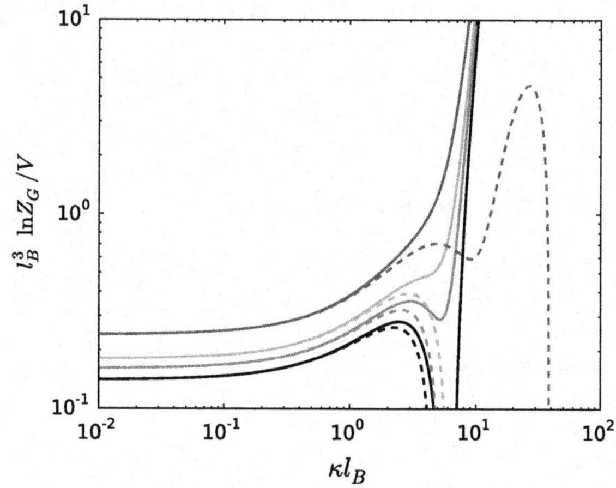

Fig. 1 Stability of the point charge model. **a** Variation of the grand potential with the inverse screening length κ for a fugacity: (i) $z_\pm l_B^3 = 0.07$ (*black*), (ii) $z_\pm l_B^3 = 0.08$ (*red*), and (iii) $z_\pm l_B^3 = 0.09$ (*green*). (iv) $z_\pm l_B^3 = 0.12$ (*green*). The *solid lines* represent the situation without a cutoff, and the *dashed lines* are for a cutoff of $\Lambda l_B = 20$

a cutoff wavevector Λ. Fluctuations of length scales less than $2\pi/\Lambda$ are neglected. The resulting expression for the grand partition function is

$$\frac{1}{V}\ln Z_G[\gamma] \geq \sum_\alpha z_\alpha \exp\left[\frac{\kappa l_B}{2}\frac{2}{\pi}\arctan\frac{\Lambda}{\kappa}\right] - \frac{\kappa^3}{24\pi}\frac{2}{\pi}\left[\arctan\frac{\Lambda}{\kappa} - \frac{\Lambda}{\kappa} + \left(\frac{\Lambda}{\kappa}\right)^3 \ln\left(1 + \left(\frac{\kappa}{\Lambda}\right)^2\right)\right] \tag{50}$$

This is plotted as the dashed lines in Fig. 1.

Suppressing the short wavelength fluctuations stabilizes the model. The maximum of $\ln Z_G$ no longer occurs for $\kappa \to \infty$; it now occurs for finite values of the inverse screening length for all values of the fugacity. At high electrolyte fugacities, the maximum in the grand partition function is no longer related to the Debye-Hückel value. In this situation, the system depends sensitively on the value of Λ. Its properties are controlled by the physics at short wavelengths. At these conditions, the point charge model is irrelevant and not applicable.

At low electrolyte fugacities, the maximum is close to the Debye-Hückel value, although it is slightly shifted; the lower the fugacity (ion concentration), the nearer this peak is to the Debye-Hückel value. For these conditions, the system is relatively insensitive to the precise value of the cutoff Λ. In this case, the short wavelength physics are irrelevant, and the point charge model is relevant.

It is also interesting to note that multiple peaks can appear in the dependence of the grand partition function on the inverse screening length. Each peak correspond to a distinct phase of the system. Phase coexistence occurs when the height of the peaks are the same. Consequently, we see that this approach is able to predict the vapor-liquid transition of the electrolyte system [19]. Note, however, the quantitative features of this transition (e.g., critical temperature and density) will depend the details of short length scale physics of the system.

4.4 Dielectric Interfaces

Variations in the dielectric constant of the background medium will affect the manner in which the electrostatic potential propagates from a charge. Here, we examine the influence of a dielectric interface on the properties of the point charge electrolyte. An electrolyte is immersed in a medium with dielectric constant ε and confined to remain in the half-space defined by $z > 0$. For $z < 0$, there is a planar body with dielectric constant ε'. A schematic diagram of the system is shown in Fig. 3.

When a charge of magnitude q is placed near a planar dielectric discontinuity, the difference in the polarizabilities of the two media leads to an induced charge distribution on the interface. The potential generated by this induced charge can be mathematically represented by a point charge of magnitude $q(\varepsilon - \varepsilon')/(\varepsilon + \varepsilon')$, located within the dielectric body (see Fig. 2 the same distance from the interface as the original charge. This is known as an image charge.

Consequently, charges are repelled from a low dielectric surface (i.e. $\varepsilon' < \varepsilon$), regardless of the sign of the charge. This leads to an electrostatic depletion effect in the case of the low dielectric bodies [13, 20, 21]. Charges are attracted to high dielectric bodies (i.e. $\varepsilon' > \varepsilon$), such as metals.

In this planar geometry, the presence of the dielectric interface shifts the Green's function [1]

$$G_0(\mathbf{r}, \mathbf{r}') = \frac{1}{\varepsilon|\mathbf{r} - \mathbf{r}'|} + \frac{\varepsilon - \varepsilon'}{\varepsilon + \varepsilon'} \frac{1}{\varepsilon\sqrt{(x - x')^2 + (y - y')^2 + (z + z')^2}}. \tag{51}$$

The first term on the right of the equation is the Green's function in a uniform dielectric (i.e. G_{free}). The second term is the influence of the dielectric interface and can be interpreted as the potential that eminates from the image charge.

The renormalized Green's function $G_{\mathcal{K}}$ of this system can be determined by solving the variational condition given in Eq. (29). It can be written as the Green's function for the uniform bulk system, where the inverse screening length is equal to κ, plus a term due to the presence of the dielectric interface:

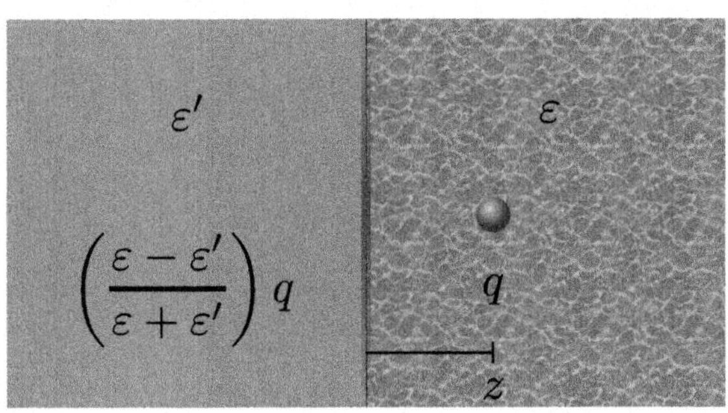

Fig. 2 Schematic diagram of a charge located a distance z from a dielectric interface

$$G_{\mathscr{K}}(\mathbf{r},\mathbf{r}') = \frac{e^{-\kappa|\mathbf{r}-\mathbf{r}'|}}{\varepsilon|\mathbf{r}-\mathbf{r}'|} + \delta G_{\mathscr{K}}(\mathbf{r},\mathbf{r}') \tag{52}$$

where $\delta G_{\mathscr{K}}(\mathbf{r},\mathbf{r}')$ is shift in the Green's function [20, 22, 23].

If we make the simplifying assumption that the inverse screening length is approximately constant for $z > 0$, we find that [20]

$$\delta G_{\mathscr{K}}(\mathbf{r},\mathbf{r}) = -\left(\frac{\eta-1}{\eta+1}\right)\frac{e^{-2\kappa z}}{2\varepsilon z} - \frac{2\eta}{\eta+1}\frac{\kappa}{\varepsilon}\int_1^\infty dx \frac{\sqrt{x^2-1}-x}{\eta\sqrt{x^2-1}+x} e^{-2x\kappa z}. \tag{53}$$

The resulting density profiles for the cations $\rho_+(\mathbf{r})$ and the anions $\rho_-(\mathbf{r})$ are given by

$$\rho_+(\mathbf{r}) = \rho_-(\mathbf{r}) = \rho_\pm \exp\left[-\frac{\beta q^2}{2}\delta G_{\mathscr{K}}(\mathbf{r},\mathbf{r})\right], \tag{54}$$

where $\rho_\pm$ is the ion concentration far from the dielectric interface (i.e. $z \to \infty$). This is plotted in Fig. 3a for a system with an electrolyte concentration $\rho_\pm l_B^3 = 0.1$ near planar interfaces of various dielectric constants. When the dielectric constant of the interface is larger that that of the solvent (i.e. $\varepsilon' > \varepsilon$), the ions absorb to the interface. When $\varepsilon' < \varepsilon$, such as for electrolytes near an air-water interface, the ions desorb from the interface. Even in the case where there is no dielectric discontinuity

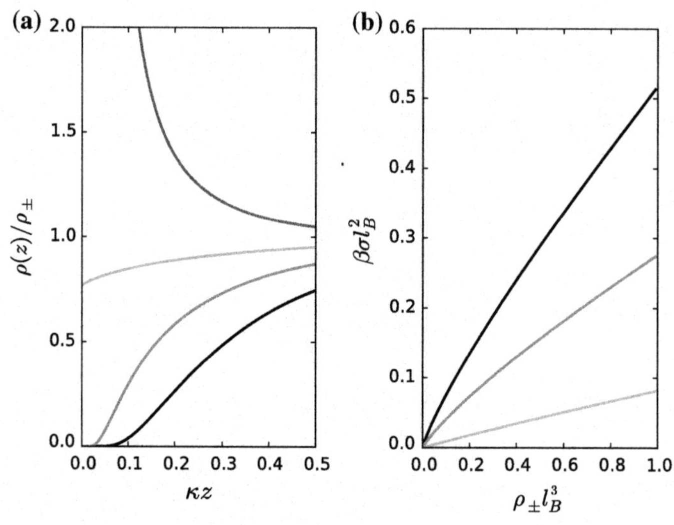

Fig. 3 **a** Ion density profile for an electrolyte solution of concentration $\rho_\pm l_B^3 = 0.1$ near a dielectric interface with (i) $\varepsilon'/\varepsilon = 0$ (*black*), (ii) $\varepsilon'/\varepsilon = 0.5$ (*red*), (iii) $\varepsilon'/\varepsilon = 1$ (*green*), and (iv) $\varepsilon'/\varepsilon = 2$ (*blue*). **b** Incremental interfacial tension

(i.e. $\varepsilon' = \varepsilon$), there is a slight desorption of ions from the interface. This is due to the exclusion of particles from $z < 0$ and the preference of charged particles to be fully surrounded by other charged particles.

Adsorption of molecules on an interface will lead to a decrease in the interfacial tension. On the other hand, desorption of solutes from an interface will increase the interfacial tension. This effect is what leads to the increase of the surface tension of water with the addition of electrolytes. For the planar geometry within constant screening approximation, the interfacial tension is given by

$$\beta\sigma = -\frac{\kappa^2}{32\pi}\left(\frac{\eta-1}{\eta+1}\right) - 2\rho_{\pm}\int dz \left[e^{-(\beta q^2/2)\delta G_{\mathscr{K}}(z,z)} - 1\right]. \tag{55}$$

where $\rho_{\pm}$ is the bulk electrolyte concentration. The dependence of the interfacial tension with the bulk electrolyte concentration is presented in Fig. 3b.

5 Conclusions

We have described a field theory approach to the description of electrolyte systems. The evaluation of this theory using the variational perturbation approximation has been discussed. The theory has been developed for particles with a rigid charge distribution. The point charge model is used to illustrate the application of the method. In this case, it was found that the mean-field approximation for this system leads to the Poisson-Boltzman theory and the first-order variational perturbation approximation leads to the Debye-Hückel theory.

The theoretical framework presented here is quite flexible and applicable to a wide variety of systems and geometries. For example, it has been applied to particles with nonspherically symmetric charge distributions, such as lines [24] and disks [8]. In these cases, the system can form liquid crystalline phases, where the orientations of the molecules are ordered, although the translational degrees of freedom are not. It has also been applied to examine ions confined within pores [13, 22, 23].

When truncated at first order, the variational perturbation approximation is able to go beyond the Poisson-Boltzmann theory, allowing it to capture phenomena like the influence of dielectric interfaces and the coupling between dispersion and electrostatic interactions (i.e. screening of the dispersion). However, it is only accurate for systems where the electrostatic interactions are relatively weak (e.g., low surface charge densities, low ion valencies, and high temperatures). This is due to the assumption that the fluctuations in a system can be well represented by a Gaussian probability distribution, which is good for long wavelength fluctuations but breaks down at short wavelengths. As a consequence, it cannot capture effects that occur at high charge densities, such as overcharging. In principle, the approximation can be extended to higher orders to increase its range of applicability to higher electrostatic coupling strengths; however, the rate of improvement of this expansion is relatively slow.

Another approach to improving the theory is to divide the long and short wavelength fluctuations and treat each with a separate approximation scheme: the variational perturbation approximation for long wavelengths and a virial expansion, for example, for the short wavelength fluctuations. This splitting strategy has been found to be successful in describing electrolyte systems over a broad range of conditions [25, 26], from the weak to the intermediate and including the strong electrostatic coupling regime. This "splitting" approach is able to quantitatively describe phenomena such as charge inversion and like-charge attraction.

References

1. J.D. Jackson, *Classical Electrodynamics* (Wiley, New York, 1975)
2. T.L. Hill, *Statistical Mechanics: Principles and Selected Applications* (McGraw-Hill, New York, 1956)
3. R.L. Stratonovich, Dokl. Akad. Nauk SSSR **115**(6), 1097 (1957)
4. J. Hubbard, Phys. Rev. Lett. **3**, 77 (1959). doi:10.1103/PhysRevLett.3.77
5. A.L. Kholodenko, A.L. Beyerlein, Phys. Rev. A **43**(4), 3309 (1986)
6. R.D. Coalson, A. Duncan, J. Chem. Phys. **97**(8), 5653 (1992). doi:10.1063/1.463950
7. R.R. Netz, Phys. Rev. E **60**(3), 3174 (1999)
8. L. Lue, Fluid Phase Equil. **241**, 236 (2006). doi:10.1016/j.fluid.2005.11.007
9. J. Mahanty, B.W. Ninham, *Dispersion Forces* (Academic Press, London, 1976)
10. M.M. Hatlo, L. Lue, Soft Matter **4**, 1582 (2008). doi:10.1039/B803783C
11. R.R. Netz, Eur. Phys. J. E **3**(2), 131 (2000)
12. R.R. Netz, Eur. Phys. J. E **5**, 189 (2001)
13. M.M. Hatlo, R.A. Curtis, L. Lue, J. Chem. Phys. **128**(16), 164717 (2008). doi:10.1063/1.2908738
14. L. Lue, N. Zoeller, D. Blankschtein, Langmuir **15**(11), 3726 (1999). doi:10.1021/la9813376
15. D.S. Dean, R.R. Horgan, Phys. Rev. E **69**(6), 061603 (2004). doi:10.1103/PhysRevE.69.061603
16. A.M. Walsh, R.D. Coalson, J. Chem. Phys. **100**(2), 1559 (1994)
17. N. Ben-Tal, R.D. Coalson, J. Chem. Phys. **101**(6), 5148 (1994). doi:10.1063/1.467371
18. H. Kleinert, *Path Integrals in Quatum Mechanics, Statistics, and Polymer Physics*, 2nd edn. (World Scientific, Singapore, 1995)
19. Y. Levin, Rep. Prog. Phys. **65**, 1577 (2002). doi:10.1088/0034-4885/65/11/201
20. R.A. Curtis, L. Lue, J. Chem. Phys. **123**, 174702 (2005). doi:10.1063/1.2102890
21. R.A. Curtis, L. Lue, Curr Opin. Colloid Interface Sci. **20**(1), 19 (2015). doi:10.1016/j.cocis.2014.12.001
22. S. Buyukdagli, M. Manghi, J. Palmeri, Phys. Rev. E **81**(4), 041601 (2010)
23. S. Buyukdagli, C.V. Achim, T. Ala-Nissila, J. Chem. Phys. **137**(10), 104902 (2012)
24. M.M. Hatlo, A. Karatrantos, L. Lue, Phys. Rev. E **80**(6), 061107 (2009). doi:10.1103/PhysRevE.80.061107
25. M.M. Hatlo, L. Lue, Soft Matter **5**, 125 (2009). doi:10.1039/b815578j
26. M.M. Hatlo, L. Lue, EPL (Europhys. Lett.) **89**(2), 25002 (2010). doi:10.1209/0295-5075/89/25002

Self-consistent Field Theory of Inhomogeneous Polymeric Systems

An-Chang Shi

1 Introduction

Inhomogeneous polymeric systems, such as polymer blends, polymer solutions and block copolymers, exhibit rich phase behavior with various spatial structures. Examples include the interfaces between different polymers and the ordered phases self-assembled from block copolymers [1]. Understanding the formation and structure of these rich morphologies demands predictive theoretical frameworks that could be used to describe phase behavior and structural properties of polymeric systems. Ideally the theory should take the molecular properties of the polymers as input and be able to predict thermodynamically stable phases, the phase transition boundaries among them, as well as the physical properties of the phases. Towards this goal, a variety of theoretical methods have been developed to study the phases and phase behavior of inhomogeneous polymeric systems.

One of the most successful theoretical frameworks for inhomogeneous polymeric systems, including polymer blends, polymer solutions and block copolymers, is the self-consistent field theory (SCFT). The polymeric SCFT has its origin in the work by Edwards in the 1960s [2]. This theoretical framework was explicitly adapted to treat block copolymers by Helfand in 1975 [3], and later important contributions to the theory were made by, among others, Hong and Noolandi [4]. The applications of SCFT to numerous polymeric systems have been made in recent years. There is a large body of literature on the SCFT of polymeric systems, including a number of valuable review articles [5], book chapters and books [6].

The most fruitful application of SCFT to polymeric systems is the study of phases and phase transitions of block copolymers. In particular, powerful methods have been developed over the last decades to obtain highly accurate solutions of the SCFT equations using numerical techniques. The earliest attempts to obtain numerical

A.-C. Shi (✉)
Department of Physics and Astronomy,
McMaster University, Hamilton, ON L8S 4M1, Canada
e-mail: shi@mcmaster.ca

J. Wu (ed.), *Variational Methods in Molecular Modeling*,
Molecular Modeling and Simulation, DOI 10.1007/978-981-10-2502-0_6

solutions of SCFT for block copolymers were made by Helfand and coworkers [7]. Later Shull [8] and Whitmore and coworkers [9] have computed phase diagrams of block copolymer melts and solutions using approximate numerical techniques. The first three-dimensional numerical solutions of block copolymer phases were obtained by Matsen and Schick [10], who utilized the crystalline symmetry of the ordered phases and provides exact numerical solutions to the SCFT equations. This technique has been applied to a variety of block copolymer systems [11]. Further development of the SCFT included the theory for Gaussian fluctuations in ordered phases [12], numerical techniques for solving the mean field theory in real space [13], and fully fluctuating field-theoretical simulations [14]. Based on the numerous studies, it can be stated that SCFT forms a powerful basis for the study of inhomogeneous polymeric systems.

The essence of the self-consistent field theory is to transform a particle-based description of a statistical-mechanical system into a field-based description. In the particle-based description, the partition function of the system is written as a sum over all the particle positions, whereas in a field-based description the partition function of the system is written as a functional integral over a set of fluctuating fields. Within the field-based description, the many-body interactions are replaced by the interaction of one particle with certain fluctuating fields. This description makes it simpler to develop approximation methods. In particular, a saddle-point approximation of the functional integral leads to the mean-field approximation of the system. The resulting mean-field equations, or SCFT equations, can be solved analytically or numerically, providing information about the structure and property of the inhomogeneous polymeric phases. This theoretical framework is flexible in that it applies to any statistical-mechanical systems.

This chapter provides a systematic description of the self-consistent field theory using a binary polymer blend as a model system. In particular, we will present a detailed derivation of the field-theoretical formulation of polymers using both canonical and grand-canonical ensembles. The theoretical development results in the free-energy functional of the system, which can be used as the starting point to study the phases, phase transitions, and the physical properties of inhomogeneous polymeric systems. We try to present a pedagogical account of the theoretical development and hope that this approach would serve as a reference source for the readers who are interested in learning and applying the SCFT to different polymeric systems.

2 Formulation of SCFT

The starting point of SCFT is the construction of a molecular model for the system under investigation. In this chapter we develop the self-consistent field theory using a binary mixture of A and B homopolymers as a model system. We formulate the theory by transforming the particle-based description of the partition function into a field-based description, in which the partition function is expressed as a functional integral over a set of fluctuating fields. This field-theoretical framework is then used as the starting point to develop SCFT equations in the canonical and grand-canonical ensembles.

2.1 Molecular Model of A/B Homopolymers

The model binary mixture is composed of two types of homopolymers, labeled as A-polymers and B-polymers, respectively. The system under investigation is composed of n_A A-polymer chains and n_B B-polymer chains contained in a volume V. Each polymer chain consists of N_α monomers of species $\alpha = A$ or $\alpha = B$. The length of the A-polymers is used as the reference, so that the degree of polymerizations can be written in the form $N_\alpha = \kappa_\alpha N$ with $\kappa_A = 1$ and $\kappa_B = \kappa$. Therefore the ratio of the two polymer lengths is $\kappa = N_B/N_A$. Each polymer is assumed to have an associated Kuhn length $b_\alpha = \sigma_\alpha b$ ($\alpha = A, B$), where b is a reference Kuhn length and σ_α describes the relative size or the conformational asymmetry of the A- and B-monomers. For simplicity, the monomers are further assumed to have the same monomer density, $\rho_{0A} = \rho_{0B} = \rho_0$, defined as the number of monomers per unit volume. Thus the volume per monomer is ρ_0^{-1}. In the final expressions all lengths are scaled by the Gaussian radius of gyration of the A-polymers, $R_g = b\sqrt{N/6}$. Furthermore, the chain arc length will be scaled by the degree of polymerization of the A-polymers N.

The partition function of a polymeric system is obtained by a summation over all the configurations of the polymers. For a continuous model of polymers such as Gaussian model, a configuration of a polymer chain is denoted by a space curve $\vec{R}_{a,i}(s)$, which specifies the position of the s-th monomer of the i-th chain of type $\alpha = A, B$ (Fig. 1). For a given chain configuration, $\left\{\vec{R}_{a,i}(s)\right\}$, the concentrations, i.e., the reduced local densities of the A and B monomers at a given spatial position $\vec{r}$ are given by,

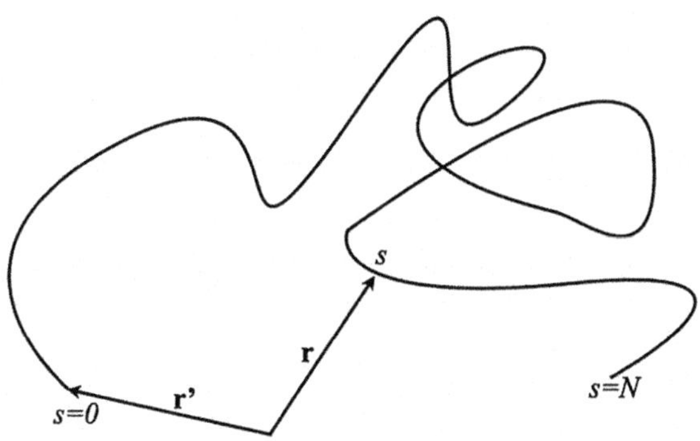

Fig. 1 In SCFT a configuration of a polymer chain is described by a space curve

$$\hat{\phi}_\alpha(\vec{r}) = \frac{1}{\rho_0} \sum_{i=1}^{n_\alpha} \int_0^{N_\alpha} ds\, \delta[\vec{r} - \vec{R}_{\alpha,i}(s)],$$

where the hat on $\hat{\phi}_\alpha(\vec{r})$ indicates that these concentrations are functionals of the chain configurations $\left\{\vec{R}_{\alpha,i}(s)\right\}$.

For simplicity, we assume that the polymer chains are flexible Gaussian chains, although other models of polymers such as semiflexible chains or rotational-isomeric-state model could be used as well. For a Gaussian chain, there is no interaction between polymer segments other than the chain connectivity represented by a harmonic potential. As a result, the probability distribution, $p_0\left[\vec{R}_{\alpha,i}(s)\right]$, for a given chain configuration is given by the standard Wiener form,

$$p_0\left[\vec{R}_{\alpha,i}(s)\right] = A_\alpha \exp\left[-\frac{3}{2b_\alpha^2}\int_0^{N_\alpha} ds \left(\frac{d\vec{R}_{\alpha,i}(s)}{ds}\right)^2\right],$$

where A_α is a normalization constant. It should be noted that in this expression the length scale is not specified. The probability $P_0\left(\left\{\vec{R}_{\alpha,i}(s)\right\}\right)$ of a given chain configuration, $\left\{\vec{R}_{\alpha,i}(s)\right\}$, of the binary mixture can be constructed from the probability of individual polymer chains,

$$P_0\left(\left\{\vec{R}(s)\right\}\right) = \left(\prod_{i=1}^{n_A} p_0\left[\vec{R}_{A,i}(s)\right]\right)\left(\prod_{i=1}^{n_B} p_0\left[\vec{R}_{B,i}(s)\right]\right).$$

It should be emphasized that the above expression is specific for a binary blend of A/B homopolymers as ideal Gaussian chains. For more complex polymers such as block copolymers, this expression should be modified according to the architecture of the polymers [5, 6].

2.2 Field-Theoretical Formulation of the Partition Function

For a statistical-mechanical system in the canonical ensemble, its thermodynamic properties are described by the Helmholtz free energy, F, of the system. The Helmholtz free energy can be obtained from the partition function, $F = -k_B T \ln \mathcal{Z}$. The partition function of the system, in turn, is given by a summation over all microstates (ν), weighted by their Boltzmann factor, $\mathcal{Z} = \sum_\nu e^{-E_\nu/k_B T}$. For the specified polymer blends under consideration, its partition function can be written in terms of the summation over the polymer configurations, which can be viewed as a functional integral over all chain configurations,

$$\mathcal{Z}_{n_A,n_B} = \frac{z_{0A}^{n_A} z_{0B}^{n_B}}{n_A! n_B!} \int \mathcal{D}\{\vec{R}(s)\} P_0(\{\vec{R}(s)\}) \prod_{\vec{r}} \delta[\hat{\phi}_A(\vec{r}) + \hat{\phi}_B(\vec{r}) - 1] e^{-W(\{\hat{\phi}\})},$$

where $z_{0\alpha}$ is the partition function of a polymer chain due to the kinetic energy of the monomers, δ is the Dirac-delta function, and $W\left(\left\{\hat{\phi}\right\}\right) = V\left(\left\{\hat{\phi}\right\}\right)/k_B T$ is the intermolecular interaction potential. For simplicity, we assume that the short-range hardcore interactions could be represented by incompressibility condition for the local polymer density, $\hat{\phi}_A(\vec{r}) + \hat{\phi}_B(\vec{r}) = 1$, and that the interaction potential has the Flory-Huggins form,

$$W\left(\left\{\hat{\phi}\right\}\right) = \rho_0 \chi \int d\vec{r}\, \hat{\phi}_A(\vec{r}) \hat{\phi}_B(\vec{r}),$$

where χ is the so-called Flory-Huggins parameter which depends on the polymers and the temperature. The above equation implies that the polymer melt is incompressible, mimicking the hardcore monomer-monomer interactions, and the delta-function is introduced in the partition function to enforce the incompressibility.

The partition function, $\mathcal{Z}_{n_A,n_B}$, contains all information about the thermodynamic properties of the system. However, a direct evaluation of $\mathcal{Z}_{n_A,n_B}$ is not possible since the integrand depends on the chain configuration through the concentration variables. This expression of the partition function presents a particle-based formulation of the theory, in the sense that the summation in the partition function is over the positions of all the particles or monomers of the system. In many applications, it is advantageous to transform the partition function into a field-based description. A standard algebraic trick to accomplish this transformation is to insert the functional identity,

$$1 = \int \mathcal{D}\{\phi_\alpha\} \prod_\alpha \prod_{\vec{r}} \delta\left[\phi_\alpha(\vec{r}) - \hat{\phi}_\alpha(\vec{r})\right],$$

into the expression of the partition function. In this expression it is important to distinguish the fields, $\phi_\alpha(\vec{r})$, which are integral variables, and the monomer concentration fields, $\hat{\phi}_\alpha(\vec{r})$, which are specified for a given polymer configuration $\left\{\vec{R}_{\alpha,i}(s)\right\}$. Using the properties of the delta-functions we can rewrite the partition function in the form,

$$\begin{aligned}\mathcal{Z}_{n_A,n_B} =& \frac{z_{0A}^{n_A} z_{0B}^{n_B}}{n_A! n_B!} \int \mathcal{D}\{\phi_\alpha\}\, e^{-W(\{\phi\})} \prod_{\vec{r}} \delta[\phi_A(\vec{r}) + \phi_B(\vec{r}) \\ & - 1] \int \mathcal{D}\left\{\vec{R}(s)\right\} P_0\left(\left\{\vec{R}(s)\right\}\right) \prod_\alpha \prod_{\vec{r}} \delta\left[\phi_\alpha(\vec{r}) - \hat{\phi}_\alpha(\vec{r})\right].\end{aligned}$$

In this expression the summation over the polymer configurations is weighted by the probability distribution $P_0\left(\left\{\vec{R}(s)\right\}\right)$ and the delta-functions, which is not straightforward to be evaluated. A further development of the theory is to introduce a set of auxiliary fields by converting the delta-function to its integral definition,

$$1 = \int \mathcal{D}\{\phi_\alpha\}\mathcal{D}\{\omega_\alpha\}\ e^{\sum_\alpha \rho_0 \int d\vec{r}\ \omega_\alpha(\vec{r})\left[\phi_\alpha(\vec{r}) - \hat{\phi}_\alpha(\vec{r})\right]},$$

where the range of the ω_α integral is along a path in the complex plane, from $-i\infty$ from to $+i\infty$, and the factor ρ_0 is introduced for convenience. Substituting this expression into the expression of the partition function, $\mathcal{Z}_{n_A,n_B}$, and rearranging the order of integrations, we obtain,

$$\mathcal{Z}_{n_A,n_B} = \frac{z_{0A}^{n_A} z_{0B}^{n_B}}{n_A!\, n_B!} \int \mathcal{D}\{\phi_\alpha\}\mathcal{D}\{\omega_\alpha\}\ e^{\sum_\alpha \rho_0 \int d\vec{r}\ \omega_\alpha(\vec{r})\phi_\alpha(\vec{r}) - W(\{\phi\})} \prod_{\vec{r}} \delta\left[\phi_A(\vec{r}) + \phi_B(\vec{r}) - 1\right] \int \mathcal{D}\left\{\vec{R}(s)\right\} P_0\left(\left\{\vec{R}(s)\right\}\right)\ e^{-\sum_\alpha \rho_0 \int d\vec{r}\ \omega_\alpha(\vec{r})\hat{\phi}_\alpha(\vec{r})}$$

The advantage of this expression is that the integration over the chain configurations can now be carried out with the relatively simpler integrand $e^{-\sum_\alpha \rho_0 \int d\vec{r}\ \omega_\alpha(\vec{r})\hat{\phi}_\alpha(\vec{r})}$. With the Gaussian probability distributions and the definition of the polymer concentrations, the integral over the chain configuration can be written as,

$$\int \mathcal{D}\left\{\vec{R}(s)\right\} P_0\left(\left\{\vec{R}(s)\right\}\right)\ e^{-\sum_\alpha \rho_0 \int d\vec{r}\ \omega_\alpha(\vec{r})\hat{\phi}_\alpha(\vec{r})}$$
$$= \prod_\alpha \left(\prod_{i=1}^{n_\alpha} \left\{ \int \mathcal{D}\left\{\vec{R}_{\alpha,i}(s)\right\} p_0\left[\vec{R}_{\alpha,i}(s)\right] \exp\left[-\int_0^{N_\alpha} ds\ \omega_\alpha\left(\vec{R}_{\alpha,i}(s)\right) \right] \right\} \right)$$
$$= \prod_\alpha \left[Q_\alpha\left(\{\omega_\alpha\}\right) V\right]^{n_\alpha},$$

where $Q_\alpha(\{\omega_\alpha\})$ is the partition function of a single polymer chain in an external fields $\omega_\alpha(\vec{r})$, which is defined by the following path integral,

$$Q_\alpha\left(\{\omega_\alpha\}\right) = \frac{1}{V} \int \mathcal{D}\left\{\vec{R}_{\alpha,i}(s)\right\} p_0\left[\vec{R}_{\alpha,i}(s)\right] \exp\left[-\int_0^{N_\alpha} ds\ \omega_\alpha\left(\vec{R}_{\alpha,i}(s)\right) \right].$$

The introduction of the single chain partition functions allows us to rewrite the partition function, $\mathcal{Z}_{n_A,n_B}$, of the binary blend of AB homopolymers as a functional integral over the concentrations fields, $\phi_\alpha(\vec{r})$, and the auxiliary fields, $\omega_\alpha(\vec{r})$,

$$\mathcal{Z}_{n_A,n_B} = \int \mathcal{D}\{\phi_\alpha\}\mathcal{D}\{\omega_\alpha\}\mathcal{D}\{\eta\}\ e^{\sum_\alpha \rho_0 \int d\vec{r}\ \omega_\alpha(\vec{r})\phi_\alpha(\vec{r}) - W(\{\phi\}) - \rho_0 \int d\vec{r}\ \eta(\vec{r})\left[\phi_A(\vec{r}) + \phi_B(\vec{r}) - 1\right]} \times \frac{(z_{0A} Q_A V)^{n_A}}{n_A!} \frac{(z_{0B} Q_B V)^{n_B}}{n_B!}.$$

In this expression the delta-function due to the incompressibility condition has been rewritten using a Fourier formulation of the delta-function, with $\eta(\vec{r})$ as a field coupled to that. With the above expression of the partition function of the system, we have transformed the particle-based theory of the polymer blends to a field-based formulation. The central quantity of the theory is the single-molecule, or single-chain, partition function, $Q_\alpha(\{\omega_\alpha\})$, which is the partition function of one polymer chain in the field $\omega_\alpha(\vec{r})$. It should be noted that in deriving this field-theoretical formulation of the partition function, we have not made any approximations. Thus the partition function of the system in the form of a functional integral over the fields, $\phi_\alpha(\vec{r})$ and $\omega_\alpha(\vec{r})$, is an exact formulation of the theory. This field-theoretical formulation of the partition function is the starting point of the SCFT.

2.3 *The Single-Chain Partition Function and the Propagators*

It is useful to explore the properties of the single-chain partition function. The single-chain or single-molecule partition functions, $Q_\alpha(\{\omega_\alpha\})$, given in the above section are obtained by summing over all the chain configurations in the presence of an external field $\omega_\alpha(\vec{r})$. These quantities contain contributions of the chain configurations to the total partition function. It is obvious that the single-chain partition function, $Q_\alpha(\{\omega_\alpha\})$, is a functional of the fields, $\omega_\alpha(\vec{r})$. In our derivation, the single-chain partition function is obtained as a path-integral or functional-integral of the form,

$$Q_\alpha(\{\omega_\alpha\}) = \frac{1}{V}\int \mathcal{D}\left\{\vec{R}_{\alpha,i}(s)\right\}\, p_0\left[\vec{R}_{\alpha,i}(s)\right] \exp\left[-\int_0^{N_\alpha} ds\, \omega_\alpha\left(\vec{R}_{\alpha,i}(s)\right)\right].$$

The expression of the single-chain partition function as a path integral is an elegant formulation, resembling its counterpart in quantum mechanics. For computational purpose, however, it is more convenient to express the single-chain partition function in terms of the chain propagators or Greens function, $Q_\alpha\left(\vec{r},\ s|\vec{r}'\right)$, of the polymers,

$$Q_\alpha(\{\omega_\alpha\}) = \frac{1}{V}\int d\vec{r}_1 d\vec{r}_2\, Q_\alpha(\vec{r}_1,\ s|\vec{r}_2).$$

where the chain propagator is defined by the following constrained path integral,

$$Q_\alpha\left(\vec{r},\ s|\vec{r}'\right) = \frac{1}{V}\int_{\vec{R}(0)=\vec{r}'}^{\vec{R}(s)=\vec{r}} \mathcal{D}\left\{\vec{R}(s)\right\}\, e^{-\int_0^{N_\alpha} ds\left[\frac{3}{2b_\alpha^2}\int_0^{N_\alpha} ds\left(\frac{d\vec{R}(s)}{ds}\right)^2 + \omega_\alpha\left(\vec{R}(s)\right)\right]}.$$

The physical meaning of the propagators is that $Q_\alpha\left(\vec{r},\ s|\vec{r}'\right)$ represents the conditional probability distribution of the sth-monomer at position $\vec{r}$, given that the 0th-monomer is at position $\vec{r}'$, in the presence of an external field $\omega_\alpha(\vec{r})$. A direct evaluation of the path integrals is not straightforward. Alternatively, and more conveniently, it can be shown that the propagators can be obtained from the following differential equations, or the modified diffusion equations,

$$\frac{\partial}{\partial s} Q_\alpha\left(\vec{r},\ s|\vec{r}'\right) = \frac{b_\alpha^2}{6}\nabla^2 Q_\alpha\left(\vec{r},\ s|\vec{r}'\right) - \omega_\alpha\left(\vec{r}\right) Q_\alpha\left(\vec{r},\ s|\vec{r}'\right),$$

with the initial conditions, $Q_\alpha\left(\vec{r},\ 0|\vec{r}'\right) = \delta\left(\vec{r} - \vec{r}'\right)$. It is interesting to observe that the theoretical framework of Gaussian chain statistics is analogous with the quantum mechanics. In particular, the modified diffusion equation can be regarded as the Schrödinger equation with a Hamiltonian $\mathcal{H} = -\frac{b_\alpha^2}{6}\nabla^2 + \omega_\alpha\left(\vec{r}\right)$ and an imaginary time. Many developments of polymer physics stem from this analogy.

In further applications, it is convenient to introduce the end-integrated propagators, $q_\alpha\left(\vec{r}, s\right)$, which is defined by,

$$q_\alpha\left(\vec{r}, s\right) = \int d\vec{r}'\, Q_\alpha\left(\vec{r}, s|\vec{r}'\right).$$

The end-integrated propagators satisfy the same differential equation as $Q_\alpha\left(\vec{r},\ s|\vec{r}'\right)$, with different initial conditions, $q_\alpha\left(\vec{r}, 0\right) = 1$. The physical meaning of $q_\alpha\left(\vec{r}, s\right)$ is that it represents the probability distribution of the sth-monomer at position $\vec{r}$, irrespective to where the end-segment at $s = 0$ is, in the presence of an external field $\omega_\alpha\left(\vec{r}\right)$. In terms of the end-integrated propagators, the single chain partition function is given by,

$$Q_\alpha\left(\{\omega_\alpha\}\right) = \frac{1}{V}\int d\vec{r}\, q_\alpha\left(\vec{r},\ N_\alpha\right).$$

In principle, for each segment at s, there are two end-integrated propagators, $q_\alpha\left(\vec{r}, s\right)$ and $q_\alpha^\dagger\left(\vec{r}, s\right)$, representing the probability of reaching to that particular segment starting from the two ends at $s = 0$ and $s = N_\alpha$, respectively. For homopolymers, these two end-integrated propagators are the same due to the symmetry of the chain. However, $q_\alpha\left(\vec{r}, s\right)$ and $q_\alpha^\dagger\left(\vec{r}, s\right)$ are in general different for more complex polymers. For example, the two ends of A-B diblock copolymers are not equivalent, thus these two propagators, $q_\alpha\left(\vec{r}, s\right)$ and $q_\alpha^\dagger\left(\vec{r}, s\right)$, are different and needed to be computed separately.

2.4 *Functional Derivatives of the Propagators*

In the derivation of the mean-field equations, functional derivatives of the single-chain partition functions are required. All these functional derivatives can be obtained from the functional derivative of the propagators with respect to the field,

$$\frac{\delta Q_\alpha\left(\vec{r}_1,\ s|\vec{r}_2\right)}{\delta\omega_\alpha\left(\vec{r}\right)}.$$

This functional derivative could be obtained using the path integral formulation. Alternatively we can use the differential equation as the starting point. Taking the functional derivative of the modified diffusion equation, we have,

$$\frac{\partial}{\partial s}\frac{\delta Q_\alpha\left(\vec{r}_1,\ s|\vec{r}_2\right)}{\delta\omega_\alpha\left(\vec{r}\right)} = \frac{b_\alpha^2}{6}\nabla_1^2\frac{\delta Q_\alpha\left(\vec{r}_1,\ s|\vec{r}_2\right)}{\delta\omega_\alpha\left(\vec{r}\right)} - \omega_\alpha\left(\vec{r}_1\right)\frac{\delta Q_\alpha\left(\vec{r}_1,\ s|\vec{r}_2\right)}{\delta\omega_\alpha\left(\vec{r}\right)} - \frac{\delta\omega_\alpha\left(\vec{r}_1\right)}{\delta\omega_\alpha\left(\vec{r}\right)}Q_\alpha\left(\vec{r}_1,\ s|\vec{r}_2\right).$$

Using the fact that $\frac{\delta\omega_\alpha(\vec{r}_1)}{\delta\omega_\alpha(\vec{r})} = \delta(\vec{r}_1 - \vec{r})$ and rearranging the terms, we obtain an equation for the required functional derivative $\frac{\delta Q_\alpha(\vec{r}_1,\ s|\vec{r}_2)}{\delta\omega_\alpha(\vec{r})}$,

$$\left[\frac{\partial}{\partial s} - \frac{b_\alpha^2}{6}\nabla_1^2 + \omega_\alpha\left(\vec{r}_1\right)\right]\frac{\delta Q_\alpha\left(\vec{r}_1,\ s|\vec{r}_2\right)}{\delta\omega_\alpha\left(\vec{r}\right)} = -\delta\left(\vec{r}_1 - \vec{r}\right)Q_\alpha\left(\vec{r},\ s|\vec{r}_2\right).$$

It is now useful to define the Green functions $G_\alpha\left(\vec{r}_1,\ s|\vec{r}_2, s'\right)$ which are solutions of the differential equations,

$$\left[\frac{\partial}{\partial s} - \frac{b_\alpha^2}{6}\nabla_1^2 + \omega_\alpha\left(\vec{r}_1\right)\right]G_\alpha\left(\vec{r}_1,\ s|\vec{r}_2, s'\right) = -\delta\left(s - s'\right)\delta\left(\vec{r}_1 - \vec{r}_2\right).$$

It is easy to show that the Green function $G_\alpha\left(\vec{r}_1,\ s|\vec{r}_2, s'\right)$ are related to the propagators,

$$G_\alpha\left(\vec{r}_1, s|\vec{r}_2, s'\right) = \theta\left(s - s'\right)Q_\alpha\left(\vec{r}_1, s - s'|\vec{r}_2\right),$$

where $\theta\left(s - s'\right)$ is the Heaviside step function. Using the Green functions, it is straightforward to show that the functional derivative $\frac{\delta Q_\alpha(\vec{r}_1,\ s|\vec{r}_2)}{\delta\omega_\alpha(\vec{r})}$ can be expressed as,

$$\begin{aligned}\frac{\delta Q_\alpha\left(\vec{r}_1,\ s|\vec{r}_2\right)}{\delta\omega_\alpha\left(\vec{r}\right)} &= -\int_{-\infty}^{\infty} ds' \int d\vec{r}' G_\alpha\left(\vec{r}_1,\ s|\vec{r}', s'\right)\delta\left(\vec{r}' - \vec{r}\right)Q_\alpha\left(\vec{r},\ s'|\vec{r}_2\right)\\ &= -\int_0^s ds' Q_\alpha\left(\vec{r}_1,\ s - s'|\vec{r}\right)Q_\alpha\left(\vec{r},\ s'|\vec{r}_2\right),\end{aligned}$$

where we have used the implicit relations that $Q_\alpha\left(\vec{r}_1,\ s|\vec{r}\right) = 0$ for $s < 0$..

3 Canonical Ensemble and Helmholtz Free Energy Functional

3.1 SCFT in Canonical Ensemble

In the canonical ensemble, the numbers of polymer chains in the volume are fixed. In this case the partition function could be written in terms of a functional integral over a Boltzmann factor weighted by a free energy functional,

$$\mathcal{Z}_{n_A,n_B} = \int \mathcal{D}\{\phi_\alpha\}\mathcal{D}\{\omega_\alpha\}\mathcal{D}\{\eta\}\ e^{\sum_\alpha \rho_0 \int d\vec{r}\ \omega_\alpha(\vec{r})\phi_\alpha(\vec{r}) - W(\{\phi\}) - \rho_0 \int d\vec{r}\ \eta(\vec{r})[\phi_A(\vec{r})+\phi_B(\vec{r})-1]}$$
$$\times \frac{(z_{0A}Q_A V)^{n_A}}{n_A!}\frac{(z_{0B}Q_B V)^{n_B}}{n_B!}$$
$$= \int \mathcal{D}\{\phi_\alpha\}\mathcal{D}\{\omega_\alpha\}\mathcal{D}\{\eta\}\ e^{-\mathcal{F}(\{\phi\},\{\omega\},\{\eta\})},$$

where the free energy functional, or, more precisely, the "Hamiltonian", of the system, $\mathcal{F}(\{\phi\},\{\omega\})$, is given by,

$$\mathcal{F}(\{\phi\},\{\omega\},\{\eta\})$$
$$= W(\{\phi\}) - \sum_\alpha \rho_0 \int d\vec{r}\ \omega_\alpha(\vec{r})\phi_\alpha(\vec{r})$$
$$+ \rho_0 \int d\vec{r}\ \eta(\vec{r})\left[\phi_A(\vec{r}) + \phi_B(\vec{r}) - 1\right] - \ln\frac{(z_{0A}Q_A V)^{n_A}}{n_A!} - \ln\frac{(z_{0B}Q_B V)^{n_B}}{n_B!}$$
$$= \rho_0\chi \int d\vec{r}\ \phi_A(\vec{r})\phi_B(\vec{r}) - \sum_\alpha \rho_0 \int d\vec{r}\ \omega_\alpha(\vec{r})\phi_\alpha(\vec{r})$$
$$+ \rho_0 \int d\vec{r}\ \eta(\vec{r})\left[\phi_A(\vec{r}) + \phi_B(\vec{r}) - 1\right] - \sum_\alpha n_\alpha \ln\left(\frac{z_{0\alpha} e Q_\alpha V}{n_\alpha}\right),$$

where the Stirling formula $n! \approx \left(\frac{n}{e}\right)^n$ has been used. The average concentrations of the polymers in the volume V are given by $\bar{\phi}_\alpha = \frac{n_\alpha N_\alpha}{\rho_0 V}$. In terms of the average polymer concentrations, the free energy functional of the system can be written as,

$$\mathcal{F}(\{\phi\},\{\omega\},\{\eta\})$$
$$= \rho_0\chi \int d\vec{r}\ \phi_A(\vec{r})\phi_B(\vec{r}) - \sum_\alpha \rho_0 \int d\vec{r}\ \omega_\alpha(\vec{r})\phi_\alpha(\vec{r})$$
$$+ \rho_0 \int d\vec{r}\ \eta(\vec{r})\left[\phi_A(\vec{r}) + \phi_B(\vec{r}) - 1\right] - \rho_0 V \sum_\alpha \frac{\bar{\phi}_\alpha}{N_\alpha} \ln\left(\frac{z_{0\alpha} e N_\alpha Q_\alpha}{\rho_0 \bar{\phi}_\alpha}\right).$$

In this expression the length scale is not specified. It is convenient to use the radius of gyration of the A-polymers, $R_g = b\sqrt{\frac{N}{6}}$, as the length scale. Rescaling all lengths by R_g, we can write,

$$
\begin{aligned}
&\mathcal{F}(\{\phi\},\{\omega\},\{\eta\}) \\
&\quad= \frac{\rho_0 R_g^3}{N}\Big\{\chi N \int d\vec{r}\, \phi_A(\vec{r})\,\phi_B(\vec{r}) - \sum_\alpha \int d\vec{r}\, N\omega_\alpha(\vec{r})\,\phi_\alpha(\vec{r}) \\
&\qquad + \int d\vec{r}\, N\eta(\vec{r})\left[\phi_A(\vec{r}) + \phi_B(\vec{r}) - 1\right] - V\sum_\alpha \frac{\bar{\phi}_\alpha}{\kappa_\alpha}\ln\left(\frac{z_{0\alpha} e N_\alpha Q_\alpha}{\rho_0 \bar{\phi}_\alpha}\right)\Big\}.
\end{aligned}
$$

In many cases it is convenient to express the free energy in terms of the free energy density,

$$
\begin{aligned}
f(\{\phi\},\{\omega\},\{\eta\}) &\equiv \frac{N\mathcal{F}(\{\phi\},\{\omega\},\{\eta\})}{\rho_0 V R_g^3} \\
&= \frac{1}{V}\int d\vec{r}\,\Big\{\chi N\phi_A(\vec{r})\,\phi_B(\vec{r}) - \sum_\alpha N\omega_\alpha(\vec{r})\,\phi_\alpha(\vec{r}) + N\eta(\vec{r})\left[\phi_A(\vec{r}) + \phi_B(\vec{r}) - 1\right]\Big\} \\
&\quad - \sum_\alpha \frac{\bar{\phi}_\alpha}{\kappa_\alpha}\ln Q_\alpha + \sum_\alpha \frac{\bar{\phi}_\alpha}{\kappa_\alpha}\ln\left(\frac{\bar{\phi}_\alpha}{e}\right) + \sum_\alpha \mu_{0\alpha}\bar{\phi}_\alpha.
\end{aligned}
$$

where $\mu_{0\alpha} \equiv \frac{1}{\kappa_\alpha}\ln\frac{\rho_0}{z_{0\alpha}N_\alpha}$ are constants, which can be viewed as a constant term in the fields $\omega_\alpha(\vec{r})$. This constant can be ignored since adding a constant to the fields does not change the physics of the system.

The modified diffusion equation for the chain propagators can be rewritten in the form,

$$
N\frac{\partial}{\partial s}Q_\alpha\left(\vec{r},\, s|\vec{r}'\right) = \frac{Nb^2\sigma_\alpha^2}{6}\nabla^2 Q_\alpha\left(\vec{r},\, s|\vec{r}'\right) - N\omega_\alpha(\vec{r})\,Q_\alpha\left(\vec{r},\, s|\vec{r}'\right).
$$

We can now define the length scale using R_g and the arc-length scale using N. Furthermore we redefine the fields by including the factor N, N, $N\omega_\alpha(\vec{r}) \to \omega_\alpha(\vec{r})$. In this scaled form the modified diffusion equation becomes,

$$
\frac{\partial}{\partial s}Q_\alpha\left(\vec{r},\, s|\vec{r}'\right) = \sigma_\alpha^2\nabla^2 Q_\alpha\left(\vec{r},\, s|\vec{r}'\right) - \omega_\alpha(\vec{r})\,Q_\alpha\left(\vec{r},\, s|\vec{r}'\right),
$$

with the initial conditions, $Q_\alpha\left(\vec{r},\, 0|\vec{r}'\right) = \delta\left(\vec{r} - \vec{r}'\right)$. The end-integrated propagators, $q_\alpha(\vec{r}, s)$, are given by,

$$
q_\alpha(\vec{r}, s) = \int d\vec{r}'\, Q_\alpha\left(\vec{r}, s|\vec{r}'\right).
$$

These end-integrated propagators satisfy the same differential equation as $Q_\alpha\left(\vec{r},\, s|\vec{r}'\right)$, with different initial conditions, $q_\alpha(\vec{r}, 0) = 1$. In terms of the end-integrated propagators, the single chain partition function is given by,

$$
Q_\alpha(\{\omega_\alpha\}) = \frac{1}{V}\int d\vec{r}\, q_\alpha(\vec{r},\, \kappa_\alpha).
$$

The formula of the functional derivative stays the same,

$$\frac{\delta Q_\alpha(\vec{r}_1,\ s|\vec{r}_2)}{\delta\omega_\alpha(\vec{r})} = -\int_0^s ds' \, Q_\alpha(\vec{r}_1,\ s-s'|\vec{r})\, Q_\alpha(\vec{r},\ s'|\vec{r}_2),$$

The free energy density now has the simple form,

$$\begin{aligned} f(\{\phi\},\{\omega\},\{\eta\}) \\ = \frac{1}{V}\int d\vec{r}\left\{\chi N\phi_A(\vec{r})\phi_B(\vec{r}) - \sum_\alpha \omega_\alpha(\vec{r})\phi_\alpha(\vec{r}) + \eta(\vec{r})\left[\phi_A(\vec{r})+\phi_B(\vec{r})-1\right]\right\} \\ - \sum_\alpha \frac{\bar{\phi}_\alpha}{\kappa_\alpha}\ln Q_\alpha + \sum_\alpha \frac{\bar{\phi}_\alpha}{\kappa_\alpha}\ln\left(\frac{\bar{\phi}_\alpha}{e}\right) + \sum_\alpha \mu_{0\alpha}\bar{\phi}_\alpha. \end{aligned}$$

The partition function of the system is then written in the form,

$$\mathcal{Z}_{n_A,n_B} = \int \mathcal{D}\{\phi_\alpha\}\mathcal{D}\{\omega_\alpha\}\mathcal{D}\{\eta\}\ e^{-F(\{\phi\},\{\omega\},\{\eta\})}.$$

This expression of the partition function is exact. However evaluation of the partition function is in general a formidable task. A variety of approximate methods have been developed to evaluate the partition function. The most fruitful method is the mean-field approximation, which amounts to evaluating the functional integral using a saddle-point technique. Technically the saddle-point approximation is obtained by demanding that the functional derivative of the integrand is zero,

$$\frac{\delta F(\{\phi\},\{\omega\},\{\eta\})}{\delta\phi_\alpha(\vec{r})} = \frac{\delta F(\{\phi\},\{\omega\},\{\eta\})}{\delta\omega_\alpha(\vec{r})} = \frac{\delta F(\{\phi\},\{\omega\},\{\eta\})}{\delta\eta(\vec{r})} = 0.$$

Carrying out these functional derivatives leads to the following mean-field equations, or SCFT equations, of the binary blends of A/B homopolymers:

$$\phi_A(\vec{r}) = \frac{\bar{\phi}_A}{\kappa_A Q_A}\int_0^{\kappa_A} ds\, q_A(\vec{r},s)\, q_A(\vec{r},\kappa_A - s),$$

$$\phi_B(\vec{r}) = \frac{\bar{\phi}_B}{\kappa_B Q_B}\int_0^{\kappa_B} ds\, q_B(\vec{r},s)\, q_B(\vec{r},\kappa_B - s),$$

$$\omega_A(\vec{r}) = \chi N\phi_B(\vec{r}) + \eta(\vec{r}),$$

$$\omega_B(\vec{r}) = \chi N\phi_A(\vec{r}) + \eta(\vec{r}),$$

$$\phi_A(\vec{r}) + \phi_B(\vec{r}) = 1.$$

The field $\eta(\vec{r})$ can also be viewed as a Lagrangian multiplier to ensure the incompressibility condition. It should be note that, with $\alpha = A, B$, there are five SCFT equations for five unknown variables, $\{\phi_A(\vec{r}), \phi_B(\vec{r}), \omega_A(\vec{r}), \omega_B(\vec{r}), \eta(\vec{r})\}$. The parameters controlling the thermodynamic properties of the system are the average polymer concentrations $\bar{\phi}_\alpha$, the molecular weights κ_α, the relative Kuhn lengths σ_α, and the interaction parameters χN.

Within the mean-field approximation, the free energy per chain of the system is obtained by inserting the solution of the SCFT equations into the free energy expression,

$$f(\{\phi\}, \{\omega\}, \{\eta\}) = \frac{1}{V}\int d\vec{r}\left\{\chi N\phi_A(\vec{r})\phi_B(\vec{r}) - \sum_\alpha \omega_\alpha(\vec{r})\phi_\alpha(\vec{r})\right\} - \sum_\alpha \frac{\bar{\phi}_\alpha}{\kappa_\alpha}\ln Q_\alpha(\{\omega_\alpha\}) + \sum_\alpha \frac{\bar{\phi}_\alpha}{\kappa_\alpha}\ln\left(\frac{\bar{\phi}_\alpha}{\bar{\phi}}\right) + \sum_\alpha \mu_{0\alpha}\bar{\phi}_\alpha.$$

It is useful to observe that adding a constant to the fields, $\omega_\alpha(\vec{r}) \to \omega_\alpha(\vec{r}) + c_\alpha$, leads to a change in the propagator of the form, $Q_\alpha\left(\vec{r}, s|\vec{r}'\right) \to e^{-c_\alpha s} Q_\alpha\left(\vec{r}, s|\vec{r}'\right)$, and, $Q_\alpha \to e^{-c_\alpha \kappa_\alpha} Q_\alpha$. These changes leave the mean-field concentrations and the free energy density invariant. Therefore we can choose the constant level of the fields based on convenience.

3.2 Homogeneous Phase

The simplest and exact solution of the SCFT equations is obtained for a homogeneous phase, in which the polymer concentrations and the mean-field potentials are constants, $\phi_\alpha(\vec{r}) = \phi_\alpha = \bar{\phi}_\alpha$, $\omega_\alpha(\vec{r}) = \omega_\alpha$, leading to the trivial solutions, $Q_\alpha\left(\vec{r}, s|\vec{r}'\right) = \frac{1}{(4\pi\sigma_\alpha^2 s)^{3/2}} e^{-\frac{(\vec{r}-\vec{r}')^2}{4\sigma_\alpha^2 s} - \omega_\alpha s}$, $q_\alpha(\vec{r}, s) = e^{-\omega_\alpha s}$, and $Q_\alpha = e^{-\omega_\alpha \kappa_\alpha}$. The mean-field equations become,

$$\phi_A = \bar{\phi}_A, \ \phi_B = \bar{\phi}_B;$$

$$\omega_A = \chi N\phi_B + \eta, \omega_B = \chi N\phi_A + \eta;$$

$$\phi_A + \phi_B = 1.$$

In this case the free energy per chain of a homogeneous phase is given by the simple expression,

$$f_H = \chi N\phi_A\phi_B + \frac{\phi_A}{\kappa_A}\ln\left(\frac{\phi_A}{e}\right) + \frac{\phi_B}{\kappa_B}\ln\left(\frac{\phi_B}{e}\right) + \mu_{0A}\phi_A + \mu_{0B}\phi_B.$$

It is noted that this expression of the free energy of a polymer blend is exactly the same as the textbook result of Flory and Higgins. The Flory-Higgins free energy is the starting point to analyze the phase behavior of polymer blends, including the construction of phase diagrams and the calculation of spinodal lines.

One useful quantity to analyze the thermodynamic property is the chemical potential. For a polymer blend, the total free energy of the system is given by, $F_H = \frac{\rho_0 V f_H}{N}$. The chemical potentials of the two polymers are defined as the free energy difference when one more polymer chain is added to the system,

$$\mu_\alpha = F_H\left(n_\alpha + 1\right) - F_H\left(n_\alpha\right) = \frac{\partial F_H}{\partial n_\alpha} = \left(\frac{\rho_0 V}{N}\right)\frac{\partial f_H}{\partial n_\alpha} + f_H \frac{\partial}{\partial n_\alpha}\left(\frac{\rho_0 V}{N}\right).$$

In a homogenous system, the number of polymer chains is related to the concentrations by $\phi_\alpha = \frac{n_\alpha \kappa_\alpha N}{\rho_0 V}$. The incompressibility condition requires, $\sum_\alpha \phi_\alpha = \frac{N}{\rho_0 V}\sum_\alpha \left(n_\alpha \kappa_\alpha\right) = 1$, therefore the volume is given by, $\frac{\rho_0 V}{N} = \sum_\alpha \left(n_\alpha \kappa_\alpha\right)$, *or* $\frac{\partial}{\partial n_\alpha}\left(\frac{\rho_0 V}{N}\right) = \kappa_\alpha$. Another useful relation is that, $\frac{\partial \phi_\beta}{\partial n_\alpha} = \frac{\kappa_\alpha N}{\rho_0 V}\left(\delta_{\alpha\beta} - \phi_\beta\right)$. Using these relations it is straightforward to derive the chemical potential per monomer of the blends,

$$\frac{\mu_\alpha}{\kappa_\alpha} = \frac{\partial f_H}{\partial \phi_\alpha} + \left(f_H - \sum_\beta \phi_\beta \frac{\partial f_H}{\partial \phi_\beta}\right).$$

Using the expression of the homogeneous free energy density, the chemical potential of the binary blends is obtained as,

$$\frac{\mu_A}{\kappa_A} = \frac{\ln \phi_A}{\kappa_A} + \chi N \phi_B - K + \mu_{0A},$$

$$\frac{\mu_B}{\kappa_B} = \frac{\ln \phi_B}{\kappa_B} + \chi N \phi_A - K + \mu_{0B},$$

where the constant K is given by $K = \chi N \phi_A \phi_B + \frac{\phi_A}{\kappa_A} + \frac{\phi_B}{\kappa_B}$. At thermal equilibrium, the chemical potential of the system must be a constant. This condition determines the equilibrium values of the bulk concentrations ϕ_α^b, or the binodal points of the blends above the critical point.

4 Grand-Canonical Ensemble and Grand Potential Functional

4.1 SCFT in Grand-Canonical Ensemble

For systems undergoing macroscopic phase separation, it is more convenient to work in the grand-canonical ensemble, in which the chemical potentials of the polymers are control parameters. In the grand canonical ensemble, the chemical potential of the polymers is given whereas the number of polymers in the system is not fixed. The thermodynamic properties of the system in the grand-canonical ensemble is given by the grand potential Ω, which is related to the grand-partition function, Ξ, via $\Omega = -k_B T \ln \Xi$. Thus in the grand-canonical formulation, we need to calculate the grand partition function Ξ. For the binary mixtures of A/B polymers, the grand-partition function is defined by a summation over all number of polymer chains,

$$
\begin{aligned}
\Xi &= \sum_{n_A=0}^{\infty}\sum_{n_B=0}^{\infty} e^{n_A\mu_A+n_B\mu_B}\, \mathcal{Z}_{n_A,n_B} \\
&= \int \mathcal{D}\{\phi_\alpha\}\mathcal{D}\{\omega_\alpha\}\mathcal{D}\{\eta\}\; e^{\sum_\alpha \frac{\rho_0}{N}\int d\vec{r}\;\omega_\alpha(\vec{r})\phi_\alpha(\vec{r}) - W(\{\phi\}) - \frac{\rho_0}{N}\int d\vec{r}\;\eta(\vec{r})[\phi_A(\vec{r})+\phi_B(\vec{r})-1]} \\
&\quad \times \prod_\alpha \left(\sum_{n_\alpha=0}^{\infty} \frac{(e^{\mu_\alpha} z_{0\alpha} Q_\alpha V)^{n_\alpha}}{n_\alpha!}\right) \\
&= \int \mathcal{D}\{\phi_\alpha\}\mathcal{D}\{\omega_\alpha\}\mathcal{D}\{\eta\} \\
&\quad \times e^{\sum_\alpha \frac{\rho_0}{N}\int d\vec{r}\;\omega_\alpha(\vec{r})\phi_\alpha(\vec{r}) - W(\{\phi\}) - \frac{\rho_0}{N}\int d\vec{r}\;\eta(\vec{r})[\phi_A(\vec{r})+\phi_B(\vec{r})-1] + e^{\mu_A} z_{0A} Q_A V + e^{\mu_B} z_{0B} Q_B V} \\
&= \int \mathcal{D}\{\phi_\alpha\}\mathcal{D}\{\omega_\alpha\}\mathcal{D}\{\eta\}\; e^{-\Omega(\{\phi\},\{\omega\},\{\eta\})},
\end{aligned}
$$

where the grand potential functional $\Omega\left(\{\phi\},\{\omega\},\{\eta\}\right)$ is given by,

$$
\begin{aligned}
\Omega\left(\{\phi\},\{\omega\},\{\eta\}\right) = &\\
&\rho_0\chi \int d\vec{r}\;\phi_A(\vec{r})\,\phi_B(\vec{r}) - \sum_\alpha \frac{\rho_0}{N}\int d\vec{r}\;\omega_\alpha(\vec{r})\,\phi_\alpha(\vec{r}) \\
&+ \frac{\rho_0}{N}\int d\vec{r}\;\eta(\vec{r})\left[\phi_A(\vec{r})+\phi_B(\vec{r})-1\right] - e^{\mu_A} z_{0A} Q_A V - e^{\mu_B} z_{0B} Q_B V.
\end{aligned}
$$

Equivalently a grand potential density can be defined as,

$$
\begin{aligned}
g\left(\{\phi\},\{\omega\},\{\eta\}\right) &\equiv \frac{N}{\rho_0 V R_g^3}\,\Omega\left(\{\phi\},\{\omega\},\{\eta\}\right) \\
&= \frac{1}{V}\int d\vec{r}\left\{\chi N \phi_A(\vec{r})\,\phi_B(\vec{r}) - \sum_\alpha \omega_\alpha(\vec{r})\,\phi_\alpha(\vec{r}) + \eta(\vec{r})\left[\phi_A(\vec{r})+\phi_B(\vec{r})-1\right]\right\} \\
&\quad - \sum_\alpha e^{\mu_\alpha-\kappa_\alpha\mu_{0\alpha}}\frac{Q_\alpha}{\kappa_\alpha},
\end{aligned}
$$

where $\mu_{0\alpha} \equiv \frac{1}{\kappa_\alpha}\ln\frac{\rho_0}{z_{0\alpha}N_\alpha}$ are constants as defined before in the case of canonical ensemble.

The mean-field equations are again obtained by demanding that the first-order functional derivatives of the free energy functional vanish,

$$\frac{\delta\Omega\left(\{\phi\},\{\omega\},\{\eta\}\right)}{\delta\phi_\alpha(\vec{r})}=\frac{\delta\Omega\left(\{\phi\},\{\omega\},\{\eta\}\right)}{\delta\omega_\alpha(\vec{r})}=\frac{\delta\Omega\left(\{\phi\},\{\omega\},\{\eta\}\right)}{\delta\eta(\vec{r})}=0.$$

Carrying out these functional derivatives leads to the following mean-field, or SCFT, equations for the polymer blends in the grand-canonical ensemble,

$$\phi_A(\vec{r})=\frac{e^{\mu_A-\kappa_A\mu_{0A}}}{\kappa_A}\int_0^{\kappa_A}ds\, q_A(\vec{r},s)\,q_A(\vec{r},\kappa_A-s),$$

$$\phi_B(\vec{r})=\frac{e^{\mu_B-\kappa_B\mu_{0B}}}{\kappa_B}\int_0^{\kappa_B}ds\, q_B(\vec{r},s)\,q_B(\vec{r},\kappa_B-s),$$

$$\omega_A(\vec{r})=\chi N\phi_B(\vec{r})+\eta(\vec{r}),$$

$$\omega_B(\vec{r})=\chi N\phi_A(\vec{r})+\eta(\vec{r}),$$

$$\phi_A(\vec{r})+\phi_B(\vec{r})=1.$$

This set of equations is similar to the canonical SCFT equations. The difference is that, in the current case, the chemical potentials are the controlling parameters. In particular, the average concentrations are given in terms of the chemical potentials as, $\bar{\phi}_\alpha=e^{\Delta\mu_\alpha}Q_\alpha$ where $\Delta\mu_\alpha=\mu_\alpha-\kappa_\alpha\mu_{0\alpha}$ is the effective chemical potential of the polymers.

Once solutions of the mean-field equations are obtained, the grand potential density can be computed using the expression,

$$g\left(\{\phi\},\{\omega\},\{\eta\}\right)=\frac{1}{V}\int d\vec{r}\left\{\chi N\phi_A(\vec{r})\,\phi_B(\vec{r})-\sum_\alpha\omega_\alpha(\vec{r})\,\phi_\alpha(\vec{r})\right\}-\sum_\alpha\frac{e^{\Delta\mu_\alpha}Q_\alpha}{\kappa_\alpha}.$$

Again in the case of grand-canonical ensemble, there are five SCFT equations for five unknown variables, $\{\phi_A(\vec{r}),\ \phi_B(\vec{r}),\ \omega_A(\vec{r}),\ \omega_B(\vec{r}),\ \eta(\vec{r})\}$. The parameters controlling the thermodynamic properties of the system are the chemical potentials of the polymers μ_α, the molecular weights κ_α, the relative Kuhn lengths σ_α, and the interaction parameters χN.

4.2 *Homogeneous Phase*

It is instructive to work out the grand potential for the homogeneous phase and related the expression with the results from the canonical ensemble. For a homogeneous or disordered phase all concentrations and fields are constants, $\phi_\alpha(\vec{r})=\phi_\alpha$, ω_α

$(\vec{r}) = \omega_\alpha$. Due to the lack of spatial variation, the solutions of the propagators are easily obtained in this case,

$$Q_\alpha\left(\vec{r},\ s|\vec{r}'\right) = \frac{1}{\left(4\pi\sigma_\alpha^2 s\right)^{3/2}} e^{-\frac{(\vec{r}-\vec{r}')^2}{4\sigma_\alpha^2 s}-\omega_\alpha s},\ q_\alpha\left(\vec{r},\ s\right) = e^{-\omega_\alpha s},\ Q_\alpha = e^{-\omega_\alpha \kappa_\alpha}.$$

Therefore the mean-field equations become,

$$\phi_A = e^{\Delta\mu_A - \omega_A\kappa_A},\ \phi_B = e^{\Delta\mu_B - \omega_B\kappa_B};$$

$$\omega_A = \chi N\phi_B + \eta,\ \omega_B = \chi N\phi_A + \eta;$$

$$\phi_A + \phi_B = 1.$$

These five equations are to be solved to obtain $\{\phi_A, \phi_B, \omega_A\ \omega_B,\ \eta\}$ as functions of the chemical potentials $\{\Delta\mu_A, \Delta\mu_B\}$. The corresponding grand potential density is now obtained by inserting these solutions to the follow expression,

$$g = \chi N\phi_A\phi_B + \sum_\alpha \omega_\alpha\phi_\alpha - \sum_\alpha \frac{e^{\Delta\mu_\alpha - \omega_\alpha\kappa_\alpha}}{\kappa_\alpha}.$$

It should be noted that, although the above expression is given in terms of the concentrations $\{\phi_A, \phi_B\}$ and the fields $\{\omega_A\ \omega_B\}$, the grand potential is a function of the chemical potential thus all the variables $\{\phi_A, \phi_B, \omega_A\ \omega_B,\ \eta\}$ should be taken as functions of the chemical potentials $\{\Delta\mu_A, \Delta\mu_B\}$.

5 Summary of SCFT

The self-consistent field theory of a binary polymer blend can now be summarized for both canonical and grand canonical ensembles in a unified formulation. The basic quantity in the theory is the chain propagators, which are solutions of the modified diffusion equations,

$$\frac{\partial}{\partial s}Q_\alpha\left(\vec{r},\ s|\vec{r}'\right) = \sigma_\alpha^2\nabla^2 Q_\alpha\left(\vec{r},\ s|\vec{r}'\right) - \omega_\alpha\left(\vec{r}\right)Q_\alpha\left(\vec{r},\ s|\vec{r}'\right),$$

with the initial conditions, $Q_\alpha\left(\vec{r},\ 0|\vec{r}'\right) = \delta\left(\vec{r} - \vec{r}'\right)$. The physical significance of the propagator is that it represents the conditional probability distribution for a polymer chain of type α in an external field $\omega_\alpha\left(\vec{r}\right)$. It is obvious that the propagators are functional of the field $\omega_\alpha\left(\vec{r}\right)$. Mathematically, the propagator can be expressed as a path integral, thus linking the mathematical structure of polymer statistics with quantum mechanics. One useful formula is the functional derivative of the propagator with respect to $\omega_\alpha(\vec{r})$,

$$\frac{\delta Q_\alpha(\vec{r}_1, s|\vec{r}_2)}{\delta\omega_\alpha(\vec{r})} = -\int_0^s ds' Q_\alpha(\vec{r}_1, s-s'|\vec{r})\, Q_\alpha(\vec{r}, s'|\vec{r}_2).$$

For numerical implementations, it is useful to work with the end-integrated propagators, $q_\alpha(\vec{r}, s)$, which are defined by,

$$q_\alpha(\vec{r}, s) = \int d\vec{r}'\, Q_\alpha(\vec{r}, s|\vec{r}').$$

These end-integrated propagators satisfy the same differential equation as $Q_\alpha(\vec{r}, s|\vec{r}')$, with different initial conditions, $q_\alpha(\vec{r}, 0) = 1$. In terms of the end-integrated propagators, the single chain partition function is given by,

$$Q_\alpha(\{\omega_\alpha\}) = \frac{1}{V}\int d\vec{r}_1 d\vec{r}_2\, Q_\alpha(\vec{r}_1, s|\vec{r}_2) = \frac{1}{V}\int d\vec{r}\, q_\alpha(\vec{r}, \kappa_\alpha).$$

It should be noted that the end-integrated propagators assume more complex form for block copolymers because their two or more ends are not equivalent. The above equations determining the propagators and the single-china partition functions when the fields, $\omega_\alpha(\vec{r})$, are specified. These equations apply to both canonical and grand canonical ensembles.

The self-consistent mean field equations are a set of nonlinear and nonlocal equations determining the concentrations $\phi_\alpha(\vec{r})$ and the fields $\omega_\alpha(\vec{r})$, as well as the field $\eta(\vec{r})$ which ensures the incompressibility condition. In the canonical ensemble, the number of polymer chains n_α or, equivalently, the average polymer concentrations $\bar{\phi}_\alpha$, are fixed control parameters. In this case the polymer concentrations are determined by,

$$\phi_\alpha(\vec{r}) = \frac{\bar{\phi}_\alpha}{\kappa_\alpha Q_\alpha}\int_0^{\kappa_\alpha} ds\, q_\alpha(\vec{r}, s)\, q_\alpha(\vec{r}, \kappa_\alpha - s).$$

In the grand canonical ensemble, the chemical potentials of the polymer chains $\Delta\mu_\alpha = \mu_\alpha - \kappa_\alpha\mu_{0\alpha}$ are fixed control parameters. In this case the polymer concentrations are determined by,

$$\phi_\alpha(\vec{r}) = \frac{e^{\Delta\mu_\alpha}}{\kappa_\alpha}\int_0^{\kappa_\alpha} ds\, q_\alpha(\vec{r}, s)\, q_\alpha(\vec{r}, \kappa_\alpha - s).$$

These two expressions are related to each other by the fact that, from the grand canonical ensemble expression, we can obtain the average concentrations as

$\bar{\phi}_\alpha = e^{\Delta\mu_\alpha} Q_\alpha$. Inserting this relation to the grand canonical ensemble expression leads to the canonical result.

The fields $\omega_\alpha(\vec{r})$ and $\eta(\vec{r})$ are determined self-consistently from the following equations, which are the same for both ensembles,

$$\begin{aligned}\omega_A(\vec{r}) &= \chi N \phi_B(\vec{r}) + \eta(\vec{r}),\\ \omega_B(\vec{r}) &= \chi N \phi_A(\vec{r}) + \eta(\vec{r}),\\ \phi_A(\vec{r}) + \phi_B(\vec{r}) &= 1.\end{aligned}$$

These SCFT equations must be solved to obtain solutions corresponding to different phases. Inserting the SCFT solutions to the expression of the free energy leads to the evaluation of the free energy of different phases. A comparison of the free energy of the phases can then be used to construct phase diagrams of the system.

The appropriate free energies for the two ensembles are the Helmholtz free energy $F(T, V, n_\alpha)$ and the grand potential $\Omega(T, V, \mu_\alpha)$. These two free energies are related by a Legend transformation, $\Omega(T, V, \mu_\alpha) = F(T, V, n_\alpha) - \sum_\alpha n_\alpha \mu_\alpha$. In terms of the free energy densities, $F = \frac{\rho_0 V}{N} f$ and $\Omega = \frac{\rho_0 V}{N} g$, we have $g(T, V, \mu_\alpha) = f(T, V, n_\alpha) - \sum\limits_\alpha \frac{\mu_\alpha \bar{\phi}_\alpha}{\kappa_\alpha}$. The free energy densities are specified in terms of the SCFT solutions (where the constants $\kappa_\alpha \mu_{0\alpha}$ have been absorbed into the chemical potential, $\mu_\alpha - \kappa_\alpha \mu_{0\alpha} \to \mu_\alpha$),

$$f(\bar{\phi}_\alpha) = \frac{1}{V}\int d\vec{r} \left\{ \chi N \phi_A(\vec{r}) \phi_B(\vec{r}) - \sum_\alpha \omega_\alpha(\vec{r}) \phi_\alpha(\vec{r}) \right\} - \sum_\alpha \frac{\bar{\phi}_\alpha}{\kappa_\alpha} \ln \frac{e Q_\alpha}{\bar{\phi}_\alpha},$$

$$g(\mu_\alpha) = \frac{1}{V}\int d\vec{r} \left\{ \chi N \phi_A(\vec{r}) \phi_B(\vec{r}) - \sum_\alpha \omega_\alpha(\vec{r}) \phi_\alpha(\vec{r}) \right\} - \sum_\alpha \frac{e^{\mu_\alpha} Q_\alpha}{\kappa_\alpha}.$$

These two expressions are equivalent and they are related by the Legend transformation. This statement can be easily proven using the relationship $\bar{\phi}_\alpha = e^{\mu_\alpha} Q_\alpha$.

For a binary polymer blend, there are two chemical potentials, μ_A and μ_B. In general these two chemical potentials could be changed independently. However for an incompressible system, these two chemical potentials are not independent because the incompressibility condition requires that, $1 = \sum_\alpha \bar{\phi}_\alpha = \sum_\alpha e^{\mu_\alpha} Q_\alpha$. In practice we could choose one chemical potential as the independent parameter, and the second one should be determined self-consistently from the theory.

6 Ginzburg-Landau Free Energy Functional of Polymer Blends

Within the mean-field approximation, the SCFT developed above provides a useful framework for the study of the thermodynamic property of inhomogeneous polymeric systems. On the other hand, the polymeric SCFT equations are a set of nonlinear and nonlocal equations, whose solutions could be obtained using various approximate methods or numerical techniques. For many applications, it might be advantageous to cast the free energy functional in terms of the density variables alone, thus paralleling the classical density functional theory. For polymer blends, a widely used free energy functional is casted in the form of a Ginzburg-Landau theory [15, 16],

$$F_{GL}\left(\{\phi_\alpha(\vec{r})\}\right) = \int d\vec{r} \left\{ f\left(\phi_\alpha(\vec{r})\right) + \sum_\alpha g\left(\phi_\alpha(\vec{r})\right)\left[\nabla\phi_\alpha(\vec{r})\right]^2 \right\},$$

where $f\left(\phi_\alpha(\vec{r})\right)$ is a local free energy density, the gradient terms provides a penalty to the inhomogeneity in the system and $g\left(\phi_\alpha(\vec{r})\right)$ includes contributions from the polymer configurations. It is noted that this particular form of the Ginzburg-Landau free energy applies to polymer blends. More complex gradient terms are required to describe the formation of ordered phases from block copolymers [17].

In principle, it is possible to obtain a free energy functional $\mathcal{F}\left(\{\phi\}\right)$ by integrating out the field variables within the SCFT framework,

$$\mathcal{Z}_{n_A,n_B} = \int \mathcal{D}\{\phi_\alpha\}\mathcal{D}\{\omega_\alpha\}\mathcal{D}\{\eta\}\; e^{-\mathcal{F}(\{\phi\},\{\omega\},\{\eta\})} = \int \mathcal{D}\{\phi_\alpha\}\; e^{-\mathcal{F}(\{\phi\})}.$$

Regrettably, performing this functional integration exactly is not possible in most cases. Various approximate methods, such as the random phase approximation [17], have been developed to evaluate this functional integral, resulting in different forms of the free energy functional. Among the many approximate methods, a simple and effective one is to carry out a mean-field or saddle-point approximation over the integral over the fields, $\frac{\delta F(\{\phi\},\{\omega\},\{\eta\})}{\delta\omega_\alpha(\vec{r})} = 0$, resulting in the following mean-field equations,

$$\phi_\alpha(\vec{r}) = \frac{\bar{\phi}_\alpha}{\kappa_\alpha Q_\alpha}\int_0^{\kappa_\alpha} ds\, q_\alpha(\vec{r},s)\, q_\alpha(\vec{r},\kappa_\alpha - s),$$

where the propagators are solutions of the modified diffusion equations,

$$\frac{\partial}{\partial s} q_\alpha(\vec{r},\, s) = \sigma_\alpha^2 \nabla^2 q_\alpha(\vec{r},\, s) - \omega_\alpha(\vec{r})\, q_\alpha(\vec{r},\, s).$$

Normally this equation is used to obtain the density profiles when the fields $\omega_\alpha(\vec{r})$, thus the propagators $q_\alpha(\vec{r},s)$, are given. In the current case, however, the task is reversed in that we would like to obtain the fields $\omega_\alpha(\vec{r})$ or the propagator $q_\alpha(\vec{r},s)$ for a given density profile $\phi_\alpha(\vec{r})$, $\omega_\alpha(\vec{r}) = \omega_\alpha(\vec{r},\left[\phi_\alpha(\vec{r})\right])$. Inserting this solution to the SCFT free energy functional,

$$F\left(\{\phi\},\{\omega\},\{\eta\}\right) = \int d\vec{r} \left\{ \chi N \phi_A(\vec{r})\,\phi_B(\vec{r}) - \sum_\alpha \left[\omega_\alpha(\vec{r}) + \frac{1}{\kappa_\alpha}\ln\frac{eQ_\alpha}{\bar{\phi}_\alpha}\right]\phi_\alpha(\vec{r}) \right\},$$

we obtain the desired density functional for the system $F(\{\phi\})$.

Even within the saddle-point approximation, inverting the mean-field equations to obtain the solution $\omega_\alpha(\vec{r}) = \omega_\alpha(\vec{r}, [\phi_\alpha(\vec{r})])$ is not straightforward. For cases where the inhomogeneity is weak, it is possible to carry out an expansion in terms of the gradients of the density [16]. In what follows we will find approximation solutions of the relationship $\omega_\alpha(\vec{r}) = \omega_\alpha(\vec{r}, [\phi_\alpha(\vec{r})])$ at the weak-segregation and strong-segregation limits.

In order to proceed, it is convenient to write the integral $\int d\vec{r}\ \omega_\alpha(\vec{r})\,\phi_\alpha(\vec{r})$ in terms of the propagators. To this end, we multiply the modified diffusions equation by $q_\alpha(\vec{r},\ \kappa_\alpha - s)$, leading to the expression,

$$q_\alpha(\vec{r},\ \kappa_\alpha - s)\frac{\partial}{\partial s}q_\alpha(\vec{r},\ s) = \sigma_\alpha^2 q_\alpha(\vec{r},\ \kappa_\alpha - s)\,\nabla^2 q_\alpha(\vec{r},\ s) - \omega_\alpha(\vec{r})\,q_\alpha(\vec{r},\ \kappa_\alpha - s)\,q_\alpha(\vec{r},\ s).$$

Using the expression of the mean-field density profile we have,

$$\omega_\alpha(\vec{r})\,\phi_\alpha(\vec{r}) = -\frac{\bar{\phi}_\alpha}{\kappa_\alpha Q_\alpha}\int_0^{\kappa_\alpha} ds\left\{q_\alpha(\vec{r},\ \kappa_\alpha - s)\frac{\partial}{\partial s}q_\alpha(\vec{r},\ s) - \sigma_\alpha^2 q_\alpha(\vec{r},\ \kappa_\alpha - s)\,\nabla^2 q_\alpha(\vec{r},\ s)\right\}.$$

Therefore the integral $\int d\vec{r}\ \omega_\alpha(\vec{r})\,\phi_\alpha(\vec{r})$ can be written as,

$$\begin{aligned}\int d\vec{r}\ \omega_\alpha(\vec{r})\,\phi_\alpha(\vec{r}) &= -\frac{\bar{\phi}_\alpha}{\kappa_\alpha Q_\alpha}\int d\vec{r}\int_0^{\kappa_\alpha} ds\left\{q_\alpha(\vec{r},\ \kappa_\alpha - s)\frac{\partial}{\partial s}q_\alpha(\vec{r},\ s)\right.\\ &\quad\left. + \sigma_\alpha^2\left[\nabla q_\alpha(\vec{r},\ \kappa_\alpha - s)\right]\left[\nabla q_\alpha(\vec{r},\ s)\right]\right\},\end{aligned}$$

where we have used integration by part to write the second terms in terms of the gradients of the propagators.

At the weak-segregation limit, the amplitude variation is small so the gradient terms could be treated as perturbations. Ignoring the gradient term in the modified diffusion equation, we obtain an analytic solution of the propagator, $q_\alpha(\vec{r},\ s) \approx e^{-\omega_\alpha(\vec{r})s}$. The density profile is then given by,

$$\phi_\alpha(\vec{r}) \approx \frac{\bar{\phi}_\alpha}{Q_\alpha}e^{-\omega_\alpha(\vec{r})\kappa_\alpha}.$$

Inverting this expression gives an approximate expression of the fields and the propagators in terms of the density profiles. In particular, the propagators are related to the density profiles by,

$$q_\alpha(\vec{r}, s) \approx \left[\frac{Q_\alpha}{\bar{\phi}_\alpha}\phi_\alpha(\vec{r})\right]^{\frac{s}{\kappa_\alpha}}.$$

Using this expression of the propagators, it is straightforward to obtain,

$$\frac{\bar{\phi}_\alpha}{\kappa_\alpha Q_\alpha}\int_0^{\kappa_\alpha} ds \left\{q_\alpha(\vec{r}, \kappa_\alpha - s)\frac{\partial}{\partial s}q_\alpha(\vec{r}, s)\right\} = \frac{\phi_\alpha(\vec{r})}{\kappa_\alpha}\ln\left[\frac{Q_\alpha}{\bar{\phi}_\alpha}\phi_\alpha(\vec{r})\right],$$

$$\frac{\bar{\phi}_\alpha}{\kappa_\alpha Q_\alpha}\int_0^{\kappa_\alpha} ds \left\{\sigma_\alpha^2\left[\nabla q_\alpha(\vec{r}, \kappa_\alpha - s)\right]\left[\nabla q_\alpha(\vec{r}, s)\right]\right\} = \frac{\sigma_\alpha^2}{6}\frac{\left[\nabla\phi_\alpha(\vec{r})\right]^2}{\phi_\alpha(\vec{r})}.$$

Inserting these results to the expression of the free energy functional, we obtain,

$$F_{GL}(\{\phi_\alpha(\vec{r})\}) = \int d\vec{r}\left\{\chi N\phi_A(\vec{r})\phi_B(\vec{r}) + \sum_\alpha \frac{\phi_\alpha(\vec{r})}{\kappa_\alpha}\ln\left[\frac{\phi_\alpha(\vec{r})}{e}\right] + \sum_\alpha \frac{\sigma_\alpha^2}{6}\frac{\left[\nabla\phi_\alpha(\vec{r})\right]^2}{\phi_\alpha(\vec{r})}\right\},$$

which can be written in the standard form,

$$F_{GL}(\{\phi_\alpha(\vec{r})\}) = \int d\vec{r}\left\{f_{FH}(\{\phi_\alpha(\vec{r})\}) + \sum_\alpha \frac{\sigma_\alpha^2}{6}\frac{\left[\nabla\phi_\alpha(\vec{r})\right]^2}{\phi_\alpha(\vec{r})}\right\},$$

where the Flory-Huggins free energy density is given by,

$$f_{FH}(\{\phi_\alpha(\vec{r})\}) = \chi N\phi_A(\vec{r})\phi_B(\vec{r}) + \sum_\alpha \frac{\phi_\alpha(\vec{r})}{\kappa_\alpha}\ln\frac{\phi_\alpha(\vec{r})}{e}.$$

This form of the free energy functional has been derived previously using random phase approximation or gradient expansion. It is the starting point to investigate the phase behavior, interfacial property and phase-separation dynamics of binary polymer blends [14, 16, 18, 19].

An approximate solution of the relationship $\omega_\alpha(\vec{r}) = \omega_\alpha(\vec{r}, [\phi_\alpha(\vec{r})])$ can also be obtained at the strong-segregation limit. At this limit the interaction between the different polymers is strong so that the phase separation is approximately complete. In this case the solution of the propagators is dominated by the ground state of the "Hamiltonian" of the chains [20],

$$q_\alpha(\vec{r}, s) \approx q_{\alpha,0}e^{-\varepsilon_{\alpha,0}s}\psi_{\alpha,0}(\vec{r}),$$

where $\psi_{\alpha,0}(\vec{r})$ is the ground-state eigenfunction of the chain Hamiltonian $\mathcal{H}_\alpha = -\sigma_\alpha^2\nabla^2 + \omega_\alpha(\vec{r})$ and $\varepsilon_{\alpha,0}$ is the corresponding eigenenergy, $\mathcal{H}_\alpha\psi_{\alpha,n}(\vec{r}) = \varepsilon_{\alpha,n}\psi_{\alpha,n}(\vec{r})$. Furthermore, $q_{\alpha,0}$ is a normalization constant. The polymer density profile corresponding to this approximate propagator is then given by,

$$\phi_\alpha(\vec{r}) \approx \frac{\bar{\phi}_\alpha}{Q_\alpha} e^{-\varepsilon_{\alpha,0}\kappa_\alpha} \left[q_{\alpha,0}\psi_{\alpha,0}(\vec{r})\right]^2.$$

Using this expression an approximate relationship between the propagator and density is obtained,

$$q_\alpha(\vec{r}, s) \approx e^{-\varepsilon_{\alpha,0}\left(s-\frac{\kappa_\alpha}{2}\right)} \left[\frac{Q_\alpha}{\bar{\phi}_\alpha}\phi_\alpha(\vec{r})\right]^{\frac{1}{2}}.$$

Inserting this expression into the integrals over the propagators leads to,

$$\frac{\bar{\phi}_\alpha}{\kappa_\alpha Q_\alpha} \int_0^{\kappa_\alpha} ds \left\{ q_\alpha(\vec{r}, \kappa_\alpha - s) \frac{\partial}{\partial s} q_\alpha(\vec{r}, s) \right\} = -\varepsilon_{\alpha,0}\phi_\alpha(\vec{r}),$$

$$\frac{\bar{\phi}_\alpha}{\kappa_\alpha Q_\alpha} \int_0^{\kappa_\alpha} ds \left\{ \sigma_\alpha^2 \left[\nabla q_\alpha(\vec{r}, \kappa_\alpha - s)\right]\left[\nabla q_\alpha(\vec{r}, s)\right] \right\} = \frac{\sigma_\alpha^2}{4} \frac{\left[\nabla\phi_\alpha(\vec{r})\right]^2}{\phi_\alpha(\vec{r})}.$$

Combining these results into the expression of the free energy results in the following Ginzburg-Landau free energy of a binary polymer blends at the strong-segregation limit,

$$F_{GL} = \int d\vec{r} \left\{ \chi N \phi_A(\vec{r}) \phi_B(\vec{r}) + \sum_\alpha \left[-\varepsilon_{\alpha,0}\phi_\alpha(\vec{r})\right] + \sum_\alpha \frac{\sigma_\alpha^2}{4} \frac{\left[\nabla\phi_\alpha(\vec{r})\right]^2}{\phi_\alpha(\vec{r})} \right\} - \sum_\alpha \frac{\bar{\phi}_\alpha}{\kappa_\alpha} \ln \frac{eQ_\alpha}{\bar{\phi}_\alpha}.$$

It is useful to observe that the terms linear in the density do not affect the thermodynamic property of the system. They could be removed by choosing proper levels of the energy. Ignoring the linear terms, we obtain,

$$F_{GL} = \int d\vec{r} \left\{ \chi N \phi_A(\vec{r}) \phi_B(\vec{r}) + \sum_\alpha \frac{\sigma_\alpha^2}{4} \frac{\left[\nabla\phi_\alpha(\vec{r})\right]^2}{\phi_\alpha(\vec{r})} \right\}.$$

It is interesting to note that this expression is similar to that derived at the weak-segregation limit. The differences are in the coefficient of the gradient term, $\frac{1}{6}$ versus $\frac{1}{4}$, and the lack of the entropic contribution at the strong-segregation limit.

7 Conclusion and Discussions

In this chapter a detailed derivation of the self-consistent field theory for inhomogeneous polymeric systems is presented using a binary blend of homopolymers as an example. The resulting SCFT equation and free energy functional could be used to study the phase diagram and interfacial structure of polymer blends. Although a binary blend of polymers is used as a model system, it should be emphasized that the theoretical methods for deriving the SCFT equations are very flexible and versatile. Extension of the theoretical framework to more complex many-body systems is straightforward. In particular, the SCFT equations developed here can be easily extended to more complex polymeric systems including multiblock copolymers [11], block copolymer blends and solutions [21], as well as semiflexible polymers [22] and charge polymers [23].

Extension of the SCFT equations to polymers with complex architectures, such as block copolymers and grafted polymers, could be carried out by constructing model polymers with appropriate architectures. In particular, the architecture of a given block copolymer could be described by specifying the constraints on the probability distribution function of the polymer configurations. For example, an AB diblock copolymer is obtained by connecting the ends of an A-polymer and a B-polymer. Thus the configurations of an AB diblock copolymer could be specified by using delta-functions to enforce the connectivity of the chains. Mathematically, the probability of find a particular configuration ($\left\{\vec{R}(s)\right\}$) of an AB diblock copolymer chain is given by,

$$P_{AB}\left(\left\{\vec{R}(s)\right\}\right) = p_0\left[\vec{R}_{A,i}(s)\right] p_0\left[\vec{R}_{B,i}(s)\right] \delta\left[\vec{R}_{A,i}\left(N_A\right) - \vec{R}_{B,i}(N_B)\right].$$

In this expression the delta-function ensures that the two ends of the A- and B-polymers are linked together, thus forming a diblock copolymers. The corresponding single-chain partition function should be modified to include contributions from both blocks,

$$Q_\alpha\left(\{\omega_\alpha\}\right) = \frac{1}{V}\int \mathcal{D}\left\{\vec{R}(s)\right\}\, P_{AB}\left(\left\{\vec{R}(s)\right\}\right)\exp\left[-\sum_{\alpha=A,B}\int_0^{N_\alpha} ds\, \omega_\alpha\left(\vec{R}_\alpha(s)\right)\right].$$

In terms of the propagators, the partition function of an AB diblock copolymer becomes,

$$Q_\alpha\left(\{\omega_\alpha\}\right) = \frac{1}{V}\int d\vec{r}_1 d\vec{r}_2 d\vec{r}_3\, Q_A\left(\vec{r}_1, N_A|\vec{r}_2\right) Q_B(\vec{r}_2, N_B|\vec{r}_3).$$

Similar expressions for more complex block copolymers can be obtained analogously.

The SCFT equations are a set of nonlinear and nonlocal equations. Analytic solutions of the SCFT equations could be obtained only for very simple cases such as the homogeneous phases. The SCFT solution of a homogeneous phase leads to a free energy expression, corresponding to that given by the Flory-Huggins theory for homogeneous polymer blends. The Flory-Huggins free energy, originally obtained using a lattice model, can be used to construct phase diagrams and calculate spinodal lines for polymer blends.

For complex ordered phases, typically formed from polymeric systems containing block copolymers, approximate analytical solutions could be obtained by using a number of approximation methods, such as the weak-segregation theory [17] or strong-segregation theory [24]. These solutions provide useful insight into the self-assembly of ordered polymeric phase. At the same time, sophisticated numerical methods to solve the SCFT equations have been developed over the last few decades [10, 13]. These solutions provide accurate description of the various ordered phases self-assembled from block copolymers [11].

The derivation of the SCFT is based on the Gaussian model of polymer chains, so that the partition function of a single polymer chain in the field can be obtained by solving the modified diffusion equations. On the other hand, the transformation from a particle-based description to a field-based description is a generic procedure, which applies to any statistical mechanical systems. For a generic molecular model, the partition function of the system will be described by the partition function of a single molecule in an external field. The evaluation of this single-particle partition function depends on the particularity of the molecular model. For example, if the molecule can be modeled as a simple particle, the single-particle partition function is given by the usual summation over the Boltzmann factor,

$$Q_\alpha(\{\omega_\alpha\}) = \frac{1}{V}\int d\vec{R}\ \exp\left[-\omega_\alpha\left(\vec{R}\right)\right].$$

On the other hand, if the molecule is more complex, such as a semiflexible chain, the single-molecule partition function has more complex expression.

Acknowledgements This work is supported by the Natural Science and Engineering Research Council (NSERC) of Canada.

References

1. F.S. Bates, G.H. Fredrickson, Phys. Today **52**, 32 (1999)
2. S. Edwards, Proc. Phys. Soc. **85**, 613 (1965)
3. E. Helfand, J. Chem. Phys. **62**, 999 (1975); Macromolecules **8**, 552 (1975)
4. K.M. Hong, J. Noolandi, Macromolecules **14**, 727 (1981)
5. M.D. Whitmore, J.D. Vavasour, Acta Polymerica **46**, 341–360 (1995); F. Schmid, J. Phys. Condens. Matter **10**, 8105–8138 (1998); M.W. Matsen, J. Condens. Matter **14**, R21–R47 (2002)
6. A.C. Shi, in *Development in Block Copolymer Science and Technology*, ed. by I.W. Hamley (Wiley, New York, 2004); M.W. Matsen, in *Soft Matter*, ed. by G. Gomper, M. Schick (Wiley, New York, 2006); G.H. Fredrickson, *The Equilibrium Theory of Inhomogeneous Polymers* (Oxford University Press, Oxford, 2006)
7. E. Helfand, Z.R. Wasserman, Macromolecules **9**, 879 (1976)
8. K.R. Shull, Macromolecules **25**, 2122 (1992)
9. J.D. Vavasour, M.W. Whitmore, Macromolecules **25**, 5477 (1992); M. Banaszak, M.W. Whitmore, Macromolecules **25**, 3406 (1992)
10. M.W. Matsen, M. Schick, Phys. Rev. Lett. **72**, 2041 (1994)
11. W. Xu, K. Jiang, P.W. Zhang, A.C. Shi, J. Phys. Chem. B **117**, 5296–5305 (2013)
12. A.-C. Shi, J. Noolandi, R.C. Desai, Macromolecules **29**, 6487 (1996); A.-C. Shi, J. Phys. Condens. Matter **11**, 10183 (1999)

13. F. Drolet, G.H. Fredrickson, Phys. Rev. Lett. **83**, 4317 (1999); Y. Bohbot-Raviv, Wang, Phys. Rev. Lett. **85**, 3428 (2000)
14. K.T. Delaney, G.H. Fredrickson, J. Phys. Chem. B **120**, 7615–7634 (2016)
15. P.G. de Gennes, J. Chem. Phys. **72**, 4756 (1980)
16. H. Tang, K.F. Freed, J. Chem. Phys. **94**, 1572 (1991)
17. L. Leibler, Macromolecules **13**, 1602 (1980); T. Ohta, K. Kawasaki, Macromolecules **19**, 2621 (1986)
18. J.F. Joanny, L. Leibler, J. Phys. (Paris) **39**, 951 (1978)
19. H. Tang, K.F. Freed, J. Chem. Phys. **94**, 6307 (1991)
20. E. Helfand, Y. Tagami, J. Chem. Phys. **56**, 3592 (1972)
21. Z. Wu, B. Li, Q. Jin, D. Ding, A.-C. Shi, J. Phys. Chem. B **114**, 15789–15798 (2010)
22. W. Song, P. Tang, H. Zhang, Y. Yang, A.-C. Shi, Macromolecules **42**, 6300–6309 (2009)
23. A.-C. Shi, J. Noolandi, Macrol. Theory Simul. **8**, 214–229 (1999)
24. A.N. Semenov, Sov. Phys. - JETP **61**, 733 (1985)

Variational Methods for Biomolecular Modeling

Guo-Wei Wei and Yongcheng Zhou

1 Introduction

Living biological systems require a constantly supply of energy to generate and maintain certain biological orders that keep the systems alive. This warrants the biophysical models that quantify the management and balance of energy in biological systems, i.e., the energy budget of metabolism. Taking cells—the building blocks of life—as an example, energy is derived from the chemical bond energy in food molecules, passed through a sequence of biochemical reactions, and is used in cells to produce activated energy carrier molecules (i.e., ATPs) for powering almost every activity of the cells, including muscle contraction, generation of electricity in nerves, and DNA replication [2]. For solvated biomolecular systems[1] discussed in this chapter, including solvated proteins, bilayer membranes, or their complexes, one can make similar energy budgets too. Various types of energies can be identified for biomolecular systems, such as

1. kinetic energies of atoms or molecules in motion;
2. potential energies for bonded atoms: potential energies characterizing the stretching, bending, torsion of the covalent bonds between atoms;

[1]Water constitutes a large percentage of cellular mass and therefore biomolecules are mostly living in an aqueous environment where various types of ions such as sodium (Na^+), potassium (K^+), calcium (Ca^{2+}), and chloride (Cl^-) present at different concentrations.

G.-W. Wei (✉)
Department of Mathematics, Michigan State University, East Lansing, MI 48824, USA
e-mail: wei@math.msu.edu

Y. Zhou
Department of Mathematics, Colorado State University, Fort Collins, CO 80523, USA
e-mail: yzhou@math.colostate.edu

J. Wu (ed.), *Variational Methods in Molecular Modeling*,
Molecular Modeling and Simulation, DOI 10.1007/978-981-10-2502-0_7

3. potential energies for unbounded atoms: electrostatic energy and van der Waals energy; and
4. kinetic and potential energy interconversions in enzymatic processes and chemical reactions.

The first three energy terms constitute the basis for the molecular dynamics (MD) simulations of non-reactive solvated biomolecular systems. Using the spatial coordinates of individual atoms as parameters, MD simulations trace the motion of each atom by using the Newton second law, where the force applied to each atom is computed as the variational of the total energy with respect to the atom's spatial coordinates [15, 20, 104, 133]. Additional forces that models temperature-dependent thermal fluctuations can be added as well, giving rise to Langevin dynamics simulations [114]. In this regard, MD simulation is indeed a classical application of the variational principle.

The large amount of solvent molecules in a molecular dynamics simulation of solvated biomolecular system can make the simulation daunting and expensive. This deficiency motivates the development of various continuum or multiscale models for part of or the entire solvated biomolecular system [16, 23, 28, 32, 46, 52, 120, 129, 147, 162]. Notably among these simplifications are implicit solvent models, which manage to replace the atomic degrees of freedom of solvent molecules with a continuum description of averaged behavior of solvent molecules while retain an atomistic description of the solute molecule [52, 120]. Accordingly, the solvent-solute interface must be identified as the boundary between the continuum solvent region and the discrete biomolecular domain. This interface is of particular importance because it is related to a range of solvent-solution interactions such as hydrogen bonding, ion-ion, ion-dipole, dipole-dipole and multipole interactions, and Debye attractions [41]. Thus the parametrization of the total energy of the system must include the geometry of this interface. Mean and Gaussian curvatures are generally involved in such parametrization because they measure the variability or non-flatness of a biomolecular surface and characterize respectively the extrinsic and intrinsic measures of the surface [76]. In these multiscale models of solvated biomolecules systems the motion of the atoms still follows the Newton's law where the force is given as the variational of the total energy with respect to the atoms' spatial coordinates, the electrostatic potential, and the interface [58, 59, 137, 147, 160]. The change in the solvent-solute interface induces variation in curvatures, whose energies might be treated as a part of the total energy functional. These curvature based or differential geometry based biomolecular models offer a manifest of mathematical analysis and computational methodologies for the dynamics of the solvent-solute interface and the equilibrium energy landscape of solvated biomolecules. In other words, one can derive dynamic partial differential equations to evolve the interface morphology, and this evolution can be mapped to the path toward the global or local minimum on the landscape of the total energy. Here in this chapter we shall present three representative applications of interface geometry based variational principles to the modeling of biomolecular interactions: (i) biomolecular electrostatics and solvation, (ii) surface microdomain formation in bilayer membranes, and (iii) curvature driven protein localization in bilayer membranes.

In the first application we consider the long-range electrostatic interactions among partially charged static atoms in the solute and the aqueous solvent with mobile ions. These interactions strongly depend on the position of solvent-solute boundary, also referred to as the molecular surface in this context, where a rapid transition of dielectric permittivity is observed. Inclusion of this interface, albeit implicitly, in the formulation of the total energy of the system facilitates the coupling of polar and nonpolar solvent-solute interactions, as well as the nonlinear solvent response, in the form of interface energy functional of surface curvature energy, electrostatic energy and van der Waals potential. Such a coupling finally gives rise to a novel variational multiscale solvation model [26, 27, 46, 47, 147]. In a more elaborated model, the solute molecule can be described in further detail by using the quantum density functional theory (DFT) in an iterative manner, which allows a more accurate account of solvent-solute interaction and response [25]. Differential geometry based solvation models have been shown to deliver superb predictions of solvation free energies for hundreds of molecules [28, 138]. This variational principle based solvation model can be further extended to describe essential biological transportation such as transmembrane ion or proton flows that depend critically on the geometry of the associated protein channels. By including the chemical potential and entropy of the diffusive ion species into the total energy functional one can obtain simultaneously the optimized channel protein surfaces as well as the corresponding I–V (current-voltage) curve [28, 149, 158].

Curvature is believed to play an important role in many biological processes, such as protein-DNA and protein-membrane interactions, including membrane curvature sensing. Classical phase field modeling of surface pattern formation in bilayer membranes contains a curvature term in its definition of the total energy [18, 24, 42, 44, 56, 102]. However, when modeling the surface pattern formation in our second application here, we show that it is the geodesic curvature rather than the curvature of pattern interfaces that plays an essential role in modulating the interface energy. Noting that this geodesic curvature is defined on a general differentiable manifold, and thus the classical phase field modeling of phase separation with specified intrinsic curvature can be regarded as a special case of this geodesic curvature model in the Euclidean spaces. By providing various intrinsic geodesic curvatures that model the geometry of the contact of different species of lipids, we are able to simulate the generation of lipid rafts as the formation and equalization of localized surface domains.

In contrast to most amphiphilic lipids whose relatively long and geometrically regular hydrophobic tails allow they to pack together, membrane proteins usually do not present in large distinct domains in membrane surfaces, although small amount of membrane proteins can compound together forming functional complexes such as ion channels or membrane transporters. Most membrane proteins have amphipathic transmembrane helices, which contain both hydrophobic and hydrophilic groups, complementing to amphiphilic lipids. Therefore, the localization of these membrane proteins in general can not be modeled using the geodesic curvature based phase separation model as described in our second application. Many membrane proteins, however, do prefer bilayer membranes with particular curvature, in the sense that they can induce particular curvature in the bilayer membrane and they tend to be localized in regions with specific curvature. Therefore, one can imagine that membrane curvature can provide a driving force for the distribution of membrane proteins in the bilayer, and thus an appropriate energy functional that represents the membrane curvature must be added to the classical electrochemical potential and entropy to describe the localization of membrane proteins.

These three applications of variational principles in biomolecular modeling are by no means exhaustive, even in the context of solvation analysis and membrane-protein interactions. There are inspiring studies of ion and water transport in membrane channels using energetic variational approaches, where the effects of surface charge density and non-uniform particle sizes can be readily included in investigations thanks to the flexibility of variational approaches [67, 69–71, 83, 89, 147, 149, 153]. Similar flexibility also enables the extension of the application of variational principles from the standard phase field modeling of bilayer membrane deformation and morphology [42, 44, 45] to multi-components membranes [86, 157], pore formation [35, 113], and double layer [38, 57]. Some of these models, particular those for bilayer membranes, share various degree of similarity to the models used for self-assembly or phase separation of polymers or co-polymers. It is this wide diversity of lipid structures and the complicated interactions between proteins and lipid bilayers in solution that makes the energetic variational modeling of bilayer membranes unique and challenging. As we shall present below, most of our efforts are concentrated on the formulation of potential energy functional of these interactions so that the variational principle can be applied and numerical solutions can be found by solving the corresponding systems of nonlinear partial differential equations (PDEs).

2 Variational Multiscale Methods for Biomolecular Electrostatics and Solvation

By definition, the solvation energy of biomolecules is the cost of free energy required to transfer the biomolecules from the vacuum to the solvent environment. It is therefore an essential quantitative characterization of the solute-solvent interactions. Electrostatic free energy, also called polar solvation free energy, is an important component of the solvation free energy since most biomolecules are charged and there are always mobile ions in the solvent under physiological conditions. Various critical applications of the electrostatic and solvation free energies can be found in chemistry, biophysics, and medicine. We refer the reader to [31, 40, 48, 52, 73, 79, 91, 92, 109, 117, 137, 142–144] for theoretical underpinning of these applications and the determination of the electrostatics and solvation free energies. Apart from electrostatic effects, the solvation free energy also involves the nonpolar energy, namely, the energy cost for creating a suitable cavity in the continuum solvent to allow the transferring of the biomolecules and for the dispersive interactions between the solvent and the biomolecule on the surface of this cavity. Implicit solvent models are particularly appearing for computing the solvation free energy since the number of solvent degrees of freedom can be dramatically reduced by a well fitted bulk dielectric permittivity while the atomistic representations of solute biomolecules can be retained to maintain a detailed modeling of the solute. The framework of implicit solvent models allows the solvation free energy to be decomposed into two components, polar solvation and nonpolar solvation [79, 81, 137]. In this approach, the electrostatic contribution can be readily computed from the solution of the Poisson-Boltzmann equation, or the Poisson equation if there is no explicit ion in the solvent [6, 7, 61, 63, 88, 101, 136]. The solution of these equation depends on the contrast of dielectric permittivity in vacuum and the solvent environments, and this contrast is concentrated at the boundary between the biomolecule and the solvent. Likewise, the calculation of nonpolar solvation free energy depends on the geometry of the biomolecular surface. The fact that both polar and nonpolar components are determined by the solvent-solute interface warrants the importance of a biophysically justifiable, mathematically well-posed, and computational feasible definition of the molecular surface or dielectric interface. In fact, the decoupling of polar and nonpolar components makes implicit solvent models conceptually convenient and computationally simple.

However, there are many structural imperfections associated with implicit solvent models. First, intrinsic thermodynamical and kinetic coupling makes it impossible to completely separate the electrostatic component from the non-electrostatic components in the solvation modeling. Additionally, a pre-prescribed solvent-solute interface, such as solvent excluded surface and van der Waals surface, decouples polar and nonpolar components. As a result, the solvation induced solute polarization and solvent response are not appropriately accounted in implicit solvent models. Moreover, implicit solvent models neglect potential solvation induced surface reconstruction and possible conformational changes. Finally, thermodynamically, the change in the Gibbs free energy of solvation can be formally decomposed into the change in internal energy, work, and entropy effect. There is no guarantee that all of these components are fully accounted in implicit solvent models. In addition to the aforementioned structural or organizational imperfections, the performance of implicit solvent models is subject to a wide range of implementation deficiencies, such as the modeling of nonpolar component, the treatment of the electrostatic component, the exclusion of high-order polarization, the exclusion of curvature, the geometric singularity of solvent-solute interface, the stability of numerical schemes and algorithms, the grid convergence of the solvation free energy, to mention only a few.

Some of the aforementioned problems have been the subjects of intensive study in the past few decades. One approach starts from improving the surface definitions, so that earlier van der Waals surface, solvent accessible surface [77], and molecular surface (MS) [111] are replaced by smooth surface expressions [22, 60–62, 159]. Geometric analysis, which combines differential geometry (DG) and differential equations, is a powerful mathematical tool for signal and image processing, data analysis, and surface construction [100, 139–141, 145]. Geometric PDEs and DG theories of surfaces provide a natural and simple description for a solvent-solute interface. The first curvature-controlled PDEs for molecular surface construction and solvation analysis was introduced in 2005 [146]. A variational solvent-solute interface, namely a minimal molecular surface (MMS), was proposed for molecular surface generation in 2006 [9, 10]. In this work, the minimization of surface free energy is equivalent to the minimization of surface area, which can be implemented via the mean curvature flow, or the Laplace-Beltrami flow, and gives rise to the MMS. The MMS approach has been used in implicit solvent models [10, 28]. Potential-driven geometric flows, which admit potential driven terms, have also been proposed for biomolecular surface construction [8]. This approach was adopted by many researchers [21, 29, 30, 154–156] for biomolecular surface identification and electrostatics/solvation modeling.

It is natural to extend DG based variational theory of the solvent-solute interface into a full solvation model by incorporating a variational formulation of the PB theory [28, 59, 116, 147] following the spirit of a similar approach by McCammon and coworkers [46, 47]. However, the formalism of McCammon and coworkers does not involve geometric flow and has a Gaussian curvature term that might lead to jumps in the energy when there are topological changes. Our DG based variational model addresses many of the aforementioned imperfections of implicit solvent models. For example, by parametrizing both polar and nonpolar components of the solvation energy using the geometry of the interface, these two components can be coupled naturally in a single free energy functional. Application of the variational principle and the equilibrium solution of the associated Laplace-Beltrami flow gives rise to an optimal biomolecular surface along with an optimized solvation energy.

2.1 Polar Solvation Free Energy

We start with the definition of polar solvation energy, which is associated with the energy difference for charging biomolecules in the vacuum and the solvent environment. Variational formulation of Poisson-Boltzmann equation was discussed in the literature [59, 116]. Here we recast this formulation in our DG based formalism. Considering a solvated biomolecular system occupying a three-dimensional (3D) domain $\Omega \in \mathbb{R}^3$, one can relate the polar solvation energy of the biomolecule to the electrostatic potential $\Phi(\mathbf{r}) : \mathbb{R}^3 \to \mathbb{R}$ by the formulation [27, 147]

$$G_p = \int_\Omega \left\{ S\left[\rho_m \Phi - \frac{1}{2}\varepsilon_m |\nabla\Phi|^2\right] - (1-S)\left[\frac{1}{2}\varepsilon_s|\nabla\Phi|^2 + k_B T \sum_{i=1}^{N_c} c_i \left(e^{-q_i\Phi/K_BT} - 1\right)\right]\right\} d\mathbf{r}, \tag{1}$$

where $S(\mathbf{r})$ and $1 - S(\mathbf{r})$ are respectively the domain indicators for the solute and the solvent domains. We set $0 \le S(\mathbf{r}) \le 1$, which is related to the widely used phase-field function $|\bar{\phi}(\mathbf{r})| \le 1$ by

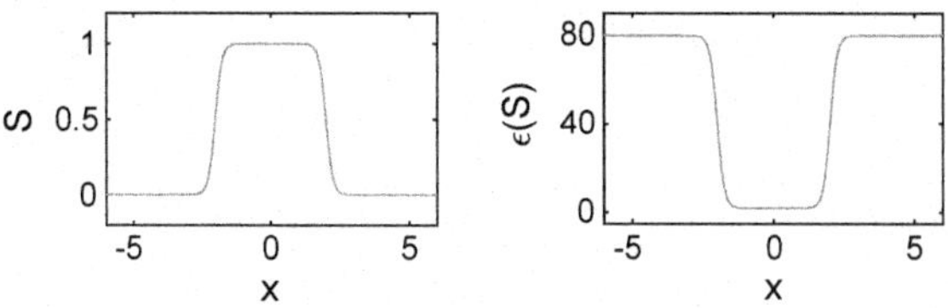

Fig. 1 *Left* A typical phase field function S changes smoothly from its value of -1 in the solvent domain to the value of 1 in the solute domain. *Right* The dielectric constant $\varepsilon(S)$ depends on the phase field function and changes smoothly from a value of 78 (or 80) in the solvent domain to a value of 2 (or 1) in the solute domain

$$S = \frac{1+\bar{\phi}}{2}, \quad 1 - S = \frac{1-\bar{\phi}}{2}. \tag{2}$$

Here S and $1 - S$ are introduced to distinguish the contributions to the total free energy from the solute region Ω_m and solvent region Ω_s. The dielectric permittivity in these two complementary subdomains of Ω are given by ε_m and ε_s, respectively. The fixed charge density ρ_m of biomolecule consists of a summation of partial charges (Q_j) from atoms

$$\rho_m(\mathbf{r}) = \sum_j Q_j \delta(\mathbf{r} - \mathbf{r}_j), \tag{3}$$

where $\mathbf{r}_j \in \mathbb{R}^3$ is the position of jth charged atom. In Eq. (1), q_i and c_i are respectively the charge and bulk concentration of the ith ion species, N_c is the number of ions species in the solvent, k_B is the Boltzmann constant, and T is the temperature.

The surface function $S(\mathbf{r})$ can be chosen initially as a smooth function to simplify the numerical implementation, as seen in the left chart of Fig. 1. We show below the classical Poisson-Boltzmann equation can be reproduced by using this energy functional when a sharp solvent-solute interface is adopted, i.e., when S becomes a Heaviside function. In the sequel we shall work on a generalized Poisson-Boltzmann equation in the sense that the transition from the solvent region to the solute region is smooth rather than discontinuous.

2.2 *Nonpolar Solvation Free Energy*

The nonpolar solvation energy involves a number of terms. The scaled-particle theory (SPT) for nonpolar solutes in aqueous solutions [105, 124] utilize a solvent-accessible surface area term [95, 127]. Solvent-accessible volume was shown to be relevant in large length scale regimes [68, 90]. It was pointed out that van der Waals (vdW) interactions near solvent-solute interface are important as well [33, 54, 55, 137]. Dzubiella et al. convert these terms into a nonpolar energy functional, which, however involves a Gaussian curvature term [46]. We modify this functional in spirit of our MMS [9, 10] to give the following nonpolar term [27, 147]

$$G_{np} = \gamma A_m + pV_m + \rho_0 \int_{\Omega_s} U^{\mathrm{att}} d\mathbf{r}. \tag{4}$$

Here the first term on the right is the surface energy given by the surface tension γ and the biomolecule's surface area A_m. This term measures the disruption of inter- and intra-molecular noncovalent bonds of solvent molecules when an internal surface is created. In our approach, the surface tension γ does not depend on Gaussian curvature so that the first term in Eq. (4) avoids possible energy jumps suggested by the Gauss-Bonnet theorem. Additionally, such a term follows our minimum surface energy functional formulation [9, 10]. The second term represents the mechanical work for expanding a volume of V_m in solvent against a hydrostatic pressure p. The last term quantifies the attractive dispersion effects near the solvent-solute interface, determined by the solvent bulk density ρ_0 and the attractive portion of the van der Waals potential U^{att} at position $\mathbf{r}$. Since the biomolecular surface is not explicitly known in the present modeling, we relate the surface area and its enclosed volume to the surface function S through

$$V_m = \int_{\Omega_m} d\mathbf{r} = \int_{\Omega} S d\mathbf{r} \tag{5}$$

and the coarea formula [147, 150]

$$A_m = \int_{\Omega} |\nabla S| d\mathbf{r}. \tag{6}$$

With these relations we can assemble the polar and nonpolar contributions to give the formulation of the total solvation free energy functional for biomolecules at equilibrium [27, 147]

$$\begin{aligned} G_{\mathrm{tot}} = \int_{\Omega} \Bigg\{ & \gamma|\nabla S| + pS + (1-S)\rho_0 U^{\mathrm{att}} + S\left[(\rho_m \Phi) - \frac{1}{2}\varepsilon_m |\nabla \Phi|^2 \right] + \\ & (1-S)\left[-\frac{1}{2}\varepsilon_s |\nabla \Phi|^2 - k_B T \sum_{i=1}^{N_c} c_i (e^{-q_i \Phi / K_B T} - 1) \right] \Bigg\} d\mathbf{r}. \end{aligned} \tag{7}$$

There are a variety of definitions of nonpolar free energies alternative to that in Eq. (4), but most of them are determined by the surface area, its enclosed volume and ver der Waals term in a similar way [79, 81, 137]. The present formulation and the variational principle introduced here are applicable to these alternative nonpolar solvation models as well.

2.3 Governing Equations

We search for the critical point of the free energy functional to obtain the optimal free energy of the biomolecular systems. By construction, the free energy functional is determined by the surface function S and the potential Φ. The latter indeed depends on the position of dielectric interface hence on the surface function S as well. Since the electrostatic potential follows the Poisson equation, it is theoretically possible to replace the electrostatic potential using the convolution of the Green's function with the change density. However, the dependence of this Green's function on the surface function S does not have an explicit representation. Consequently, it is practically impossible to represent the total energy as the functional of the surface function only and compute its variation. In our investigations we shall compute the critical point by evolving the gradient flow of the free energy functional to a steady state; while the electrostatic potential defined by the vanishing variation $\frac{\delta G_{\text{tot}}}{\delta \Phi}$ is used as a constraint during the evolution. These two variations are

$$\frac{\delta G_{\text{tot}}}{\delta \Phi} = S\rho_m + \nabla \cdot ((1-S)\varepsilon_s + S\varepsilon_m)\nabla\Phi) + (1-S)\sum_{i=1}^{N_c} c_i q_i e^{-q_i\Phi/K_B T}, \quad (8)$$

$$\frac{\delta G_{\text{tot}}}{\delta S} = -\nabla \cdot \left(\gamma \frac{\nabla S}{|\nabla S|}\right) + p - \rho_0 U^{\text{att}} + \rho_m \Phi + \frac{1}{2}(\varepsilon_s - \varepsilon_m)|\nabla\Phi|^2$$
$$+ k_B T \sum_{i=1}^{N_c} c_i (e^{-q_i\Phi/K_B T} - 1). \quad (9)$$

The vanishing variation in Eq. (8) gives rise to a generalized Poisson-Boltzmann equation (GPBE) [27, 147]

$$-\nabla \cdot (\varepsilon(S)\nabla\Phi) = S\rho_m + (1-S)\sum_{i=1}^{N_c} c_i q_i e^{-q_i\Phi/K_B T}. \quad (10)$$

where the dielectric function

$$\varepsilon(S) = (1-S)\varepsilon_s + S\varepsilon_m, \quad (11)$$

is also plotted in the right chart in Fig. 1. The gradient flow for the surface function S follows a generalized Laplace-Beltrami equation [27, 147]

$$\frac{\partial S}{\partial t} = -|\nabla S| \frac{\delta G_{\text{tot}}}{\delta S} = |\nabla S| \left[\nabla \cdot \left(\gamma \frac{\nabla S}{|\nabla S|}\right) + V\right], \quad (12)$$

where a generalized potential function V collects the relevant terms in Eq. (9) as

$$V = -p + \rho_0 U^{\mathrm{att}} - \rho_m \Phi + \frac{1}{2}(\varepsilon_m - \varepsilon_s)|\nabla \Phi|^2 - k_B T \sum_{i=1}^{N_c} c_i (e^{-q_i \Phi / K_B T} - 1), \tag{13}$$

and $|\nabla S|$ is added to the front of the variation to introduce the local curvature of the molecular surface to adjust the rate at which the surface function evolves toward its steady configuration. In this sense Eq. (12) is a generalized geometric flow equation. Note that the time in Eq. (12) is artificial.

We expect that the GPBE with smooth S converges to its sharp interface limit when S becomes a Heaviside function with a discontinuity located at the dielectric interface Γ, in that case the GPBE can be written as the following two elliptic equations

$$-\varepsilon_m \nabla^2 \Phi_m = \rho_m, \qquad \mathbf{r} \in \Omega_m, \tag{14}$$

$$-\varepsilon_s \nabla^2 \Phi_s = \sum_{i=1}^{N_c} c_i q_i e^{-q_i \Phi_s / K_B T}, \ \mathbf{r} \in \Omega_s. \tag{15}$$

These two equations are coupled through the interface conditions on Γ. In this case, to make the above two equations well posed, one has to introduce two interface jump conditions,

$$\Phi_s = \Phi_m, \quad \varepsilon_m \nabla \Phi_m \cdot \bar{\mathbf{n}} = \varepsilon_s \nabla \Phi_s \cdot \bar{\mathbf{n}}, \quad \mathbf{r} \in \Gamma \tag{16}$$

where Φ_m, Φ_s are the limit values of the electrostatic potential from solution domains Ω_m and Ω_s, respectively, and $\bar{\mathbf{n}}(\mathbf{r})$ is the unit normal vector on Γ.

2.4 Computational Simulations and Summary

A second-order finite difference scheme was designed to solve the coupled generalized Poisson-Boltzmann equation (10) and the Laplace-Beltrami equation (12). Most of physical parameters involved in Eq. (12) are taken from the references [81, 99] and the CHARMM force field. A constant surface tension γ is chosen in our investigation whose value shall vary for different molecular surfaces [81, 99]. In particular, γ is implemented as a fitting parameter so that the optimized solvation free energy ΔG from our computational studies can match the experimental measurements. By definition,

$$\Delta G = G_{\mathrm{tot}} - G_0, \tag{17}$$

where G_{tot} is defined in Eq. (7) and G_0 is the total energy of the solvent molecules in vacuum with $\varepsilon_s = \varepsilon_m = 1$ and without nonpolar energy. To facilitate the fitting of γ we rewrite Eq. (12) as

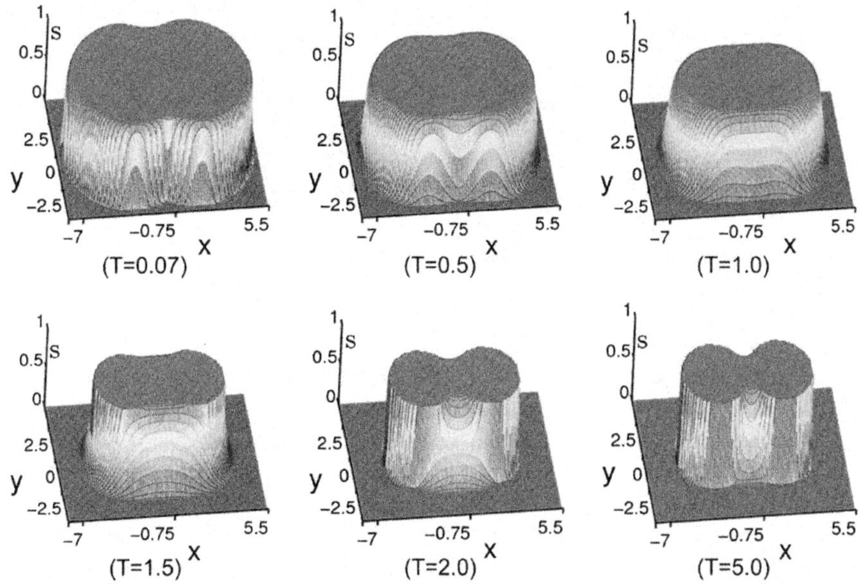

Fig. 2 The phase field function evolves from its initial configuration to the final state where the surface $S = 0.0$ fits the molecular surface for a diatomic system. Here we show only the profiles of S at the cross section $(x, y, 0.05)$ sampled at six moments during the evolution

$$\frac{\partial S}{\partial t} = \gamma |\nabla S| \left[\nabla \cdot \left(\frac{\nabla S}{|\nabla S|} \right) + \frac{V}{\gamma} \right]. \tag{18}$$

More details of the numerical methods for solving the coupled partial differential equations can be found in [27]. In Fig. 2 we show a simulation where the initial surface function is set such that the target diatomic system is well contained in the region $S = 1$. The surface function evolves from the initial profile toward the final configuration that fits the molecular surface of a diatomic system, reaching a state where the total solvation energy is optimized. A more realistic simulation on the protein (PDB ID: 1frd) is shown in Fig. 3, where isosurfaces defined by different S are plotted along with the electrostatic potential Φ on the surface. While $S = \frac{1}{2}$ is usually chosen as the molecular surface, the three surfaces are very close due to the high resolution of the numerical method. The availability of the surface position and surface potential could significantly facilitate the analysis of binding affinity of protein-protein or protein-ligand systems, of which the electrostatic potential is an important component [5, 34, 63, 87, 108, 119].

Numerically, this model can be computed by using both the Eulerian formulation, in which the solute boundary is embedded in the 3D Euclidean space so evaluation of the electrostatic potential can be carried out directly [27], and the Lagrangian formulation, wherein the solvent-solute interface is extracted as a sharp surface and subsequently used in solving the GPB equation for the electrostatic potential [26].

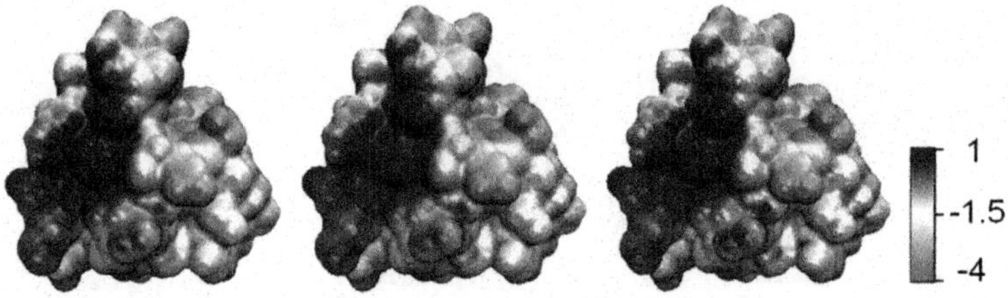

Fig. 3 Electrostatic potential on molecular surfaces with different values of S. *Left* $S = 0.25$; *Middle* $S = 0.5$; *Right* $S = 0.75$

Lagrangian formulation requires direct tracking of the sampling points on the molecular surface, which is convenient for the surface visualization, the mapping of the surface electrostatic potential field, and the enforcement of the van der Waals radii in constraint. However, it suffers from the development of singularities while evolving molecular surface and the difficulty of handling the change of topology. In contrast, the Eulerian representation gets around of the explicit tracking of sampling points by modeling the solvent-solute interface either a smooth 3D density profile or as a specific level set of the smooth profile. The dynamics of the solvent-solute interface can be obtained by evolving this 3D density profile following the Laplace-Beltrami flow of the energy functional. The Eulerian representation is therefore capable of reproducing complicated dynamics of surface topology. As we shall introduce below, it also greatly facilitates the computation of a number of geometric quantities that are otherwise difficult to compute in the Lagrangian representation, such as the area of entire surface and surface enclosed volume.

The parametrization of solvation energy using the surface function S allows one to track the molecular surface by following the isosurface $S = 0.5$ during the evolution of S. This formulation is referred to the Eulerian formulation. Alternatively, one can explicitly define a molecular surface Γ to separate the solvent and solute domains, and to use this surface to parametrize the solvation energy. Denote such an energy functional as $G_{\text{tot}}(\Gamma)$. Similar to the optimization procedure presented above, the total energy is optimized by evolving Γ following the gradient flow of the energy, and in this case, the energy variation is with respect to the spatial coordinates of this explicitly defined surface Γ. Numerically, this can be achieved by discretizing Γ into a collection of surface elements or surface vectors $\{\hat{S}_j\}$, each element parametrized by a local coordinate system (x_1, x_2), and thus $G_{\text{tot}}(\Gamma)$ becomes $G_{\text{tot}}(\hat{S}_j)$. Furthermore, we can constrain the motion of Γ to the normal direction $\mathbf{n}(x_1, x_2)$ only, for that a tangential displacement of Γ does not change the surface configuration and the solvation energy. A scalar displacement field $\psi(x_1, x_2)$ in the normal direction can be defined through

$$\hat{S}_j^{\sigma}(x_1, x_2) = \hat{S}(x_1, x_2) + \sigma\psi(x_1, x_2)\mathbf{n}(x_1, x_2), \tag{19}$$

which states that the surface element $\hat{S}_j$ is updated from its original position by $\sigma\psi(x_1, x_2)$ along the normal direction to the new position $\hat{S}_j^\sigma$, where σ is a number to scale the normal displacement field $\psi(x_1, x_2)$. The optimization of the total energy at a particular molecular surface Γ means that any normal displacement will violate the nature of optimum at this point, indicating

$$\left.\frac{\partial \hat{S}_j^\sigma}{\partial \sigma}\right|_{\sigma=0} = 0. \tag{20}$$

Now we can observe the transition of the independent variables in calculating the energy variation:

$$\frac{\delta}{\delta \Gamma} \to \frac{\partial}{\partial \hat{S}_j^\sigma} \to \frac{\partial}{\partial \sigma}, \tag{21}$$

as a result of replacing the motion of the explicit surface Γ using the scaled normal motion of a collection of surface elements. The readers are referred to [26] for the detailed calculation of the energy variation, the derivation of the equation governing the gradient flow, and the numerical techniques for solving the equation. This investigation also shows that the optimized solvation energy and molecular surface are well matching those generated by the Eulerian formulation if there is no topological change in Γ during its evolution. Notice that a single point on $\hat{S}_j$ may evolves to two distinct points, or two distinct points in two different surface elements may converge to a single point when there is a topological change during the evolution of Γ. This intrinsic singularity in handling the topological change limits the applications of the Lagrangian formulation to complex biomolecular systems, for which it is impossible to set an initial surface Γ that is topologically equivalent to the final optimized molecular surface. The Eulerian formulation is hence suggested for the investigations of the solvation energy and molecular surfaces of general biomolecular systems.

Recently, differential geometry based implicit solvent model has been tested extensively via solvation analysis [26–28, 39, 132, 138]. The differential geometry based nonpolar model was found to deliver some of the best nonpolar solvation predictions [28]. However, for general molecules with a significant polar component, our initial predictions were not up to the state of the art [26, 27]. It turns out that both the generalized Laplace-Beltrami equation and the generalized Poisson-Boltzmann equation can be easily solved individually. However, when these equations are coupled, there is a stability problem [155, 156]. Essentially, when S admits unphysical values beyond its physical definition $0 \le S \le 1$, the dielectric function (11) will adopt unphysical (negative) values as well, which gives rise to an instability in updating the Laplace-Beltrami equation (12). This issue hinders the performance of DG based solvation models for molecules with significant polar component. To address this problem, a convex optimization algorithm [138] has been developed to ensure the stability in solving coupled PDEs (10) and (12). As a result, the differential geometry based solvation model is found to deliver some of the most accurate prediction of experimental solvation free energies for more than 100 molecules of both polar and nonpolar types [138].

Most recently, Wei and coworkers have taken a different treatment of non-electrostatic interactions between the solvent and solute in the DG based solvation models so that the resulting total energy functional and PB equations are consistent with more detailed descriptions of solvent densities at equilibrium [148, 149]. To account for solute response to solvent polarization, a quantum mechanical (QM) treatment of solute charges was introduced to the DG-based solvation models using the Kohn-Sham density functional theory (DFT) [25]. This multiscale approach self-consistently computes the solute charge density distribution which simultaneously minimizes both the DFT energy as well as the solvation energy contributions.

Currently, efforts are invested to improve the accuracy and robustness of DG based solvation models by combining physical models with knowledge based models, namely, machine learning approaches. Additionally, DG based solvation models and machine learning approaches are utilized for accurate predictions of the protein binding energies and ligand binding affinities over a wide range of conformational states. Furthermore, it is worth noting that the method depends only on the representation of the solvent-solute interfaces, and this representation is independent of the atomic or coarse-grained description of the biomolecules. It is therefore possible to adopt this method to compute the potential of mean force of coarse-grained biomolecular structures along selected coordinate, and the results can be utilized for parametrization the force field for coarse-grained molecular systems as well. Finally, we would like to point out that many critical applications to biophysics, chemistry, and medicine mostly remain unexplored.

3 Variational Methods for Pattern Formation in Bilayer Membranes

As one of the most important biomolecular systems, the lipid bilayer membranes sustain the regular functions of cell and subcelluar compartments by regulating the transmembrane ion or molecular flows and by providing platforms for various essential biochemical processes [2, 123]. These critical functions of bilayer membranes are determined by their lipid compositions, the specific membrane proteins, and their dynamical arrangement in the bilayers during the course of membrane morphology change as a result of various membrane-solvent, membrane-membrane, or membrane-protein interactions. Applications of the variational principle for bilayer membrane modeling have been mostly focused on four types of problems: (i) mean-curvature dependent membrane morphology [37, 42, 45, 96], (ii) ionic or proton flows in protein channels [153, 158], (iii) lateral diffusion on membrane surfaces [161], and (iv) pattern formation in bilayer membranes [17, 43, 151]. Here in this section we focus on the local pattern formation in bilayer membranes, for that there are many controversial investigations concerning the biophysical underpinning of these patterns, their spatial and temporal distributions, and their roles in modulating relevant biochemical processes [3, 106, 128, 134]. These patterns are called lipid rafts, which are small (10–200 nm), heterogeneous, highly dynamic, sterol- and sphingolipid-enriched domains that compartmentalize cellular processes [118]. Lipids move laterally within the domains mostly rather than over the entire membrane surface [4]. Classical phase separation models manage to minimize the total area of the domain boundaries and large domains appear at the end of the minimization; this process is usually referred to as coarsening. When these classical models are directly extended to model surface phase separation, the total arc length of the domain boundaries on the surface is minimized to generate large domains, which do not match the measured sizes of lipid rafts [17, 43, 151].

3.1 Classical Phase Field Models

We first examine the classical phase separation model for binary systems. Consider two species of particles in $\mathbb{R}^3$ with respective mass or volume fractions $m_1, m_2 \in [0, 1]$. The interactions between particles of the same species are favorable while the interactions between different species are unfavorable. This preference can be modeled by defining a phase field function

$$\bar{\phi} = \frac{m_1 - m_2}{m_1 + m_2}, \tag{22}$$

where $\bar{\phi}(\mathbf{r}) \in [-1, 1]$, $\mathbf{r} \in \mathbb{R}^3$ and minimizing the Ginzburg-Landau free energy functional in $\Omega \in \mathbb{R}^3$

$$G(\bar{\phi}) = \int_{\Omega} \left(f(\bar{\phi}) + \frac{\sigma}{2} |\nabla \bar{\phi}|^2 \right) d\mathbf{r}, \tag{23}$$

where $f(\bar{\phi})$ is a double well potential that has two minimums at $\bar{\phi} = \pm 1$. A typical choice is

$$f(\bar{\phi}) = \frac{\bar{\phi}^4}{4} - \frac{\bar{\phi}^2}{2} \tag{24}$$

which has two symmetric potential wells of the same depth at $\bar{\phi} = \pm 1$. It is apparent that a complete phase separation with $\bar{\phi}$ changing discontinuously between 1 and -1 is favorable by $f(\bar{\phi})$ when $G(\bar{\phi})$ is minimized. Such an unphysical distribution of $\bar{\phi}$ is to be penalized by the term $\frac{\sigma}{2}|\nabla \bar{\phi}|^2$ that regulates the transitional gradient of $\bar{\phi}$ between 1 and -1.

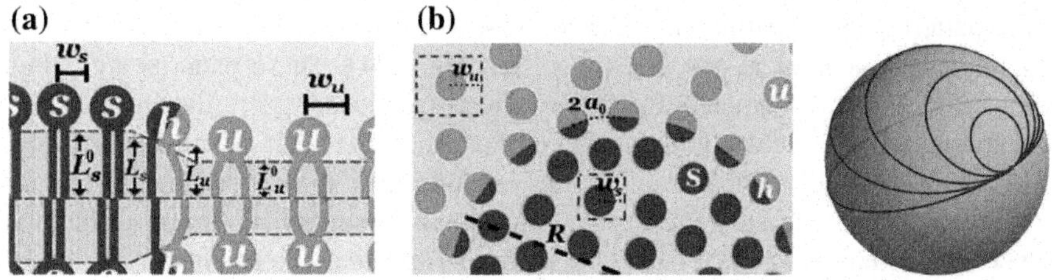

Fig. 4 *Left* Schematic illustration of the mismatch of the lipid structures at the interface that induces a transitional hybrid region between two lipid domains [14]. *Middle* Within the transitional hybrid layer the otherwise regular lattices of the lipids in either domain relax to match each other, causing a bending interface [14]. *Right* Circles on a sphere have constant geodesic curvatures. The great circle, i.e., the lowest circle, has a vanishing geodesic curvature in particular

3.2 Geodesic Curvature Based Membrane Models

3.2.1 Lagrangian Formulation

Our variational model is motivated by the recent theoretical studies of the hybrid lipids saturation at the interface between saturated and unsaturated of lipids with geometrical and molecular mechanical mismatch [14]. As illustrated in Fig. 4, two species of lipids at their interface have different intermolecular interactions that are determined by their structures. Otherwise, the regular lattice of either species of lipids has to be relaxed in a way such that the intermolecular interactions in the transitional region near the interface will fit the different lattice structure of other species. This relaxation generates curved interface between two species of lipids in a manner similar to the generation of surface tension. Since the domain boundary is a curve on a two-dimensional (2D) surface embedded in $\mathbb{R}^3$, it is the geodesic curvature of the interface rather than the interface curvature that determines the intermolecular interactions between two species of lipids near the interface.[2] The geodesic curvature of the interface measures how far the interface curve is from being a geodesic. We define the curvature energy of the microdomain boundary by a one-dimensional (1D) on-curve integration

$$G = \int_C k(H - H_0)^2 ds, \tag{25}$$

where C is the domain boundary contour embedded in $\mathbb{R}^3$, H is the geodesic curvature, H_0 is the spontaneous geodesic curvature of the lipid mixture to be separated,

[2] In Sect. 2, we use S to denote the surface function, which is a domain indicator, and use Φ to denote the electrostatic potential following the traditional usage in the studies of biomolecular electrostatics. Here in Sects. 3 and 4 the models do not involve electrostatics, and we denote ϕ the phase field function, while use S to denote the 2D surface embedded in $\mathbb{R}^3$ when applicable. An interface in Sect. 2 refers to solvent-solute boundary region, whereas in Sects. 3 and 4, it refers a boundary curve on a given surface.

and k is the geodesic curvature energy coefficient. The spontaneous geodesic curvature H_0 is an intrinsic property of the combination of any two species of lipids in the bilayer membrane that will be separated to form local microdomains as a result of geometric and molecular mechanical mismatch. In the transitional region near the interface two species of lipids are arranged in a hybrid state rather than the regular lattice structure. Indeed a recent theoretical study adopted a free energy for the hybrid packing of two species of lipids (denoted by the subscript 1 and 2 below) at the interface [13, 14]:

$$\mathscr{F} = k_s(L_1 - L_1^0)^2 + k_u(L_2 - L_2^0)^2 + \gamma(L_1 - L_2)^2, \tag{26}$$

where L_i is the length of the lipid chains in the transitional region and L_i^0 is the length of the equilibrium chain in the bulk. Parameters k_s and k_u are the free energetic costs of mismatch between two species and their hybrids at the interface, respectively and similarly, γ is the energy cost of mismatch between two chains of the hybrid. Furthermore, the following relations are identified to related the domain curvature and lipid geometrical properties:

$$V_i = L_i a_0 w_i \left(1 \pm \frac{w_i H}{2}\right), \quad i = 1, 2, \tag{27}$$

where V_i is the molecular volume of the lipid chains, w_i is the length that characterizes the molecular spacing of the lipid head groups, and $a_0 = (w_1 + w_2)/2$ is the headgroup spacing of the hybrids along the interface. Here the subtraction sign is chosen if the species is included in the microdomain, otherwise the addition sign is used. The chain length in the equilibrium bulk state, L_i^0, can be computed from the molecular volume divided by the head group area in the bulk state

$$L_i^0 = \frac{V_i}{w_i^2}. \tag{28}$$

Equations (26–27) represent the interface bending energy $\mathscr{F}$ as a function of it geodesic curvature H. The minimizer H_0 can be analytically calculated to the linear order:

$$H_0 = \frac{1}{w_T}\left[\frac{(1-2B)w_d}{(1+2B)w_T} + \frac{2BV_d}{(1+2B)V_T}\right], \tag{29}$$

where B is a constant characterizing the free energetic cost of lipid mismatch at the interface, $w_T = (w_1 + w_2)/2$, $w_d = w_1 - w_2$, $V_T = (V_1 + V_2)/2$, and $V_d = V_1 - V_2$. By truncating the Taylor series approximation of $\mathscr{F}(H)$ with respect to H_0 to the second order we get an energy functional in the form of Eq. (25).

3.2.2 Eulerian Formulation

It has been seen in Sect. 2 that the parametrization of solvation energy using the surface function allows one to implicitly track the molecular surface by following the iso-surface extraction during the evolution of the surface function, which is referred as to the Eulerian formulation. We could also evolve a phase field function to minimize the energy in Eq. (25) and to obtain the configuration of microdomains. This is achieved by using the following 2D Eulerian formulation of the microdomain geodesic curvature energy defined on the entire membrane surface S:

$$G(\phi) = \int_S \frac{k\varepsilon}{2}\left(\Delta_{\mathbf{x}}\phi + \frac{1}{\varepsilon^2}(\phi + H_c\varepsilon)(1-\phi^2)\right)^2 d\mathbf{x} \tag{30}$$

where $H_c = \sqrt{2}H_0$ and ε is a small positive parameter that characterizes the width of the transitional layer from $\phi(\mathbf{x}) = -1$ to $\phi(\mathbf{x}) = 1$. Here S is a surface embedded in $\mathbb{R}^3$, $\mathbf{x} = (x_1, x_2)$ and $d\mathbf{x}$ is an infinitesimal surface element. The equivalence of this Eulerian formulation (30) to the Lagrangian formulation (25) is analogous to the equivalence between the Canham-Helfrich-Evans curvature energy and the membrane elastic energy [1, 42]. In particular, if the phase field function is defined by

$$\phi(\mathbf{x}) = \tanh\left(\frac{d(\mathbf{x})}{\sqrt{2}\varepsilon}\right) \tag{31}$$

with $d(\mathbf{x})$ being the signed geodesic distance at the surface point $\mathbf{x}$ to the interface contour C where $\phi = 0$, then

$$\nabla_{\mathbf{x}}\phi = \frac{1}{\varepsilon}q'(d(\mathbf{x}))\nabla_{\mathbf{x}}d, \quad \Delta_{\mathbf{x}}\phi = \frac{1}{\varepsilon}q''(d(\mathbf{x}))|\nabla_{\mathbf{x}}d|^2 + \frac{1}{\varepsilon}q'(d(\mathbf{x}))\Delta_{\mathbf{x}}d,$$

where

$$q(x) = \tanh\left(\frac{x}{\sqrt{2}\varepsilon}\right), q'(\mathbf{x}) = \frac{1}{\sqrt{2}}\left[1 - \tanh^2\left(\frac{x}{\sqrt{2}\varepsilon}\right)\right],$$

$$q''(\mathbf{x}) = -\frac{1}{\varepsilon}\tanh\left(\frac{x}{\sqrt{2}\varepsilon}\right)\operatorname{sech}^2\left(\frac{x}{\sqrt{2}\varepsilon}\right),$$

and $\nabla_{\mathbf{x}}$, $\nabla_{\mathbf{x}}\cdot$ are surface gradient and surface divergence operators, respectively. The geodesic curvature of a contour is given by

$$H = \nabla_{\mathbf{x}} \cdot \mathbf{n}, \tag{32}$$

where $\mathbf{n}$ is the normal vector to the contour C. Since $\mathbf{n} = \nabla_{\mathbf{x}}d$ we have $H = \nabla_{\mathbf{x}} \cdot \nabla_{\mathbf{x}}d = \Delta_{\mathbf{x}}d$ and

$$\Delta_{\mathbf{x}} d = \frac{\varepsilon}{q'} \Delta_{\mathbf{x}} \phi - \frac{q''}{q'} |\nabla_{\mathbf{x}} d|^2, \quad \nabla_{\mathbf{x}} d = \frac{\varepsilon}{q'} \nabla_{\mathbf{x}} \phi.$$

Therefore, one has

$$\Delta_{\mathbf{x}} d = \frac{\varepsilon}{q'} \Delta_{\mathbf{x}} \phi - \frac{q''}{q'} \left| \frac{\varepsilon}{q'} \nabla_{\mathbf{x}} \phi \right|^2 .$$

Writing $q'(\mathbf{x})$ and $q''(\mathbf{x})$ in terms of $q(\mathbf{x})$ we can convert the above representation to

$$\Delta_{\mathbf{x}} d = \frac{\sqrt{2}\varepsilon}{1-q^2} \left(\Delta_{\mathbf{x}} \phi + \frac{2q}{1-q^2} |\nabla_{\mathbf{x}} \phi|^2 \right),$$

which is the geodesic curvature $H = \Delta_{\mathbf{x}} d$. Replacing $q(\mathbf{x})$ with ϕ one obtains the final form of H as

$$\begin{aligned} H &= \frac{\sqrt{2}\varepsilon}{1-\phi^2} \left(\Delta_{\mathbf{x}} \phi + \frac{2\phi}{1-\phi^2} |\nabla_{\mathbf{x}} \phi|^2 \right) \\ &= \frac{\sqrt{2}\varepsilon}{1-\phi^2} \left(\Delta_{\mathbf{x}} \phi + \frac{1}{\varepsilon^2} (1-\phi^2)\phi \right), \end{aligned} \tag{33}$$

where we assume $\|\mathbf{n}\| = 1$ in the last step of derivation. When minimizing the curvature energy in Eq. (30) the following constraint

$$A(\phi) = \int_S \phi(\mathbf{x}) d\mathbf{x} = \text{constant} \tag{34}$$

must be enforced such that quantities of both species of lipids are conserved.

To derive the equation of the geometric flow for the energy $G(\phi)$ we compute its first variation with respect to ϕ:

$$\frac{\delta G}{\delta \phi} = k \left[\Delta_{\mathbf{x}} W - \frac{1}{\varepsilon^2} (3\phi^2 + 2H_c \varepsilon \phi - 1) W \right] \tag{35}$$

where

$$W = \varepsilon \Delta_{\mathbf{x}} \phi - \frac{1}{\varepsilon} (\phi + H_c \varepsilon)(\phi^2 - 1).$$

We then split the linear and nonlinear components (W_L and W_N) of W to facilitate the numerical treatments. They are given respectively by

$$W_L = \varepsilon \Delta_{\mathbf{x}} \phi + \frac{1}{\varepsilon} \phi + H_c, \quad W_N = -\frac{1}{\varepsilon} \phi^3 - H_c \phi^2.$$

We then have the full expansion of the variation

$$\begin{aligned}\frac{\delta G}{\delta \phi} &= k\Delta_{\mathbf{x}} W_L + \frac{k}{\varepsilon^2} W_L + k\Delta_{\mathbf{x}} W_N - \frac{k}{\varepsilon^2}(3\phi^2 + 2H_c\varepsilon\phi)(W_N + W_L) + \frac{k}{\varepsilon^2} W_N \\ &= k\varepsilon\Delta_{\mathbf{x}}^2\phi + \frac{k}{\varepsilon}\left(2 - 6\phi^2 - 4kH_c\varepsilon\right)\Delta_{\mathbf{x}}\phi - \left(\frac{6k}{\varepsilon}\phi + 2kH_c\right)|\nabla_{\mathbf{x}}\phi|^2 \\ &\quad + k\left(-\frac{2H_c^2}{\varepsilon} + \frac{1}{\varepsilon^3}\right)\phi - \frac{3kH_c}{\varepsilon^2}\phi^2 - k\left(\frac{4}{\varepsilon^3} - \frac{2H_c^2}{\varepsilon}\right)\phi^3 + \frac{5kH_c}{\varepsilon^2}\phi^4 + \frac{3k}{\varepsilon^3}\phi^5 \\ &\quad + \frac{kH_c}{\varepsilon^2}.\end{aligned} \tag{36}$$

Also note that the variation of the mass conservation constraint is

$$\frac{\delta A}{\delta \phi} = 1. \tag{37}$$

The appearance of fourth order derivative in the variation $\delta G/\delta\phi$ motivates us to adopt the following equation of the geometric flow with an artificial time for ϕ:

$$\frac{\partial \phi}{\partial t} = -\frac{\delta G}{\delta \phi} + \lambda\frac{\delta A}{\delta \phi}, \tag{38}$$

where λ is a Lagrangian multiplier used to ensure the conservation of ϕ. We can derive a representation of λ by integrating Eq. (38) and noting that $\int_S \frac{\partial \phi}{\partial t} d\mathbf{x} = 0$, hence

$$0 = -\int_S \frac{\delta G}{\delta \phi} d\mathbf{x} + \int_S \lambda d\mathbf{x},$$

and consequently

$$\lambda = \frac{1}{|S|}\int_S \frac{\delta G}{\delta \phi} d\mathbf{x},$$

which yields

$$\frac{\partial \phi}{\partial t} = -\frac{\delta G}{\delta \phi} + \frac{1}{|S|}\int_S \frac{\delta G}{\delta \phi} d\mathbf{x}. \tag{39}$$

Equation (39) is a fourth-order nonlinear surface diffusion equation. Alternatively, one could derive a Cahn-Hilliard equation for the surface phase field function ϕ as

$$\frac{\partial \phi}{\partial t} = \Delta_{\mathbf{x}}\left(\frac{\delta G}{\delta \phi}\right), \tag{40}$$

which guarantees the conservation of ϕ and thus does not need a Lagrangian multiplier. However, it involves a sixth order surface derivative and thus is more complicated when the equation is to be solved numerically on a discretized surface S.

To simplify the exposition of numerical treatments we adopt $\lambda = \frac{1}{|S|}\int_S \frac{\delta A}{\delta \phi} d\mathbf{x}$ and define $g = \frac{\delta G}{\delta \phi}$. Then we write Eq. (39) as

$$\phi_t = -g + \lambda. \tag{41}$$

To implement the time discretization we average the nonlinear function $g(\phi)$ over the current and next time steps ϕ_n, ϕ_{n+1} to implement a Crank-Nicolson approximation

$$\frac{\phi_{n+1} - \phi_n}{\Delta t} + g(\phi_{n+1}, \phi_n) - \lambda(\phi_n) = 0, \tag{42}$$

where the averaged function is defined by

$$\begin{aligned} g(\phi_{n+1}, \phi_n) &= \frac{k}{2}\Delta_{\mathbf{x}}(f_c(\phi_{n+1}) + f_c(\phi_n)) \\ &\quad - \frac{k}{2\varepsilon^2}(\phi_{n+1}^2 + \phi_{n+1}\phi_n + \phi_n^2 + \varepsilon H_c(\phi_{n+1} + \phi_n) - 1)(f_c(\phi_{n+1}) + f_c(\phi_n)), \end{aligned}$$

and

$$f_c(\phi) = k\left(\varepsilon \Delta_{\mathbf{x}}\phi - (\frac{1}{\varepsilon} + \varepsilon H_c)(\phi^2 - 1)\right).$$

To numerically solve Eq. (42) which is an implicit scheme for ϕ_{n+1}, we define an interior iteration for computing ψ_m such that $\psi_m \to \phi_{n+1}$ as $m \to \infty$. The equation for ψ_m reads as

$$\frac{\psi_{m+1} - \phi_n}{\Delta t} + g(\psi_{m+1}, \psi_m, \phi_n) - \lambda(\psi_m) = 0, \tag{43}$$

where new averaged functions are defined by

$$\begin{aligned} g(\psi_{m+1}, \psi_m, \phi_n) &= \frac{k}{2}\Delta_{\mathbf{x}}\tilde{f}_c(\psi_{m+1}, \psi_m, \phi_n) \\ &\quad - \frac{k}{2\varepsilon^2}(\psi_m^2 + \psi_m\phi_n + \phi_n^2 + \varepsilon H_c(\psi_m + \phi_n) - 1)(f_c(\psi_m) + f_c(\phi_n)), \\ \tilde{f}_c(\psi_{m+1}, \psi_m, \phi_n) &= \frac{\varepsilon}{2}\Delta_{\mathbf{x}}(\psi_{m+1} + \phi_n) - \frac{1}{4\varepsilon}(\psi_m^2 + \phi_n^2 - 2)(\psi_m + \phi_n + 2\varepsilon H_c). \end{aligned}$$

Convergent ψ_m is obtained by iterating over the interior index m, usually up to a tolerance $\|\psi_{m+1} - \psi_m\| \leq \varepsilon_\psi$ for some small $\varepsilon_\psi > 0$. This convergent ψ_m is assigned to ϕ_{n+1}, and computation is advanced to the next time step. The linear and nonlinear components of ψ_{m+1} in Eq. (43) are further split. The nonlinear components are updated slower than the linear components, allowing an efficient numerical solution. The spatial approximation of the equation is obtained by a newly developed a C^0 interior penalty surface finite element method [1, 12].

3.3 Computational Simulations and Summary

We apply the geodesic curvature driven phase separation model to simulate the microdomain formation on surfaces. We present four simulations on different surfaces or with different spontaneous geodesic curvatures. The energetic histogram and the dynamics of the domain formation in each simulation are compared to those generated by the Allen-Cahn equation obtained by the direct extension of the Ginzburg-Landau energy based a classical phase separation model on surfaces [43]. We also compute the radii of the microdomains which are expected to approximate the reciprocal of the given spontaneous geodesic curvature.

In the first simulation (#1) on unit sphere with 3963 approximately uniformly distributed nodes, we choose $\varepsilon = 0.1$, $H_c = \frac{1}{0.3}$, $k = 0.01$ and $\Delta t = 0.001$. A random field is initialized on the surface such that $\int_S \phi ds = 0$. The results are compared side by side with those of the classical Allen-Cahn equation in Fig. 6. Using a K-means clustering method we are able to identify a number of microdomains whose radii are then calculated. The radius associated with each microdomain is approximately 0.23. This means the curvature is approximately $\frac{1}{0.23}$, close to the specified spontaneous geodesic curvature.

The total energies for the geodesic curvature model and the classical Allen-Cahn model are plotted in Fig. 5. Both converge as time evolves. The number of iterations is large because of the small Δt, which is constrained by the stability of our numerical method for the fourth-order nonlinear partial differential equation.

In the second simulation (#2) on the unit sphere as shown in Fig. 7, we choose $\varepsilon = 0.1$, $H_c = \frac{1}{0.40}$, $k = 0.01$ and $\Delta t = 0.002$. This spontaneous curvature matches the reported spontaneous curvature for DOPE/DOPS mixture [53]. A coarser while quasi-uniform mesh with 984 nodes is deployed on the unit sphere. The radius associated with the each microdomain is approximately 0.37, indicating a curvature approximately $\frac{1}{0.37}$. The convergence of the energies of the geodesic curvature model and the classical Allen-Cahn mode are plotted in Fig. 5 as well. The lower resolution

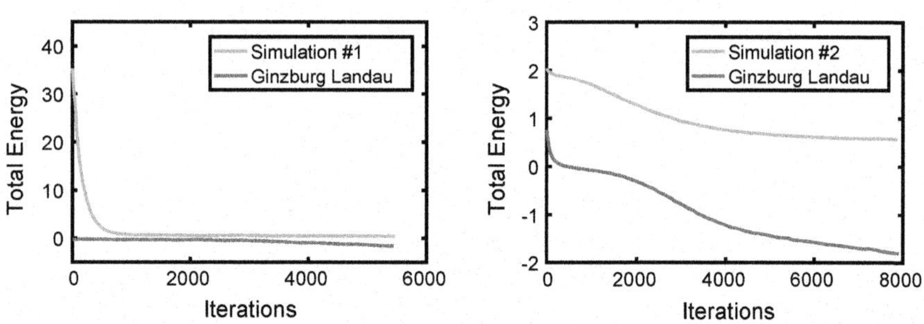

Fig. 5 Minimization of the geodesic curvature total energy and the Ginzburg-Landau Energy. *Left* Simulation #1 on unit sphere with 3963 nodes and $H_c = \frac{1}{0.3}$. *Right* Simulation #2 on unite sphere with 984 nodes and $H_c = 1/0.4$

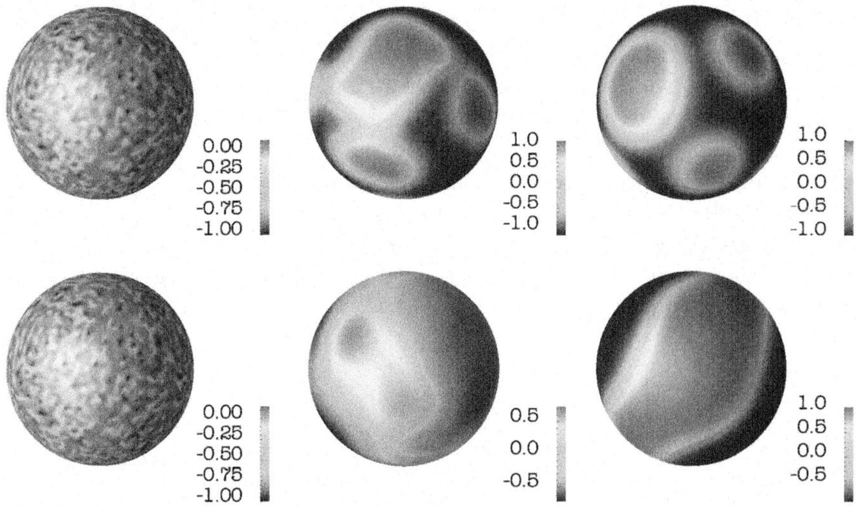

Fig. 6 Simulation #1. Formation of local microdomains simulated by the geodesic curvature energy (*top row*) and domain separation simulated by the classical Ginzburg-Landau energy (*bottom row*) from the same initial random field (*left column*) on the unite sphere with 3963 nodes. Sampling time from *left* to *right* is: $t = 0, 3$, and 7

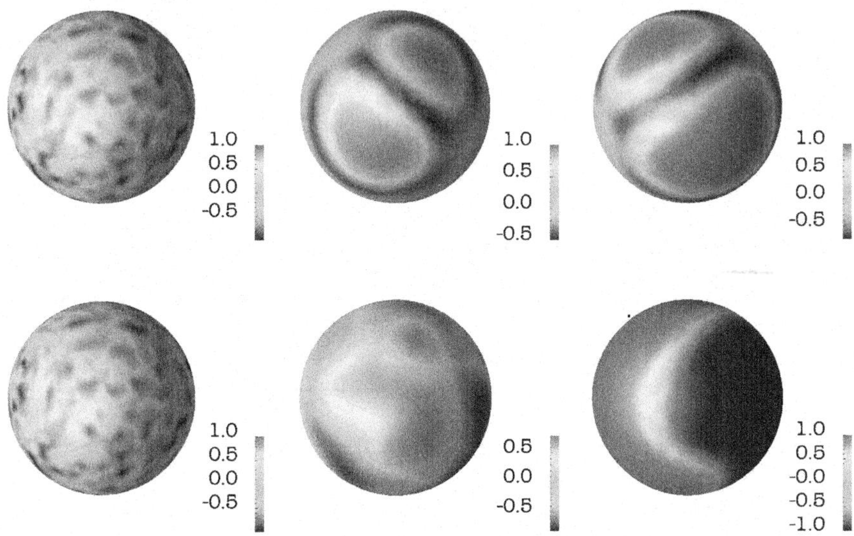

Fig. 7 Simulation #2. Formation of local microdomains simulated by the geodesic curvature energy (*top row*) and domain separation simulated by the classical Ginzburg-Landau energy (*bottom row*) from the same initial random field (*left column*) on unit sphere with 984 nodes. Sampling time from *left* to *right* is: $t = 0, 3$, and 7

resulting from the coarser mesh in the second simulation can be seen in the larger spots in the initial field and the wider transitional layers between different domains.

The third simulation (#3) is conducted on a more complicated surface as shown in Fig. 8. We choose the molecular surface of three particles of unit radius respectively centered at (0, 1, 0), (−0.864, −0.5, 0) and (0.864, −0.5, 0). The surface is quasi-uniformly meshed with 2974 nodes and we set $\varepsilon = 0.1$, $H_c = \frac{1}{0.4}$, $k = 0.01$ and $\Delta t = 0.001$. Starting with a random initial field we finally identified six microdomains using the K-mean clustering method at the equilibrium state, whose radii are estimated. As seen in Fig. 9, the radii of the microdomains approximate the given spontaneous geodesic curvatures.

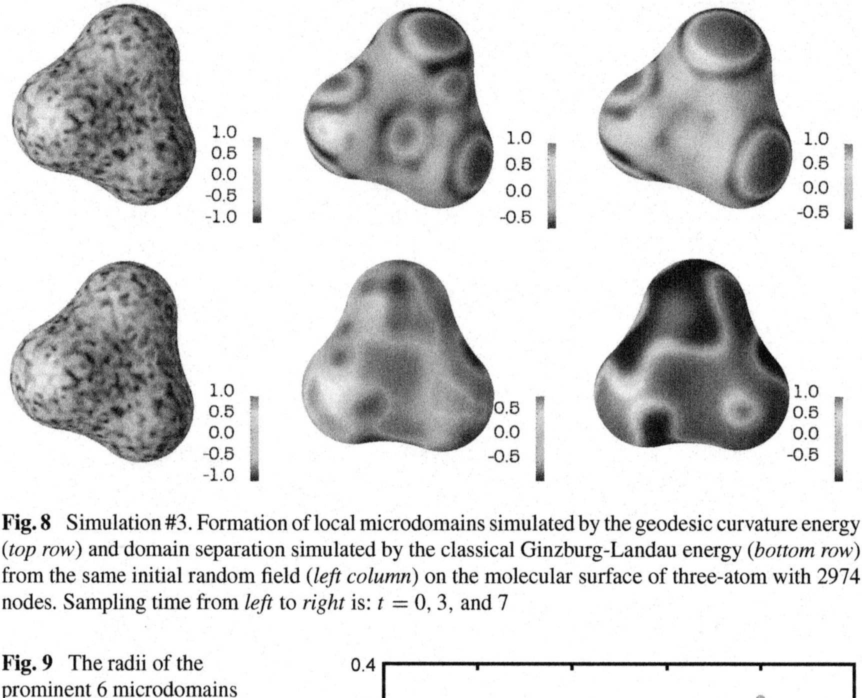

Fig. 8 Simulation #3. Formation of local microdomains simulated by the geodesic curvature energy (*top row*) and domain separation simulated by the classical Ginzburg-Landau energy (*bottom row*) from the same initial random field (*left column*) on the molecular surface of three-atom with 2974 nodes. Sampling time from *left* to *right* is: $t = 0$, 3, and 7

Fig. 9 The radii of the prominent 6 microdomains produced in Simulation #3

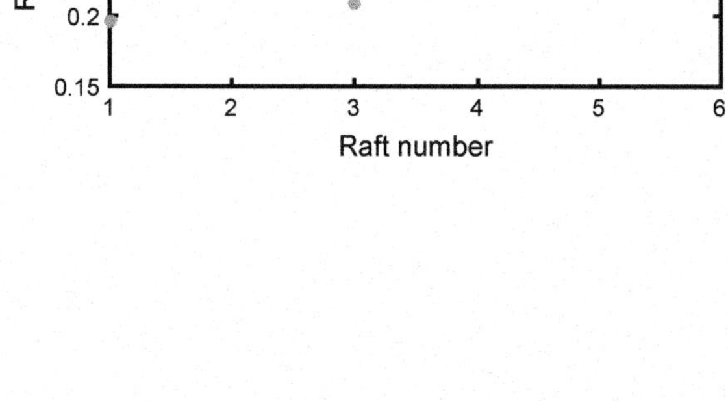

In the last simulation (#4) we choose the molecular surface of six particles of unit radius respectively centered at $(1, 0, 0)$, $(-1, 0, 0)$, $(0, 1, 0)$, $(0, -1, 0)$, $(0, 0, 1)$ and $(0, 0, -1)$. The quai-uniform surface mesh has 3903 nodes and we set $\varepsilon = 0.1$, $H_c = \frac{1}{0.4}$, $k = 0.01$ and $\Delta t = 0.001$ for the simulation. One can see from Fig. 10 that the largest raft radius obtained by the simulation is about 0.35 which means the curvature of that raft is about $\frac{1}{0.35}$, a value close to given spontaneous geodesic curvature (Fig. 11)

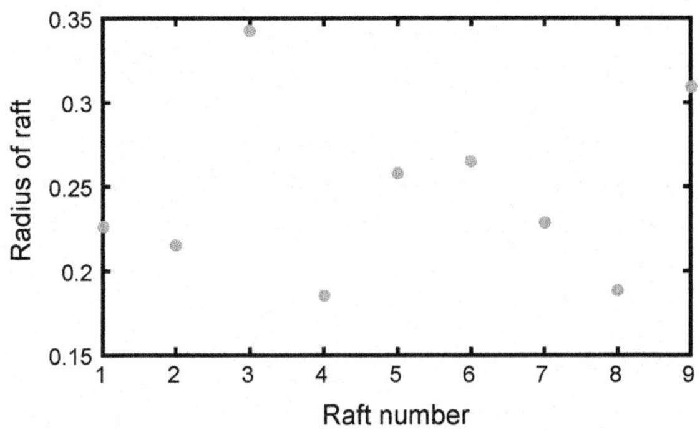

Fig. 10 The radii of the prominent 9 rafts produced by Simulation #4

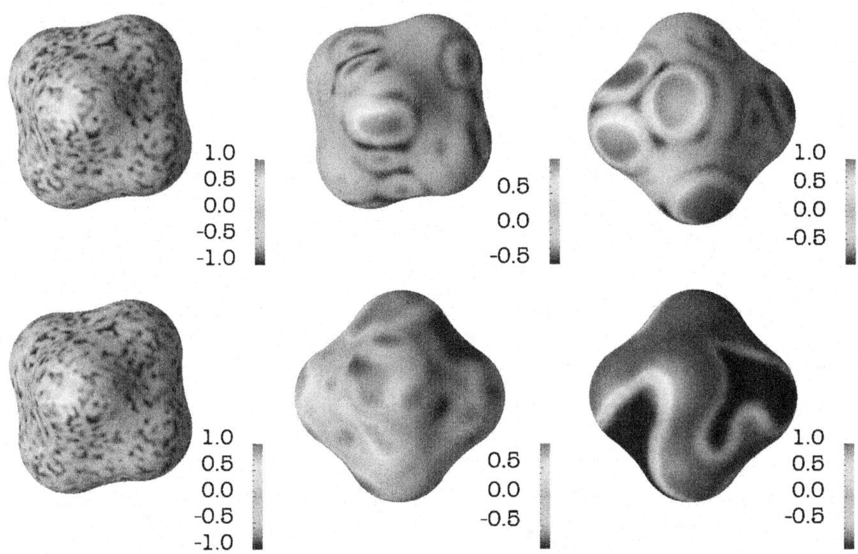

Fig. 11 Simulation #4. Formation of local microdomains simulated by the geodesic curvature energy (*top row*) and domain separation simulated by the classical Ginzburg-Landau energy (*bottom row*) from the same initial random field (*left column*) on the molecular surface of six-atom with 3903 nodes. Sampling time from *left* to *right* is: $t = 0, 3$, and 7

The radii of the microdomains generated in our simulations are not exactly the given spontaneous geodesic curvature. Rather they are distributed around the given curvature. Apart from the numerical error in simulation and in K-means clustering and radii estimate, this non-uniform distribution of domain radii is mostly related to the total quantity of the lipid phases in the initial random field. The initial quantity may not exactly cover an integer number of microdomains with the given radius. However, the overall distribution of radii around the given radius of curvature demonstrated that our geodesic curvature model is capable of predicting the formation of microdomains that are caused by the geometrical and molecular mechanical mismatch of lipid mixtures. The predicted microdomains can be compared to the observed lipid rafts, and the boundaries of these microdomains can be identified to provide locations where specific proteins can aggregate. Coupling of our model of geodesic curvature driven microdomains formation to the localization of proteins will provide a very useful quantitative technique for studying the crucial roles of these proteins in high-fidelity signal transduction in cells [66, 85].

4 Variational Methods for Curvature Induced Protein Localization in Bilayer Membranes

Rather than forming distinct domains in a way similar to lipids as modeled in Sect. 3, many membrane proteins do not form distinct domains in membranes.[3] Given the fact that their distribution on bilayer membranes is not uniform, molecular mechanisms need to be identified to quantitatively investigate this distribution and its biological consequences. On the one hand, approximately 30–90 % of all membrane proteins can freely diffuse along the membrane [50, 74, 94, 107]. On the other hand, insertion or tethering of the membrane proteins to bilayer membrane will cause membrane curvature [64, 110, 163]. For instances, the rigid proteins such as those in the BAR (Bin/Amphiphysin/Rvs) domain family can act as a scaffold to the membrane. These proteins have an intrinsic curvature and, upon attaching, the membrane bends to match the protein curvature [98]. In a similar fashion, several proteins can oligomerize to create a rigid shape and bend the membrane. Protein coats such as clathrin, COPI (COat Protein I) and COPII (COat Protein II) are examples of this type [51, 75]. Other proteins may insert themselves into the membrane. Membrane curvature is also induced when there is a difference between the length of the hydrophobic region of a membrane protein and the thickness of the hydrophobic core of the lipid bilayer in which it is embedded [103]. Epsin proteins do this by forming an alpha-helix known as H0 upon binding to the membrane, and this helix inserts itself into

[3] A protein unit consisting of several segments such as most ion channel proteins or G-protein-coupled receptors (GPCRs) is not taken as a distinct domain in this study. The whole unit is considered as a single protein instead.

the membrane [11]. Moreover, local crowding of peripheral proteins can cause membrane bending by creating an asymmetry of the monolayer areas and thereby curling the membrane away from the side on which the crowding occurred. This effect is experimentally demonstrated in [122]. Further illustrating the importance of proteins in membranes, Schmidt et al. showed that the M2 protein plays an essential role in generating regions of high curvature in the influenza A virus membrane [115]. This specific protein accumulates in regions of negative Gaussian curvature and can generate curvature in the membrane itself, allowing the replicated virus to be wrapped and released from the infected cells. While these examples should provide sufficient motivation to include proteins to the model, we note that all endocytosis and exocytosis processes are promoted in one way or another by proteins. Therefore, any viral replication process requires proteins. Antagonizing the curvature effects of proteins is a viable antiviral strategy [115]. This motivates the necessity for a model coupling membrane curvature and lateral diffusion of proteins. We shall observe below that the final governing equation for this curvature-driven lateral transportation appears a drift-diffusion equation in its essential form. This mechanism is different from the transportation of surfactants on interfaces moving with the fluid flow as investigated in the literature [130, 131, 152].

4.1 Lagrangian Formulation

Modeling generation of membrane curvature using energetic variational principle has been well established in the past few decades [19, 45, 49, 65]. These research have been inspirational to our work. However, the focus of our discussion in this section is on the curvature driven protein localization. We sketch the framework of the integration of these two components. The numerical implementation is computationally intensive because of the coupling of dynamical membrane morphology and the varying surface concentration of proteins. Consider a membrane with $(m+1)$ distinct lipid species with concentrations $\rho_l^{\text{lip}}, l = 0, \ldots, m$ and a single type of diffusive membrane proteins with a concentration ρ^{pro}. A closed membrane is modeled as a structureless surface S contained in a 3D domain $\Omega \in \mathbb{R}^3$ and separated Ω into two subdomains, one inside the membrane and the other outside. The total energy of the system is composed of the membrane curvature energy and the entropic energy from the lipids and proteins

$$G_{\text{tot}} = G_{\text{mem}} + G_{\text{ent}}, \tag{44}$$

where the membrane curvature energy is given in the classical Canham-Helfrich-Evans form

$$G_{\text{mem}} = \int_C k(H - H_0(\rho_l^{\text{lip}}, \rho^{\text{pro}}))^2 ds, \tag{45}$$

and the entropic energy for the membrane with membrane protein attachments is

$$G_{\rm ent} = \frac{1}{\beta}\int_C \left(\sum_{l=0}^{m} \rho_l^{\rm lip}\left[\ln(\rho_l^{\rm lip}(a_l^{\rm lip})^2) - 1\right] + \rho^{\rm pro}\left[\ln(\rho^{\rm pro}(a^{\rm pro})^2) - 1\right] \right) ds, \tag{46}$$

Here H is the membrane mean curvature and H_0 is the spontaneous membrane curvature, k is a curvature energy coefficient, and $\beta = 1/(k_B T)$ is the inverse of thermal energy. The effective sizes of lipids and proteins are respectively given by $a_l^{\rm lip}$ and $a^{\rm pro}$. By modeling lipids and proteins as hard disks, the occupied surface areas in the membranes are taken as $(a_l^{\rm lip})^2$ and $(a^{\rm pro})^2$, respectively. The essential feature of our model is seen in the dependence of the membrane spontaneous curvature on the local lipid composition $\rho_l^{\rm lip}$ and the protein concentration $\rho^{\rm pro}$. This dependence is justifiable considering that (i) each lipid species l has its own spontaneous curvature [93] therefore the membrane spontaneous curvature must be a function of the local lipid composition, and (ii) membrane proteins will induce membrane curvature so that the observed spontaneous curvature must be a function of the local protein concentration [72, 103, 115, 126, 135]. We define the membrane curvature induced by a single membrane protein as the spontaneous (membrane) curvature of the protein. Here we define H_0 as the average spontaneous curvature of lipids and proteins weight by their respective surface coverage fraction:

$$H_0 = \sqrt{2}\frac{\sum_{l=0}^{m} C_0^l (a_l^{\rm lip})^2 \rho_l^{\rm lip} + C_0^{\rm pro}(a^{\rm pro})^2 \rho^{\rm pro}}{\sum_{l=0}^{m} (a_l^{\rm lip})^2 \rho_l^{\rm lip} + (a^{\rm pro})^2 \rho^{\rm pro}}, \tag{47}$$

where C_0^l and $C_0^{\rm pro}$ are the spontaneous curvature of the lth species of lipids and proteins, respectively. Considering that the membrane surface is completely covered by the lipids and proteins, the following saturation constraint holds true:

$$\sum_{l=0}^{m} (a_l^{\rm lip})^2 \rho_l^{\rm lip} + (a^{\rm pro})^2 \rho^{\rm pro} = 1. \tag{48}$$

With this constraint we can write the spontaneous curvature in Eq. (47) as

$$H_0 = \sqrt{2}\left(\sum_{l=0}^{m} C_0^l (a_l^{\rm lip})^2 \rho_l^{\rm lip} + C_0^{\rm pro}(a^{\rm pro})^2 \rho^{\rm pro} \right) \tag{49}$$

and the membrane entropic energy as

$$G_{\rm ent} = \frac{1}{\beta}\int_C \left\{ \frac{1}{(a_0^{\rm lip})^2}\left(1-\rho^{\rm pro}(a^{\rm pro})^2-\sum_{l=1}^{m}\rho_l^{\rm lip}(a_l^{\rm lip})^2\right)\right.$$
$$\times\left[\ln\left(1-\rho^{\rm pro}(a^{\rm pro})^2-\sum_{l=1}^{m}\rho_l^{\rm lip}(a_l^{\rm lip})^2\right)-1\right]+$$
$$\left.\sum_{l=1}^{m}\rho_l^{\rm lip}\left[\ln(\rho_l^{\rm lip}(a_l^{\rm lip})^2)-1\right]+\rho^{\rm pro}\left(\ln(\rho^{\rm pro}(a^{\rm pro})^2)-1\right)\right\}ds. \quad (50)$$

To obtain the dynamics of the membrane morphology, one can calculate the variation of the total energy $G_{\rm tot}$ in Eq. (44) and solve the resulting equation for the gradient flow of ϕ. This process is routine and can be found in the studies of spontaneous curvature effects of pure or multi-component membranes without proteins [42, 45]. Since our interest here is to investigate the protein localization on membrane surfaces, we choose to fix the membrane morphology, i.e., H_0 is a time-independent function. We then only need to calculate the variation of the total energy with respect to the membrane protein concentration, which turns out to be

$$\frac{\delta G_{\rm tot}}{\delta\rho^{\rm pro}} = \frac{\delta G_{\rm mem}}{\delta\rho^{\rm pro}}+\frac{\delta G_{\rm ent}}{\delta\rho^{\rm pro}}$$
$$= k_BT\left[-\left(\frac{a^{\rm pro}}{a_0^{\rm lip}}\right)^2\ln\left(1-\rho^{\rm pro}(a^{\rm pro})^2-\sum_{l=1}^{m}\rho_l^{\rm lip}(a_l^{\rm lip})^2\right)+\ln(\rho^{\rm pro}(a^{\rm pro})^2)\right]$$
$$+2C_0^{\rm pro}(a^{\rm pro})^2(H-H_0). \quad (51)$$

4.2 Eulerian Formulation

While we are working on the membrane with fixed morphology, the formulation of the curvature driven protein localization is expected to interface with dynamical morphology where the membrane surface is not a prior known. For that purpose one could trace the position of membrane implicitly by evolving a phase field function $\phi(\mathbf{x})$ on surface S embedded in $\Omega \in \mathbb{R}^3$, where ϕ takes the value of -1 in the exterior of the membrane enclosure and 1 inside [42, 45]. The membrane mean curvature at $\phi = 0$ can be computed as a function of ϕ following

$$H = \frac{\sqrt{2}\varepsilon}{2(1-\phi^2)}\left(\Delta_{\mathbf{x}}\phi+\frac{1}{\varepsilon^2}(1-\phi^2)\phi\right), \quad (52)$$

where $\varepsilon > 0$ is a small parameter that adjust the transition of ϕ from -1 to 1 near the membrane as in Eq. (30). We then identify three components of the chemical potential defined by the variation in Eq. (51)

$$L^{\rm pro} = \ln(\rho^{\rm pro}(a^{\rm pro})^2), \tag{53}$$

$$R^{\rm pro} = -\left(\frac{a^{\rm pro}}{a_0^{\rm lip}}\right)^2 \ln\left(1 - \rho^{\rm pro}(a^{\rm pro})^2 - \sum_{j=1}^{m} \rho_j^{\rm lip}(a_j^{\rm lip})^2\right), \tag{54}$$

$$P^{\rm pro} = \frac{\varepsilon}{\sqrt{2}(1-\phi^2)}\left(\Delta_{\mathbf{x}}\phi + \frac{1}{\varepsilon^2}\phi(1-\phi^2)\right) - H_0 \tag{55}$$

to write this chemical potential as

$$\mu^{\rm pro} = \frac{\delta G_{\rm tot}}{\delta \rho^{\rm pro}} = k_B T(L^{\rm pro} + R^{\rm pro}) + 2C_0^{\rm pro}(a^{\rm pro})^2 \nabla_{\mathbf{x}} P^{\rm pro}. \tag{56}$$

This chemical potential allows us to define the diffusion flux vector and the transportation equation. Two options are available for the definition of the transportation equation. One could extract the membrane surface S from the phase field function ϕ and solve a surface transportation on S. This involves the dynamic meshing or mesh deformation if ϕ is evolving in time, and singularity will arise if there is topological change in S as ϕ evolves.

Alternatively, one could formally define a 3D transportation equation in the entire domain Ω but practically restrict the transportation of membrane proteins to a very small neighborhood near the membrane surface S. This is accomplished by introducing to the flux vector

$$\mathbf{J}^{\rm pro}(\mathbf{r}) = -D^{\rm pro}\delta_S \beta \rho^{\rm pro}(\mathbf{r})\nabla \mu^{\rm pro} \tag{57}$$

a function δ_S which is concentrated at the membrane S where $\phi = 0$. Various choices of such functions are available and their numerical properties differ subtly [78]. We choose

$$\delta_S = \begin{cases} \tanh(10(\phi+1)), & -1 \le \phi \le 0, \\ -\tanh(10(\phi-1)), & 0 \le \phi \le 1, \end{cases} \tag{58}$$

so that effective domain near $\phi = 0$ can be automatically identified as ϕ evolves. The general transportation equation for membrane proteins reads

$$\frac{\partial \rho^{\rm pro}(\mathbf{r})}{\partial t} + \nabla \cdot (\mathbf{v}\nabla \rho^{\rm pro}(\mathbf{r})) = -\nabla \cdot \mathbf{J}^{\rm pro}(\mathbf{r}), \tag{59}$$

where $\mathbf{v}$ is the velocity of the membrane in which the membrane proteins move. Although this velocity is taken to be zero in our computations simulations to be presented here, it can be computed if the membrane moves with the evolving phase field function. The nature of the equation can be seen if the size effects of lipids and membrane proteins are not considered, i.e., $a_l^{\rm lip} = a^{\rm pro} = 0$. In this case $R^{\rm pro} = 0$ and

$$\frac{\partial \rho^{\rm pro}}{\partial t} = \nabla \cdot (D^{\rm pro}\delta_S \nabla \rho^{\rm pro} + 2k_B T D^{\rm pro} C_0^{\rm pro}(a^{\rm pro})^2 \delta_S \rho^{\rm pro} \nabla P^{\rm pro}), \tag{60}$$

which is a drift-diffusion equation with a potential P^{pro}. The mean curvature of the membrane therefore appears a potential that drives the transportation of membrane proteins to membrane surfaces where its mean curvature well fits the spontaneous membrane curvature of proteins. To numerically solve the equation, we separate the linear and nonlinear components of the equation, which are then treated using an implicit-explicit splitting interaction methods similar to the treatment of Eq. (40) presented in Sect. 3. The spatial approximation of the equation is obtained by using the Fourier spectral method, and a change of variable is necessary to convert the equation with variable diffusion coefficient $D\delta_S$ to a constant diffusion coefficient so that the Fourier spectral method is applicable [36, 125].

4.3 Computational Simulations and Summary

To demonstrate the curvature preference of protein localization we consider in the domain $\Omega = (-4, 4)^3$ a torus because it has regions with positive and negative mean curvatures where the proteins may populate or not depending on their spontaneous curvature. The torus surface is given by

$$(R - \sqrt{x^2 + y^2})^2 + z^2 = r^2, \tag{61}$$

where R and r are the major and minor radii, respectively. Its alternative parametrization

$$(x, y, z) = ((R + r\cos\theta)\cos\varphi, (R + r\cos\theta)\sin\varphi, r\sin\varphi) \tag{62}$$

can be handy when computing the curvature. Here $0 \le \theta \le 2\pi$ is the angle made from the surface around the center of the tube, known as the poloidal angle, and $0 \le \varphi \le 2\pi$ is the angle made from the surface to the positive x-axis (projected on the xy-plane), known as the toroidal angle. When $R > r$, one gets the so-called ring torus. Here we choose $R = 2$ and $r = 1.1$. The phase field function ϕ is set as the signed distance function with this torus surface. We consider only one species of diffusion proteins and one species of lipids. The saturation condition (48) then indicates that we only need to model the distribution of proteins only. The membrane proteins are initially concentrated near the highest point of the positive y-axis, smoothly distributed along the surface, and because of the adoption of phase field function which expands the transportation domain from the surface to a small neighborhood in the vicinity of the surface, smoothly distributed from the surface to the bulk:

$$\rho = \rho_0 e^{-\sqrt{x^2+(y-R)^2+z^2}} e^{-2\left(r-\sqrt{(x-c_x)^2+(y-c_y)^2+z^2}\right)}, \tag{63}$$

where $r = \sqrt{x^2 + y^2 + z^2}$ and $(c_x, c_y, 0)$ is the center of the torus tube on the same plane of which locates the point (x, y, z). The scaling constant ρ_0 is chosen such that the maximum of the concentration is 1 on the torus surface.

We first set the spontaneous curvature of membrane proteins and lipids to be $C_0^{\rm pro} = 0.5$, $C_0^{\rm lip} = -0.1$, respectively. Notice that the mean curvature of a torus is given by

$$H_{\rm torus} = \frac{R + 2r\cos\theta}{2r(R + r\cos\theta)}, \tag{64}$$

which gives a mean curvature $H_{\rm torus} \approx 0.6158$ for the chosen values of R, r at the outer ring of the torus where $\theta = 0$ and $H_{\rm torus} \approx -0.1$ at the inner ring of the torus where $\theta = \pi$. With this first choice of $C_0^{\rm pro}$, $C_0^{\rm lip}$ we expect that the membrane proteins will populate near the outer ring where the mean curvature is close to the specified spontaneous curvature of membrane proteins. Our expectation is verified by Fig. 12, where the plots of the concentrations of the membrane proteins on the membrane

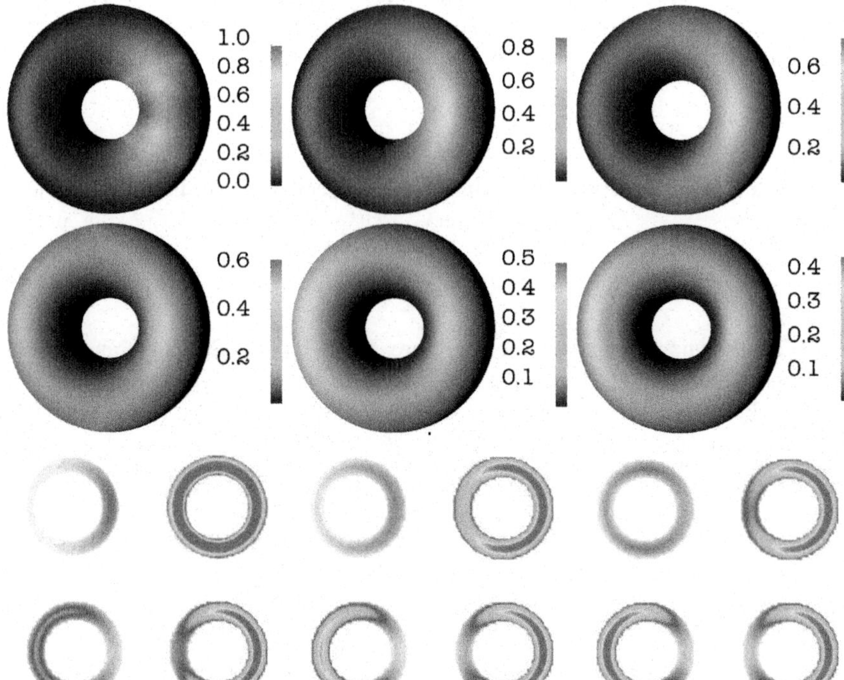

Fig. 12 Simulated localization of the membrane proteins from its initial position to the outer ring of the torus on a 128^3 uniform mesh. $\varepsilon = 0.1$. Time increment $\Delta t = 10^{-3}$. Spontaneous curvatures $C_0^{\rm pro} = 0.5$, $C_0^{\rm lip} = -0.1$, and sampling moments are $t = 0, 0.1, 0.25, 0.5, 1.0, 5.0$. Color is scaled by the maximum concentration in each plot

$\phi = 0$ and the cross section $y = 0$ at six sampling moments show the transportation of membrane proteins from its initial position to the outer ring of the torus.

In the second simulation we start with same initial condition as in the first simulation but switch the spontaneous curvatures to $C_0^{\text{pro}} = -0.1$ and $C_0^{\text{lip}} = 0.5$. It is expected that the membrane proteins will finally populate at the inner ring of the torus, and this is verified by the snapshots of concentrations in Fig. 13.

These two computational simulations demonstrate the successful modeling of the curvature driven membrane protein localization using the drift-diffusion equation (60). Full version of Eq. (59) can also be considered to include the effects of finite sizes of effects of lipids and proteins, and multiple species of lipids. Our choice of small time increment ($\Delta = 10^{-3}$) is restricted by the stability of the implicit-explicit splitting method used for integrating the nonlinear equation. We expect the development of more efficient numerical methods for the integration of the equation, in particular when it is to be coupled with the dynamic phase field function ϕ, in that case a membrane velocity shall be added to Eq. (60) to make it an advection-drift-diffusion equation. Such coupling reveals the positive feedback of membrane

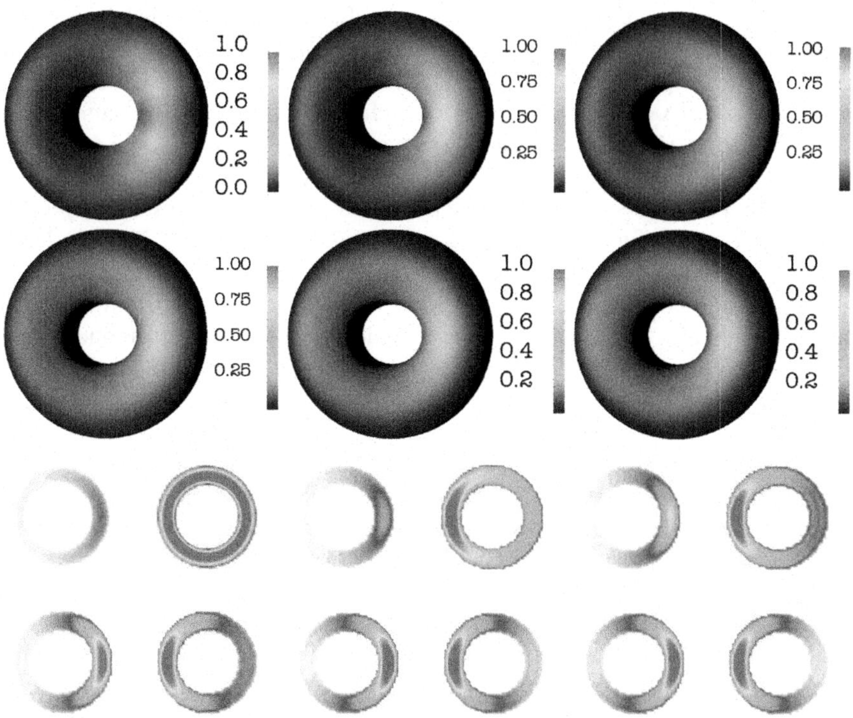

Fig. 13 Simulated localization of the membrane proteins from its initial position to the outer ring of the torus on a 128^3 uniform mesh. $\varepsilon = 0.1$. Time increment is $\Delta t = 10^{-3}$. Spontaneous curvatures are $C_0^{\text{pro}} = -0.1$ and $C_0^{\text{lip}} = 0.5$, and sampling moments are $t = 0, 0.1, 0.25, 0.5, 1.0, 5.0$. Color is scaled by the maximum concentration in each plot

curvature accumulation to membrane protein localization. On the other hand, the number of major membrane proteins involved in the membrane fusion, budding, endocytosis, or exocytosis is not a constant over the entire time course because there is continuous intracellular protein transport. Proteins may be recruited from the solution to membrane at specific regions of the membrane and meanwhile they are released from the membrane to the solution [112, 121]. The model presented here can be extended by adding a reaction term that models the dynamic exchange of membrane proteins between the membrane and the solution. Indeed, it is shown that some membrane budding proteins such as influenza virus hemagglutinin (HA) and neuraminidase (NA) are associated with raft-like microdomains, while some are not [80]. An integration of the curvature driven localization and local clustering within the microdomains will help elucidate the competing or collaborative effects of these membrane proteins in the same biophysical process.

5 Conclusions

Energetic variational principle constitutes a tangible link between multiscale theory and the experimental observation of biomolecular structure, function, and dynamics, aided by computational simulations. Although the applications of variational principle have been well established for research in various areas of mechanics, classical and modern physics, and material sciences, novel insights are offered by this principle when it is applied to the biomolecular systems. Among the progresses achieved in recent years, a significant step forward has been made using the geometry of the molecular interface to parametrize the total energy [1, 26, 27, 71, 82, 84, 97, 149]. This unified representation allows the investigators to focus on the identification of energies that characterize various molecular interactions at multiple spatial and temporal scales. The flexibility of the analytical and computational framework of the variational principle ensures that the critical states and dynamics of the biomolecular system can be tracked with confidence by evolving the total energy. Furthermore, by introducing a phase field function we can implicitly define and track the molecular interface which may subject to large deformation and topological change. The three topics presented here demonstrated the desirable flexibilities of formulating the total energy, of parametrizing the energy using phase field function, and of simulating the equilibrium state and dynamics of the system though the numerical solutions of the nonlinear partial differential equations (PDEs) for the geometric flow of the total energy.

The geometrically parametrized total energy obtained by the energetic variational principles entails a rich body of features for mathematical and numerical analysis, including the stability of its critical points, the coarsening dynamics, the solution periodicity, and the conservative discretization of the resulting PDEs, while most of them remain open as long as the applications to biomolecular problems are concerned. More broad usefulness of the methodology outlined in the present three topics are expected to be established in chemistry, biophysics, and medicine through interdisciplinary research and collaboration.

References

1. M. Adkins. Modeling local pattern formation on membrane surfaces using nonlocal interactions. Ph.D. thesis, Colorado State University, 2015
2. B. Alberts, A. Johnson, J. Lewis, K. Roberts, P. Walter, *Molecular Biology of the Cell* (Garland, Martin Raff, 2002)
3. R.G.W. Anderson, K. Jacobson, A role for lipid shells in targeting proteins to caveolae, rafts, and other lipid domains. Science **296**, 1821–1825 (2002)
4. T. Apajalahti, P. Niemela, P.N. Govindan, M.S. Miettinen, E. Salonen, S.-J. Marrink, I. Vattulainen, Concerted diffusion of lipids in raft-like membranes. Faraday Discuss. **144**, 411–430 (2010)
5. G. Archontis, T. Simonson, M. Karplus, Binding free energies and free energy components from molecular dynamics and Poisson-Boltzmann calculations. application to amino acid recognition by Aspartyl-tRNA synthetase. J. Mol. Biol. **306**(2), 307–327 (2001)
6. C. Azuara, E. Lindahl, P. Koehl, H. Orland, M. Delarue, PDB_Hydro: incorporating dipolar solvents with variable density in the Poisson-Boltzmann treatment of macromolecule electrostatics. Nucleic Acids Res. **34**(Web-Server-Issue), 38–42 (2006)
7. N.A. Baker, Biomolecular applications of Poisson-Boltzmann methods. Rev. Comput. Chem. **21**, 349–379 (2005)
8. P.W. Bates, Z. Chen, Y.H. Sun, G.W. Wei, S. Zhao, Geometric and potential driving formation and evolution of biomolecular surfaces. J. Math. Biol. **59**, 193–231 (2009)
9. P.W. Bates, G.W. Wei, S. Zhao, The minimal molecular surface (2006). arXiv:q-bio/0610038v1, [q-bio.BM]
10. P.W. Bates, G.W. Wei, S. Zhao, Minimal molecular surfaces and their applications. J. Comput. Chem. **29**(3), 380–91 (2008)
11. T. Baumgart, B.R. Capraro, C. Zhu, S.L. Das, Thermodynamics and mechanics of membrane curvature generation and sensing by proteins and lipids. Ann. Rev. Phys. Chem. **62**, 483–506 (2011)
12. S.C. Brenner, G. Shiyuan, T. Gudi, L.-Y. Sung, A quadratic c0 interior penalty method for linear fourth order boundary value problems with boundary conditions of the Cahn-Hilliard type. SIAM J. Numer. Anal. **50**(4), 2088–2110 (2012)
13. R. Brewster, P.A. Pincus, S.A. Safran, Hybrid lipids as a biological surface-active component. Biophys. J. **97**, 1087–1094 (2009)
14. R. Brewster, S.A. Safran, Line active hybrid lipids determine domain size in phase separation of saturated and unsaturated lipids. Biophys. J. **98**(6), L21–L23 (2010)
15. B.R. Brooks, R.E. Bruccoleri, B.D. Olafson, D.J. States, S. Swaminathan, M. Karplus, CHARMM: a program for macromolecular energy, minimization, ànd dynamics calculations. J. Comput. Chem. **4**, 187–217 (1983)
16. K.M. Callenberg, N.R. Latorraca, M. Grabe, Membrane bending is critical for the stability of voltage sensor segments in the membrane. J. Gen. Physiol. **140**, 55–68 (2012)
17. A.B. Camley, F.L.H. Brown, Dynamic simulations of multicomponent lipid membranes over long length and time scales. Phys. Rev. Lett. **105**, 148102 (2010)
18. F. Campelo, A. Hernandez-Machado, Shape instabilities in vesicles: a phase-field model. Euro. Phys. J. Spec. Top. **143**(1), 101–108 (2007)
19. P.B. Canham, The minimum energy of bending as a possible explanation of the biconcave shape of the human red blood cell. J. Math. Biol. **26**, 61–81 (1970)
20. D.A. Case, D.A. Pearlman, J.W. Caldwell, T.E. Cheatham, W.S. Ross, C.L. Simmerling, T.A. Darden, K.M. Merz, R.V. Stanton, A.L. Cheng, J.J. Vincent, M. Crowley, D.M. Ferguson, V. Tsui, R.J. Radmer, Y. Duan, J. Pitera, I. Massova, G.L. Seibel, U.C. Singh, P.K. Weiner, P.A. Kollman, Amber 7.0 (University of California, San Francisco, CA, 2002)
21. J. Che, J. Dzubiella, B. Li, J.A. McCammon, Electrostatic free energy and its variations in implicit solvent models. J. Phys. Chem. B **112**, 3058–3069 (2008)
22. M. Chen, T. Bin, Benzhuo Lu, Triangulated manifold meshing method preserving molecular surface topology. J. Mole. Graph. Model. **38**, 411–418 (2012)

23. X. Chen, Q. Cui, Computational molecular biomechanics: a hierarchical multiscale framework with applications to gating of mechanosensitive channels of large conductance, in *Trends in Computational Nanomechanics*, vol. 9, Challenges and Advances in Computational Chemistry and Physics, ed. by Traian Dumitrica (Springer, Netherlands, 2010), pp. 535–556
24. X. Chen, Spectrum for the allen-cahn, cahn-hilliard, and phase-field equations for generic interfaces. Commun. Partial Differ. Equ. **19**, 1371–1395 (1994)
25. Z. Chen, G.W. Wei, Differential geometry based solvation models III: Quantum formulation. J. Chem. Phys. **135**, 194108 (2011)
26. Z. Chen, N.A. Baker, G.W. Wei, Differential geometry based solvation model II: Lagrangian formulation. J. Math. Biol. **63**(6), 1139–1200 (2011)
27. Z. Chen, N.A. Baker, G.W. Wei, Differential geometry based solvation model I: Eulerian formulation. J. Comput. Phys. **229**(22), 8231–8258 (2010)
28. Z. Chen, S. Zhao, J. Chun, D.G. Thomas, N.A. Baker, P.W. Bates, G.W. Wei, Variational approach for nonpolar solvation analysis. J. Chem. Phys. **137**(8) (2012)
29. L.T. Cheng, J. Dzubiella, A.J. McCammon, B. Li, Application of the level-set method to the implicit solvation of nonpolar molecules. J. Chem. Phys. **127**(8) (2007)
30. L.-T. Cheng, Y. Xie, J. Dzubiella, J.A. McCammon, J. Che, B. Li, Coupling the level-set method with molecular mechanics for variational implicit solvation of nonpolar molecules. J. Chem. Theor. Comput. **5**(2), 257–266 (2009). PMID: 20150952
31. C. Chipot, Free energy calculations in biological systems. How useful are they in practice?, in *New Algorithms for Macromolecular Simulation*, ed. by B. Leimkuhler, C. Chipot, R. Elber, A. Laaksonen, A. Mark, T. Schlick, C. Schütte, R. Skeel (Springer, Berlin, Heidelberg, 2006), pp. 185–211
32. S. Choe, K.A. Hecht, M. Grabe, A continuum method for determining membrane protein insertion energies and the problem of charged residues. J. Gen. Physiol. **131**, 563–6573 (2008)
33. N. Choudhury, B.M. Pettitt, On the mechanism of hydrophobic association of nanoscopic solutes. J. Am. Chem. Soc. **127**(10), 3556–3567 (2005)
34. V.B. Chu, Y. Bai, J. Lipfert, D. Herschlag, S. Doniach, Evaluation of ion binding to DNA duplexes using a size-modified Poisson-Boltzmann theory. Biophys. J. **93**, 3202–3209 (2007)
35. F.S. Cohen, R. Eisenberg, R.J. Ryham, A dynamic model of open vesicles in fluids. Commun. Math. Sci. **10**, 1273–1285 (2012)
36. P. Concus, G.H. Golub, Use of fast direct methods for the efficient numerical solution of nonseparable elliptic equations. SIAM J. Numer. Anal. **10**(6), 1103–1120 (1973)
37. I.R. Cooke, M. Deserno, Coupling between lipid shape and membrane curvature. Biophys. J. **91**(2), 487–495 (2006)
38. S. Dai, K. Promislow, Geometric evolution of bilayers under the functionalized Cahn-Hilliard equation. Proc. R. Soc. A **469**, 20120505 (2013)
39. M. Daily, J. Chun, A. Heredia-Langner, G.W. Wei, N.A. Baker, Origin of parameter degeneracy and molecular shape relationships in geometric-flow calculations of solvation free energies. J. Chem. Phys. **139**, 204108 (2013)
40. M.E. Davis, J.A. McCammon, Electrostatics in biomolecular structure and dynamics. Chem. Rev. **90**(3), 509–521 (1990)
41. K.A. Dill, S. Bromberg, *Molecular Driving Forces: Statistical Thermodynamics in Chemistry and Biology*, 2 edn. (Garland Science, 2010)
42. Q. Du, C. Liu, R. Ryham, X. Wang, Modeling the spontaneous curvature effects in static cell membrane deformations by a phase field formulation. Commun. Pure Appl. Anal. **4**, 537–548 (2005)
43. Q. Du, L. Ju, L. Tian, Finite element approximation of the Cahn-Hilliard equation on surfaces. Comput. Methods Appl. Mech. Eng. **200**(29–32), 2458–2470 (2011)
44. Q. Du, C. Liu, R. Ryham, X. Wang, A phase field formulation of the Willmore problem. Nonlinearity **18**, 1249 (2005)
45. Q. Du, C. Liu, X. Wang, Simulating the deformation of vesicle membranes under elastic bending energy in three dimensions. J. Comput. Phys. **212**(2), 757–777 (2006)

46. J. Dzubiella, J.M.J. Swanson, J.A. McCammon, Coupling hydrophobicity, dispersion, and electrostatics in continuum solvent models. Phys. Rev. Lett. **96**, 087802 (2006)
47. J. Dzubiella, J.M.J. Swanson, J.A. McCammon, Coupling nonpolar and polar solvation free energies in implicit solvent models. J. Chem. Phys. **124**(8) (2006)
48. H. Edelsbrunner, P. Koehl, The geometry of biomolecular solvation, in *Combinatorial and Computational Geometry*, pp. 243–276 (2005)
49. E.A. Evans, Bending resistance and chemically induced moments in membrane bilayers. Biophys. J. **14**, 923–931 (1974)
50. J. Faraudo, Diffusion equation on curved surfaces. I. Theory and application to biological membranes. J. Chem. Phys. **116**, 5831 (2002)
51. K. Farsad, P.D. Camilli, Mechanisms of membrane deformation. Curr. Opin. Cell Biol. **15**, 372–381 (2003)
52. M. Feig, C.L. Brooks, Recent advances in the development and application of implicit solvent models in biomolecule simulations. Curr. Opin. Struct. Biol. **14**(2), 217–224 (2004)
53. N. Fuller, C.R. Benatti, R. Peter Rand, Curvature and bending constants for phosphatidylserine-containing membranes. Biophys. J. **85**, 1667–1674 (2003)
54. E. Gallicchio, R.M. Levy, AGBNP: an analytic implicit solvent model suitable for molecular dynamics simulations and high-resolution modeling. J. Comput. Chem. **25**(4), 479–499 (2004)
55. E. Gallicchio, L.Y. Zhang, R.M. Levy, The SGB/NP hydration free energy model based on the surface generalized Born solvent reaction field and novel nonpolar hydration free energy estimators. J. Comput. Chem. **23**(5), 517–529 (2002)
56. L.-T. Gao, X.-Q. Feng, H. Gao, A phase field method for simulating morphological evolution of vesicles in electric fields. J. Comput. Phys. **228**(11), 4162–4181 (2009)
57. N. Gavish, G. Hayrapetyan, K. Promislow, L. Yang, Curvature driven flow of bi-layer interfaces. Phys. D: Nonlinear Phenom. **240**(7), 675–693 (2011)
58. W. Geng, G.W. Wei, Multiscale molecular dynamics using the matched interface and boundary method. J. Comput. Phys. **230**(2), 435–457 (2011)
59. M.K. Gilson, M.E. Davis, B.A. Luty, J.A. McCammon, Computation of electrostatic forces on solvated molecules using the Poisson-Boltzmann equation. J. Phys. Chem. **97**(14), 3591–3600 (1993)
60. J.A. Grant, B.T. Pickup, A Gaussian description of molecular shape. J. Phys. Chem. **99**, 3503–3510 (1995)
61. J.A. Grant, B.T. Pickup, A. Nicholls, A smooth permittivity function for Poisson-Boltzmann solvation methods. J. Comput. Chem. **22**(6), 608–640 (2001)
62. J.A. Grant, B.T. Pickup, M.T. Sykes, C.A. Kitchen, A. Nicholls, The Gaussian Generalized Born model: application to small molecules. Phys. Chem. Chem. Phys. **9**, 4913–22 (2007)
63. P. Grochowski, J. Trylska, Continuum molecular electrostatics, salt effects, and counterion binding—a review of the Poisson-Boltzmann theory and its modifications. Biopolymers **89**, 93–113 (2008)
64. J. Gumbart, B. Roux, Determination of membrane-insertion free energies by molecular dynamics simulations. Biophys. J. **102**, 795–801 (2012)
65. W. Helfrich, Elastic properties of lipid bilayers: theory and possible experiments. Z. Naturforsch **28c**, 693–703 (1973)
66. H.J. Hocker, K.-J. Cho, C.-Y.K. Chen, N. Rambahal, S.R. Sagineedu, K. Shaari, J. Stanslas, J.F. Hancock, A.A. Gorfe, Andrographolide derivatives inhibit guanine nucleotide exchange and abrogate oncogenic Ras function. Proc. Natl. Acad. Sci. U.S.A. **110**(25), 10201–10206 (2013)
67. T.-L. Horng, T.-C. Lin, C. Liu, B. Eisenberg, PNP equations with steric effects: a model of ion flow through channels. J. Phys. Chem. B **116**(37), 11422–11441 (2012)
68. M.D. Huang, D. Chandler, Temperature and length scale dependence of hydrophobic effects and their possible implications for protein folding. Proc. Natl. Acad. Sci. **97**(15), 8324–8327 (2000)
69. Y. Hyon, B. Eisenberg, C. Liu, A mathematical model for the hard sphere repulsion in ionic solutions. Commun. Math. Sci. **9**, 459–475 (2011)

70. Y. Hyon, J.E. Fonseca, B. Eisenberg, C. Liu, Energy variational approach to study charge inversion (layering) near charged walls. Discrete Contin. Dyn. Syst. Ser. B **17**, 2725–2743 (2012)
71. Y.K. Hyon, B. Eisenberg, C. Liu, An energetic variational approach to ion channel dynamics. Math. Methods Appl. Sci. **37**(7), 952–961 (2014)
72. A. Ivankin, I. Kuzmenko, D. Gidalevitz, Cholesterol mediates membrane curvature during fusion events. Phys. Rev. Lett. **108**, 238103 (2012)
73. A. Jaramillo, S.J. Wodak, Computational protein design is a challenge for implicit solvation models. Biophys. J. **88**, 156–171 (2005)
74. K.S. Kim, J. Neu, G. Oster, Curvature-mediated interactions between membrane proteins. Biophys. J. **75**, 2274–2291 (1998)
75. T. Kirchhausen, Bending membranes. Nat. Cell Biol. **14**, 906–908 (2012)
76. W. Kuhnel, *Differential Geometry: Curves, Surfaces, Manifolds* (AMS, 2006)
77. B. Lee, F.M. Richards, The interpretation of protein structures: estimation of static accessibility. J. Mol. Biol. **55**(3), 379–400 (1971)
78. H.G. Lee, J. Kim, Regularized dirac delta functions for phase field models. Int. J. Numer. Methods Eng. **91**(3), 269–288 (2012)
79. M.S. Lee, M.A. Olson, Comparison of volume and surface area nonpolar solvation free energy terms for implicit solvent simulations. J. Chem. Phys. **139**(4) (2013)
80. G.P. Leser, R.A. Lamb, Influenza virus assembly and budding in raft-derived microdomains: a quantitative analysis of the surface distribution of HA, NA and M2 proteins. Virology **342**(2), 215–227 (2005)
81. R.M. Levy, L.Y. Zhang, E. Gallicchio, A.K. Felts, On the nonpolar hydration free energy of proteins: surface area and continuum solvent models for the solute-solvent interaction energy. J. Am. Chem. Soc. **125**, 9523–9530 (2003)
82. B. Li, Minimization of electrostatic free energy and the Poisson-Boltzmann equation for molecular solvation with implicit solvent. SIAM J. Math. Anal. **40**, 2536–2566 (2009)
83. B. Li, Continuum electrostatics for ionic solutions with non-uniform ionic sizes. Nonlinearity **22**, 811 (2009)
84. B. Li, X. Cheng, Z. Zhou, Dielectric boundary force in molecular solvation with the Poisson-Boltzmann free energy: a shape derivative approach. SIAM J. Appl. Math. **71**, 2093–2111 (2011)
85. H. Li, A.A. Gorfe, Aggregation of lipid-anchored full-length H-Ras in lipid bilayers: simulations with the MARTINI force field. PLoS ONE **8**(7), e71018, 07 (2013)
86. S. Li, J. Lowengrub, A. Voigt, Locomotion, wrinkling, and budding of a multicomponent vesicle in viscous fluids. Commun. Math. Sci. **10**, 645–670 (2012)
87. H.Y. Lin, X. Zou, Electrostatics of ligand binding: parametrization of the generalized Born model and comparison with the Poisson-Boltzmann approach. J. Phys. Chem. B **110**, 9304–9313 (2006)
88. B. Lu, Y.C. Zhou, M. Holst, J.A. McCammon, Recent progress in numerical solution of the Poisson-Boltzmann equation for biophysical applications. Commun. Comput. Phys. **3**, 973–1009 (2008)
89. B. Lu, Y.C. Zhou, Poisson-Nernst-Planck equations for simulating biomolecular diffusion-reaction processes II. Biophys. J. **100**, 2475–2485 (2011)
90. K. Lum, D. Chandler, J.D. Weeks, Hydrophobicity at small and large length scales. J. Phys. Chem. B **103**(22), 4570–4577 (1999)
91. V. Luzhkov, A. Warshel, Microscopic models for quantum-mechanical calculations of chemical processes in solutions—ld/ampac and scaas/ampac calculations of solvation energies. J. Comput. Chem. **13**(2), 199–213 (1992)
92. J.D. Madura, Y. Nakajima, R.M. Hamilton, A. Wierzbicki, A. Warshel, Calculations of the electrostatic free energy contributions to the binding free energy of sulfonamides to carbonic anhydrase. Struct. Chem. **7**(2), 131–138 (1996)
93. D. Marsh, Lateral pressure profile, spontaneous curvature frustration, and the incorporation and conformation of proteins in membranes. Biophys. J. **93**, 3884–3899 (2007)

94. A. Martinire, I. Lavagi, G. Nageswaran, D.J. Rolfe, L. Maneta-Peyret, D.-T. Luu, S.W. Botchway, S.E.D. Webb, S. Mongrand, C. Maurel, M.L. Martin-Fernandez, J. Kleine-Vehn, J. Friml, P. Moreau, J. Runions, Cell wall constrains lateral diffusion of plant plasma-membrane proteins. Proc. Natl. Acad. Sci. U.S.A. **109**(31), 12805–12810 (2012)
95. I. Massova, A. Kollman, Combined molecular mechanical and continuum solvent approach (MM-PBSA/GBSA) to predict ligand binding. Perspect. Drug Discov. Des. **18**(1), 113–135 (2000)
96. H.T. McMahon, J.L. Gallop, Membrane curvature and mechanisms of dynamic cell membrane remodelling. Nature **438**, 590–596 (2005)
97. M. Mikucki, Electromechanical and curvature-driven molecular flows for lipid membranes. Ph.D. thesis, Colorado State University, 2015
98. C. Mim, V.M. Unger, Membrane curvature and its generation by BAR proteins. Trends Biochem. Sci. **37**(12), 526–533 (2012)
99. A. Nicholls, D.L. Mobley, J. Peter Guthrie, J.D. Chodera, C.I. Bayly, M.D. Cooper, V.S. Pande, Predicting small-molecule solvation free energies: an informal blind test for computational chemistry. J. Med. Chem. **51**(4), 769–779 (2008). PMID: 18215013
100. S. Osher, J.A. Sethian, Fronts propagating with curvature-dependent speed: algorithms based on Hamilton-Jacobi formulations. J. Chem. Phys. **79**(1), 12–49 (1988)
101. I. Park, Y.H. Jang, S. Hwang, D.S. Chung, Poisson-Boltzmann continuum solvation models for nonaqueous solvents I. 1-octanol. Chem. Lett. **32**(4), 376–377 (2003)
102. R. Parthasarathy, Y. Cheng-Han, J.T. Groves, Curvature-modulated phase separation in lipid bilayer membranes. Langmuir **22**(11), 5095–5099 (2006)
103. D.L. Parton, J.W. Klingelhoefer, S.P. Mark, Aggregation of model membrane proteins, modulated by hydrophobic mismatch, membrane curvature, and protein class. Biophys. J. **101**(3), 691–699 (2011)
104. J.C. Phillips, R. Braun, W. Wang, J. Gumbart, E. Tajkhorshid, E. Villa, C. Chipot, R.D. Skeel, L. Kal, K. Schulten, Scalable molecular dynamics with NAMD. J. Comput. Chem. **26**(16), 1781–1802 (2005)
105. R.A. Pierotti, A scaled particle theory of aqueous and nonaqueous solutions. Chem. Rev. **76**(6), 717–726 (1976)
106. L.J. Pike, Lipid rafts: bringing order to chaos. J. Lipid Res. **44**, 655–667 (2003)
107. S. Ramadurai, A. Holt, V. Krasnikov, G. van den Bogaart, J. Antoinette Killian, B. Poolman, Lateral diffusion of membrane proteins. J. Am. Chem. Soc. **131**, 12650–12656 (2009)
108. M. Rami Reddy, C. Ravikumar Reddy, R.S. Rathore, M.D. Erion, P. Aparoy, R. Nageswara Reddy, P. Reddanna, Free energy calculations to estimate ligand-binding affinities in structure-based drug design. Curr. Pharm. Des. **20**, 3323–3337 (2014)
109. P. Ren, J. Chun, D.G. Thomas, M.J. Schnieders, M. Marucho, J. Zhang, N.A. Baker, Biomolecular electrostatics and solvation: a computational perspective. Quart. Rev. Biophys. **45**(11), 427–491 (2012)
110. B.J. Reynwar, G. Illya, V.A. Harmandaris, M.M. Muller, K. Kremer, M. Deserno, Aggregation and vesiculation of membrane proteins by curvature-mediated interactions. Nature **447**, 461–464 (2006)
111. F.M. Richards, Areas, volumes, packing, and protein structure. Annu. Rev. Biophys. Bioeng. **6**(1), 151–176 (1977)
112. J.E. Rothman, Mechanisms of intracellular protein transport. Nature **372**, 55–63 (1994)
113. R.J. Ryham, M.A. Ward, F.S. Cohen, Teardrop shapes minimize bending energy of fusion pores connecting planar bilayers. Phys. Rev. E **88**, 062701 (2013)
114. T. Schlick, *Molecular Modeling and Simulation: An Interdisciplinary Guide* (Springer, 2002)
115. N.W. Schmidt, A. Mishra, J. Wang, W.F. DeGrado, C.L. Gerard, Wong. Influenza virus A M2 protein generates negative Gaussian membrane curvature necessary for budding and scission. J. Am. Chem. Soc. **135**(37), 13710 (2013)
116. K.A. Sharp, B. Honig, Calculating total electrostatic energies with the nonlinear Poisson-Boltzmann equation. J. Phys. Chem. **94**, 7684–7692 (1990)

117. M.R. Shirts, D.L. Mobley, S.P. Brown, Free-energy calculations in structure-based drug design, in *Drug Design: Structure- and Ligand-Based Approaches* (Cambridge University Press, 2010)
118. K. Simons, D. Toomre, Lift rafts and signal transduction. Nat. Rev. Mol. Biol. **1**, 31–40 (2000)
119. T. Simonson, G. Archontis, M. Karplus, Free energy simulations come of age: protein-ligand recognition. Accounts Chem. Res. **35**(6), 430–437 (2002)
120. R.E. Skyner, J.L. McDonagh, C.R. Groom, T. van Mourik, J.B.O. Mitchell, A review of methods for the calculation of solution free energies and the modelling of systems in solution. Phys. Chem. Chem. Phys. **17**, 6174–6191 (2015)
121. T. Sollner, M.K. Bennett, S.W. Whiteheart, R.H. Scheller, J.E. Rothman, A protein assembly-disassembly pathway in vitro that may correspond to sequential steps of synaptic vesicle docking, activation, and fusion. Cell **75**(3), 409–418 (1993)
122. J.C. Stachowiak, C.C. Hayden, D.Y. Sasaki, Steric confinement of proteins on lipid membranes can drive curvature and tubulation. Proc. Natl. Acad. Sci. U.S.A. **107**(17), 7781–7786 (2010)
123. O. Sten-Knudsen, *Biological Membranes: Theory of Transport, Potential and Electric Impulses* (Cambridge University Press, Cambridge, 2002)
124. F.H. Stillinger, Structure in aqueous solutions of nonpolar solutes from the standpoint of scaled-particle theory. J. Solut. Chem. **2**, 141–158 (1973)
125. J. Strain, Fast spectrally-accurate solution of variable-coefficient elliptic problems. Proc. Am. Math. Soc. **122**, 843–850 (1994)
126. S. Svetina, Curvature-dependent protein-lipid bilayer interaction and cell mechanosensitivity. Eur. Bio. Phys. J. 1–7 (2015)
127. J.M.J. Swanson, R.H. Henchman, J.A. McCammon, Revisiting free energy calculations: a theoretical connection to MM/PBSA and direct calculation of the association free energy. Biophys. J. **86**(1), 67–74 (2004)
128. T. Takahashi, T. Suzuki, Function of membrane rafts in viral life cycles and host cellular response. Biochem. Res. Int. **2011**, 245090 (2011)
129. Y. Tang, G. Cao, X. Chen, J. Yoo, A. Yethiraj, Q. Cui, A finite element framework for studying the mechanical response of macromolecules: application to the gating of machanosensitive channel MscL. Biophys. J. **91**, 1248–1263 (2006)
130. K.E. Teigen, X. Li, J. Lowengrub, F. Wang, A. Voigt, A diffuse-interface approach for modelling transport, diffusion and adsorption/desorption of material quantities on a deformable interface. Commun. Math. Sci. **7**, 1009–1037 (2009)
131. K.E. Teigen, P. Song, J. Lowengrub, A. Voigt, A diffuse-interface method for two-phase flows with soluble surfactants. J. Comput. Phys. **230**, 375–393 (2011)
132. D.G. Thomas, J. Chun, Z. Chen, G.W. Wei, N.A. Baker, Parameterization of a geometric flow implicit solvation model. J. Comput. Chem. **24**, 687–695 (2013)
133. D. Van Der Spoel, E. Lindahl, B. Hess, G. Groenhof, A.E. Mark, J.C.H. Berendsen, GROMACS: fast, flexible, and free. J. Comput. Chem. **26**(16), 1701–1718 (2005)
134. F.S. Vieira, G. Correa, M. Einicker-Lamas, R. Coutinho-Silva, Host-cell lipid rafts: a safe door for micro-organisms? Biol. Cell **102**(7), 391–407 (2010)
135. A. Voth, M.S. Gregory, Membrane tension controls the assembly of curvature-generating proteins. Nat. Commun. **6**, 7219 (2015)
136. J. Wagoner, N.A. Baker, Solvation forces on biomolecular structures: a comparison of explicit solvent and Poisson-Boltzmann models. J. Comput. Chem. **25**(13), 1623–1629 (2004)
137. J.A. Wagoner, N.A. Baker, Assessing implicit models for nonpolar mean solvation forces: the importance of dispersion and volume terms. Proc. Natl. Acad. Sci. U.S.A. **103**(22), 8331–8336 (2006)
138. B. Wang, G.W. Wei, Parameter optimization in differential geometry based solvation models. J. Chem. Phys. **143**, 134119 (2015)
139. Y. Wang, G.W. Wei, S.-Y. Yang, Partial differential equation transform—variational formulation and Fourier analysis. Int. J. Numer. Methods Biomed. Eng. **27**, 1996–2020 (2011)
140. Y. Wang, G.W. Wei, S.-Y. Yang, Iterative filtering decomposition based on local spectral evolution kernel. J. Sci. Comput. **50**, 629–664 (2012)

141. Y. Wang, G.W. Wei, S.-Y. Yang, Mode decomposition evolution equations. J. Sci. Comput. **50**, 495–518 (2012)
142. A. Warshel. Consistent calculations of electrostatic energies in proteins and solutions. Abstr. Pap. Am. Chem. Soc. **214**, 191–COMP (1997)
143. A. Warshel, Z.T. Chu, Calculations of solvation free-energies in chemistry and biology, in *ACS Symposium Series*, vol. J9, no. 568, pp. 71–94 (1994)
144. A. Warshel, P.K. Sharma, M. Kato, W.W. Parson, Modeling electrostatic effects in proteins. BBA-Proteins Proteomics **1764**, 1647–1676 (2006)
145. G.W. Wei, Generalized Perona-Malik equation for image restoration. IEEE Signal Process. Lett. **6**(7), 165–167 (1999)
146. G.W. Wei, Y.H. Sun, Y.C. Zhou, M. Feig, Molecular multiresolution surfaces, pp. 1–11 (2005). arXiv:math-ph/0511001v1
147. G.-W. Wei, Differential geometry based multiscale models. Bull. Math. Biol. **72**(6), 1562–1622 (2010)
148. G.-W. Wei, Multiscale, multiphysics and multidomain models I: basic theory. J. Chem. Theor. Comput. **12**(8), 1341006 (2013)
149. G.-W. Wei, Q. Zheng, Z. Chen, K. Xia, Variational multiscale models for charge transport. SIAM Rev. **54**(4), 699–754 (2012)
150. H. Whitney, *Geometric Integration Theory* (Dover, 2005)
151. T. Witkowski, R. Backofen, A. Voigt, The influence of membrane bound proteins on phase separation and coarsening in cell membranes. Phys. Chem. Chem. Phys. **14**, 14509–14515 (2012)
152. J.J. Xu, Y. Yang, J. Lowengrub, A level-set continuum method for two-phase flows with insoluble surfactant. J. Comput. Phys. **231**(17), 5897–5909 (2012)
153. S. Xu, P. Sheng, C. Liu, An energetic variational approach for ion transport. Commun. Math. Sci. **12**, 779–789 (2014)
154. Z.Y. Yu, C. Bajaj, Computational approaches for automatic structural analysis of large biomolecular complexes. IEEE/ACM Trans. Comput. Biol. Bioinform. **5**, 568–582 (2008)
155. S. Zhao, Pseudo-time-coupled nonlinear models for biomolecular surface representation and solvation analysis. Int. J. Numer. Methods Biomed. Eng. **27**, 1964–1981 (2011)
156. S. Zhao, Operator splitting ADI schemes for pseudo-time coupled nonlinear solvation simulations. J. Comput. Phys. **257**, 1000–1021 (2014)
157. Y. Zhao, D. Qiang, Diffuse interface model of multicomponent vesicle adhesion and fusion. Phys. Rev. E **84**, 011903 (2011)
158. Q. Zheng, D. Chen, G.W. Wei, Second-order Poisson-Nernst-Planck solver for ion transport. J. Comput. Phys. **230**, 5239–5262 (2011)
159. Q. Zheng, S.Y. Yang, G.W. Wei, Molecular surface generation using PDE transform. Int. J. Numer. Methods Biomed. Eng. **28**, 291–316 (2012)
160. S. Zhou, L.-T. Cheng, J. Dzubiella, B. Li, J.A. McCammon, Variational implicit solvation with Poisson–Boltzmann theory. J. Chem. Theor. Comput. **10**(4), 1454–1467 (2014). PMID: 24803864
161. Y.C. Zhou, Electrodiffusion of lipids on membrane surfaces. J. Chem. Phys. **136**, 205103 (2012)
162. Y.C. Zhou, B. Lu, A.A. Gorfe, Continuum electromechanical modeling of protein-membrane interactions. Phys. Rev. E **82**(4), 041923 (2010)
163. J. Zimmerberg, M.M. Kozlov, How proteins produce cellular membrane curvature. Nat. Rev. Mol. Cell Biol. **7**, 9–19 (2006)

A Theoretician's Approach to Nematic Liquid Crystals and Their Applications

Apala Majumdar and Alexander H. Lewis

1 Introduction

Liquid crystals are classical examples of mesogenic materials or partially ordered materials that are intermediate in character between the conventional solid and liquid phases [5, 13, 22, 32]. Liquid crystals are a flourishing sub-field of *Soft Matter*, both from a purely scientific and an applications point of view. Liquid crystal research is inherently interdisciplinary and is a perfect meeting point for physics, chemistry, biology, engineering and, more recently, mathematics. This chapter is a self-contained but not exhaustive account of the different modelling approaches to *nematic liquid crystals*—the simplest but most widely used class of liquid crystals with examples of how to apply mathematical tools to prototype confined nematic systems yielding quantitative predictions. These examples hopefully illustrate the rich mathematical content of liquid crystal theory and, reciprocally, the use of mathematics in an interdisciplinary context.

We often grow up with the notion that matter typically exists in three forms: solid, liquid and gas. However, materials science is far more complex than this simple classification. There are everyday examples of materials that are neither fluid nor solid, such as toothpaste, butter, gels, soaps and the very human skin and cell membranes themselves [32]! These materials, to some extent, combine the flexibility and mobility of liquids with a degree of order characteristic of solids and hence, they are partially ordered materials.

A. Majumdar (✉)
Department of Mathematical Sciences, University of Bath,
Claverton Down, Bath BA2 7AY, UK
e-mail: a.majumdar@bath.ac.uk

A.H. Lewis
Mathematical Institute, University of Oxford, Andrew Wiles Building
Radcliffe Observatory Quarter, Woodstock Road, Oxford OX2 6GG, UK
e-mail: alex.lewis@maths.ox.ac.uk

J. Wu (ed.), *Variational Methods in Molecular Modeling*,
Molecular Modeling and Simulation, DOI 10.1007/978-981-10-2502-0_8

Liquid crystals are one of the most widely studied classes of partially ordered materials. Indeed, they are the only examples of thermodynamically stable soft matter phases, metastable states being common among other soft matter classes [32]. Liquid crystals were discovered accidentally in 1888 by Friedrich Reinitzer, a plant physiologist at the Charles University, in Prague, when he discovered two melting points for cholesteryl benzoate [1]. The first melting point corresponded to the temperature-induced transition from the solid to cholesteric liquid crystal phase and the second melting point corresponded to the transition from a liquid crystalline to liquid phase. Between 1889–1908, European scientists had discovered more than 200 compounds with liquid crystalline properties and by the early 1920s, the work of Georges Friedel and Otto Lehmann resulted in a basic classification scheme for liquid crystals into nematic, smectic and cholesteric phases [1]. In the 1960s, scientists became aware of the electro-optic potential of liquid crystals, notably, their sensitivity to light and external electric fields. Pioneering research at the RCA Laboratories in Princeton, in particular the contributions of Richard Williams and George Heilmeier, led to the successful commercialization of liquid crystal displays (LCDs), and liquid crystal research then accelerated throughout Europe and North America, particularly in the 70's with special reference to the works of Wolfgang Helfrich, Martin Schadt and James Fergason, to generate the multi-billion dollar LCD market in vogue today.

Nematic liquid crystals are the simplest type of liquid crystals, with constituent anisotropic rod-like molecules of length between 10–30 nm [5, 32]. These rod-like molecules tend to move freely as in a conventional liquid but whilst moving, they align along certain locally preferred directions. In other words, nematics are *anisotropic liquids* or liquids with special directions and long-range orientational order [22]. The theory of nematic liquid crystals is largely devoted to describing the concept of partial orientational order and how this order manifests in macroscopic and experimental phenomena. As with most complex materials, one can describe the partial orientational order in terms of molecular interactions at a microscopic level, in terms of a probability distribution function for the molecular orientations at a mean-field level and in terms of a macroscopic order parameter, which is typically defined in terms of a macroscopic signature of the material anisotropy, at the macroscopic level [7, 37, 43].

Nematic liquid crystals are the working material of choice for the multi-billion dollar LCD industry [5]. They have two key properties that render them useful for a range of electro-optic applications: (i) they are optically anisotropic or birefringent materials, i.e., the optical response, measured in terms of the refractive index, is typically different along the distinguished direction of molecular alignment compared to other directions, and (ii) they are sensitive to external electric fields, the sensitivity measured in terms of the dielectric anisotropy which quantifies the difference in the nematic response to electric fields along the preferred directions compared to the non-preferred directions [22]. Consequently, it is relatively straightforward to tune nematic properties with incident light, external fields and mechanical effects, making them hugely popular working materials for phones, TVs, laptops, calculators, watches and even thermometers [5, 32].

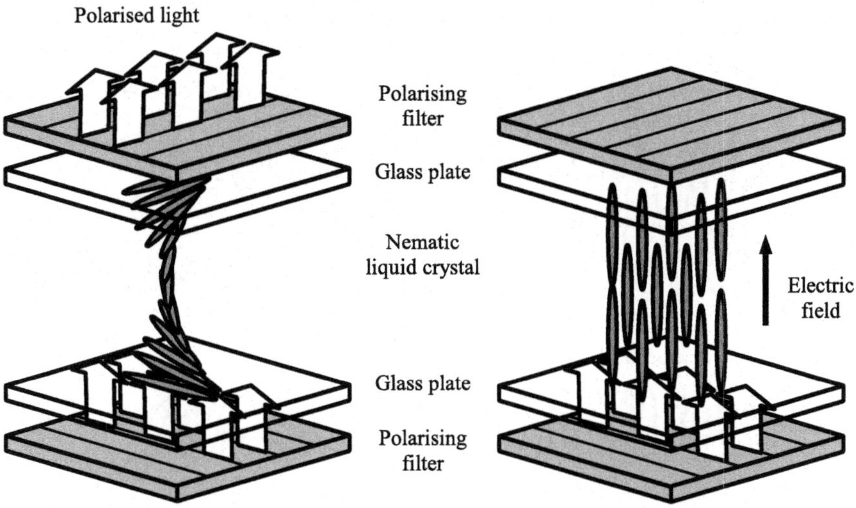

Fig. 1 Sketch of the twisted nematic device, with no external field (*left*) and applied field (*right*). Copyright 2004 From The static and dynamic continuum theory of liquid crystals: a mathematical introduction by Iain W. Stewart [44]. Reproduced by permission of Taylor and Francis Group, LLC, a division of Informa plc

The classical Twisted Nematic Display is illustrated in Fig. 1. The basic construction is straightforward: a layer of nematic material, a few microns thick, sandwiched between two flat plates. The molecular orientation is fixed on both plates; this can be easily achieved in experiments [5]. However, the molecular orientations on the two plates are conflicting, as can be seen from Fig. 1. In the absence of an electric field, the nematic molecules naturally twist between the two plates and the resulting twisted configuration is transparent to incident light [44]. When we apply an electric field along the height of the cell, the nematic molecules prefer to align with the electric field (if the material has positive dielectric anisotropy) and consequently, the nematic molecules are largely oriented in field direction throughout the cell, losing the twisted structure. The resulting field-aligned nematic configuration is opaque to incident light and on the removal of the electric field, the nematic configuration naturally relaxes to the twisted transparent state. This effect has been widely studied in the literature in terms of the *classical Fredericks transition* and we refer the interested reader to [3, 22] for quantitative details.

There are, of course, many different kinds of liquid crystals, and some of the major phases are shown in Fig. 2. For example, smectics are examples of layered liquid crystals for which molecules tend to arrange themselves in well-spaced layers and the molecules tend to align with each other inside each layer [22]. There are several variants of the smectic phase, depending on the relative orientation between the molecular orientations represented by a unit-vector **n**, referred to as *director*, and the layer normals [22]. Equally common are cholesteric liquid crystals or chiral nematic liquid crystals for which the nematic molecules organize themselves into layers but unlike smectics, there is no positional ordering within layers but rather the director field rotates between the layers with a helical profile [14, 22]. In fact, cholesterics

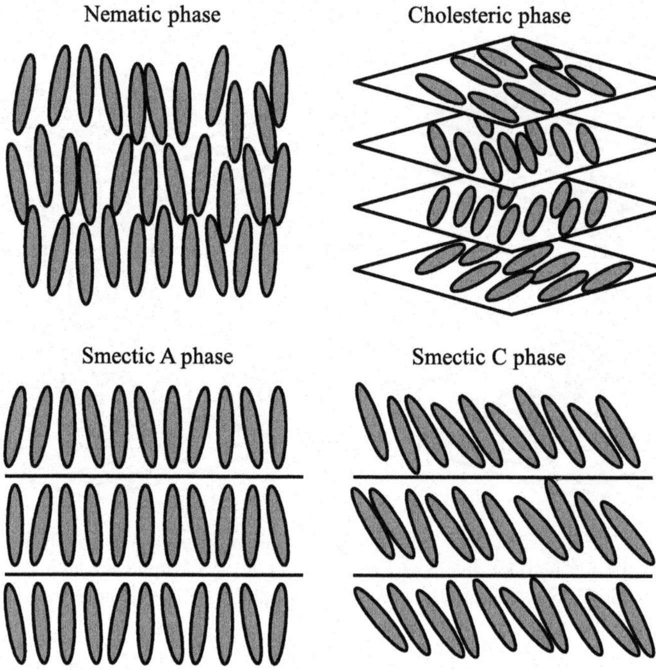

Fig. 2 Sketches of the major liquid crystal phases for rod or disk shaped particles. Copyright 2004 From The static and dynamic continuum theory of liquid crystals: a mathematical introduction by Iain W. Stewart [44]. Reproduced by permission of Taylor and Francis Group, LLC, a division of Informa plc

have also been successfully used as working materials for LCDs, examples of which include flexible ePaper, called i2R based on ChLCD technology [14].

Liquid crystal research has now moved far beyond the conventional LCD technology and scientists are keen to find new ways to exploit the symmetries and material properties of liquid crystals in diverse areas such as nanostructured materials, biosensors, pharmaceutical sciences, drug/nutrient delivery methods and even the liquid crystalline nature of DNA for synthetic biology [32]. This chapter is a modest review of nematic liquid crystals, the widely used continuum modelling approaches: the Landau-de Gennes theory, the Ericksen theory and the Oseen-Frank theory followed by a case study on the mathematical modelling of a planar bistable nematic device, reported in [45]. The case study is based on published work [29, 35, 39] and unpublished private communication [34].

2 Continuum Theories for Nematic Liquid Crystals

There are a number of theoretical approaches for nematic liquid crystals, ranging from atomistic, molecular theories to mean-field theories and fully macroscopic continuum theories [33, 36, 37, 47]. This chapter focuses on continuum theories for nematic liquid crystals and such theories are largely based on the underpinning assumption that the structural details are varying slowly on the molecular length scale. These theories do have limitations and can fail outside certain temperature regimes, near defects or interfaces or are inadequate for describing critical phenomena [40, 47]. Nevertheless, they have been used with great success for modelling generic confined nematic systems, including LCDs and new microfluidic experiments, and, given their rich mathematical landscapes, such theories are the perfect meeting point for mathematicians and practitioners in the field.

Nematic liquid crystals are broadly classified into three categories according to their type and degree of orientational ordering. Nematics are said to be in the *isotropic phase* when the constituent molecules are disordered and we cannot define distinguished directions of molecular alignment. The most widely studied nematic phase is the *uniaxial phase* wherein the constituent molecules have a single distinguished direction of molecular alignment i.e. the molecules either prefer to align along this direction or in the plane orthogonal to it. For example, if we deal with a system of rod-like molecules in an uniaxial phase, then the preferred direction might be the average of the local axes of the rods [22, 46]. Similarly, for disc-shaped nematic molecules in an uniaxial phase, the preferred direction may be the average of the unit normal to the discs. The preferred direction of uniaxial alignment is typically denoted by a unit-vector, $\mathbf{n} \in S^2$, where S^2 is the unit sphere in three dimensions. It is also useful to measure the degree of uniaxial ordering about $\mathbf{n}$; this is conventionally measured in terms of a scalar order parameter S, often defined to be

$$S = \int_0^{\pi} \left(1 - \frac{3}{2}\sin^2\theta\right) P(\theta)\sin\theta \, d\theta, \tag{1}$$

where $P(\theta)\sin\theta d\theta$ is the fraction of molecules that make an angle between θ and $\theta + d\theta$ relative to the director, $\mathbf{n}$ [4]. In particular, this definition ensures that

$$-\frac{1}{2} \leq S \leq 1, \tag{2}$$

with three characteristic values: $S = -1/2$ for which molecules are identically oriented perpendicular to $\mathbf{n}$, $S = 0$ which describes the disordered isotropic phase and $S = 1$ which describes a state of perfect molecular alignment with $\mathbf{n}$. The bounds (2) are not necessarily respected in continuum approaches [40], particularly for low temperatures, as will be detailed below and one would argue that the limiting values $S = -1/2$ and $S = 1$ are unrealistic or unattainable in a physical situation. In [7], we propose an alternative formulation that couples probabilistic approaches as in (1) and (2) and the continuum Landau-de Gennes theory, to prescribe a new *Ball-Majumdar potential*; similar approaches have also been adopted by Katriel et al. [26]. This new approach is outside the scope of this chapter and we focus on traditional continuum models.

Nematics are said to be in the biaxial phase if they have a primary and secondary preferred direction of molecular alignment, typically denoted by two unit-vectors

$\mathbf{n}, \mathbf{m} \in S^2$ and two order parameters that measure the orientational ordering about these directions, referred to as S and R respectively.

In Sect. 2.1, we describe the celebrated Landau-de Gennes theory; in Sect. 2.2, we briefly overview Ericksen's theory for uniaxial nematics with variable order and in Sect. 2.3, we outline the simplest and hugely popular Oseen-Frank theory for uniaxial nematics with constant order.

2.1 The Landau-de Gennes Theory

The Landau-de Gennes (LdG) theory is one of the most celebrated continuum theories for nematic liquid crystals [22, 37, 46]. In fact, Pierre-Gilles de Gennes was awarded the Nobel prize in Physics in 1991 for discovering that "methods developed for studying order phenomena in simple systems can be generalized to more complex forms of matter, in particular to liquid crystals and polymers" and the theory is named after him and Lev Landau of superconductivity fame.

The LdG theory describes the state of a nematic liquid crystal by a macroscopic order parameter—the $\mathbf{Q}$-tensor order parameter, which is defined in terms of macroscopic quantities such as the magnetic susceptibility [22, 46]. From an applications point of view, the LdG $\mathbf{Q}$-tensor measures the deviation of the ordered nematic phase from the isotropic disordered phase, so that we recover the isotropic fluid state when $\mathbf{Q} = 0$. In the LdG framework, the $\mathbf{Q}$ tensor is a symmetric, traceless 3×3 matrix with no a priori bounds on the eigenvalues; in particular, the eigenvalues are not constrained by the inequalities (2). The LdG $\mathbf{Q}$-tensor necessarily has five degrees of freedom and from the spectral decomposition theorem, can be written as [42]

$$\mathbf{Q} = \lambda_1 \mathbf{n} \otimes \mathbf{n} + \lambda_2 \mathbf{m} \otimes \mathbf{m} + \lambda_3 \mathbf{p} \otimes \mathbf{p}, \tag{3}$$

where $\mathbf{n}, \mathbf{m}, \mathbf{p}$ are the three eigenvectors of $\mathbf{Q}$, λ_1, λ_2 and λ_3 are the corresponding eigenvalues with $\lambda_3 = -(\lambda_1 + \lambda_2)$, from the tracelessness constraint. Informally speaking, the eigenvectors represent the preferred directions of molecular alignment and the eigenvalues measure the degree of orientational order about these directions. In the biaxial phase, $\lambda_1 \neq \lambda_2 \neq \lambda_3$ and $\mathbf{Q}$ can be concisely written as

$$\mathbf{Q} = S\left(\mathbf{n} \otimes \mathbf{n} - \frac{\mathbf{I}}{3}\right) + R\left(\mathbf{m} \otimes \mathbf{m} - \frac{\mathbf{I}}{3}\right), \tag{4}$$

where $\mathbf{I}$ is the 3×3 identity matrix, $S = (2\lambda_1 + \lambda_2)$ and $R = (2\lambda_2 + \lambda_1)$. In the uniaxial case, the LdG $\mathbf{Q}$ tensor has a pair of degenerate non-zero eigenvalues, $(\lambda_1, \lambda_2, \lambda_3) = (\lambda, \lambda, -2\lambda)$, so that $\mathbf{Q}$ can be written as

$$\mathbf{Q} = S\left(\mathbf{n} \otimes \mathbf{n} - \frac{\mathbf{I}}{3}\right), \tag{5}$$

where $\mathbf{n}$ is the distinguished eigenvector, referred to as the *director*, with the non-degenerate eigenvalue and the scalar order parameter, $S = -3\lambda$.

In the absence of any external fields or surface energies, the LdG energy functional has a particularly simple form [18, 22, 42, 46]:

$$I[\mathbf{Q}] := \int\int\int_{\Omega} \left[f_B\left(\mathbf{Q}\right) + f_E\left(\mathbf{Q}, \nabla\mathbf{Q}\right)\right] \, dV, \tag{6}$$

where $\Omega \subset \mathbb{R}^3$ is the computational domain, f_B is a bulk potential that determines the degree of orientational order as a function of temperature and f_E is an elastic energy density that penalizes spatial inhomogeneities.

The bulk potential, f_B, is a polynomial function of the eigenvalues of the $\mathbf{Q}$-tensor and drives the isotropic-nematic transition as a function of the temperature [22, 40, 46]. It is quite common to work with a quartic form of f_B as shown below:

$$f_B(\mathbf{Q}) = \frac{A}{2}\mathrm{tr}\mathbf{Q}^2 - \frac{B}{3}\mathrm{tr}\mathbf{Q}^3 + \frac{C}{4}\left(\mathrm{tr}\mathbf{Q}^2\right)^2, \tag{7}$$

where $A = \alpha(T - T^*)$ is the re-scaled temperature, $\alpha, B, C > 0$ are positive material constants, $\mathrm{tr}\mathbf{Q}^2 = \sum_{i=1}^{3} \lambda_i^2$ and $\mathrm{tr}\mathbf{Q}^3 = \sum_{i=1}^{3} \lambda_i^3$. The minimizers of f_B determine the *bulk preferred state* in the absence of any inhomogeneities or external effects and the bulk preferred state depends on A as shown below.

Proposition 1 ([6, 40]) *The critical points of the bulk energy density, f_B in (7), are given by either uniaxial or isotropic $\mathbf{Q}$-tensors of the form (5) for all admissible values of A and all material constants $\alpha, B, C > 0$.*

Proof For a symmetric, traceless matrix $\mathbf{Q}$ of the form (3), $\mathrm{tr}\mathbf{Q}^n = \sum_{i=1}^{3} \lambda_i^n$ subject to the tracelessness condition so that the bulk energy density f_B only depends on the eigenvalues λ_1, λ_2 and λ_3. Then the critical points of f_B are given by the stationary points of the function $f : \mathbb{R}^3 \to \mathbb{R}$ defined by

$$f\left(\lambda_1, \lambda_2, \lambda_3\right) = \frac{A}{2}\sum_{i=1}^{3}\lambda_i^2 - \frac{B}{3}\sum_{i=1}^{3}\lambda_i^3 + \frac{C}{4}\left(\sum_{i=1}^{3}\lambda_i^2\right)^2 - 2\delta\sum_{i=1}^{3}\lambda_i, \tag{8}$$

where we have recast f_B in terms of the eigenvalues and introduced a Lagrange multiplier δ for the tracelessness condition.

The equilibrium equations are given by a system of three algebraic equations

$$\frac{\partial f}{\partial \lambda_i} = 0 \Leftrightarrow A\lambda_i - B\lambda_i^2 + C\left(\sum_{k=1}^{3}\lambda_k^2\right)\lambda_i = 2\delta, \quad \text{for } i = 1 \ldots 3, \tag{9}$$

together with the tracelessness condition $\sum_i \lambda_i = 0$. The system (9) is equivalent to

$$\left(\lambda_i - \lambda_j\right)\left[A - B\left(\lambda_i + \lambda_j\right) + C\sum_{k=1}^{3}\lambda_k^2\right] = 0; \quad 1 \le i < j \le 3. \tag{10}$$

Let $\{\lambda_i\}$ be a solution of the system (9) with three distinct eigenvalues $\lambda_1 \neq \lambda_2 \neq \lambda_3$. We consider Eq. (10) for the pairs (λ_1, λ_2) and (λ_1, λ_3). This yields two equations

$$A - B\left(\lambda_1 + \lambda_2\right) + C\sum_{k=1}^{3}\lambda_k^2 = 0,$$
$$A - B\left(\lambda_1 + \lambda_3\right) + C\sum_{k=1}^{3}\lambda_k^2 = 0, \tag{11}$$

from which we obtain

$$-B\left(\lambda_2 - \lambda_3\right) = 0, \tag{12}$$

contradicting our initial hypothesis $\lambda_2 \neq \lambda_3$. We, thus, conclude that a stationary point of f_B must have at least two equal eigenvalues and therefore correspond to either an uniaxial or isotropic $\mathbf{Q}$-tensor. In particular, there are no biaxial stationary points for the particular choice of f_B in (7).

Therefore, we can compute the *bulk preferred state* in terms of minimizers of the bulk potential f_B in (7), which are necessarily either uniaxial or isotropic. In particular, one can check that the isotropic state, $\mathbf{Q} = 0$, is always a critical point (not necessarily a minimizer) of f_B for all values of A. A standard computation shows that there are three characteristic values of A in this framework: (i) $A = 0$ such that the isotropic critical point loses stability for $A < 0$, (ii) the nematic-isotropic transition temperature, $A = \frac{B^2}{27C}$, for which f_B has two families of minimizers: the isotropic state with $\mathbf{Q} = 0$ and a continuum of uniaxial nematic states given by

$$\mathbf{Q}_{\min} = \left\{ s_+\left(\mathbf{n}\otimes\mathbf{n} - \frac{\mathbf{I}}{3}\right); \mathbf{n} \in S^2 \text{ arbitrary}\right\},$$
$$s_+ = \frac{B + \sqrt{B^2 - 24AC}}{4C}, \tag{13}$$

and (iii) $A = \frac{B^2}{24C}$ so that $\mathbf{Q} = 0$ is the unique critical point of f_B for $A > \frac{B^2}{24C}$. To summarize, the bulk preferred state is the isotropic critical point for $A > \frac{B^2}{27C}$ and the bulk preferred state is defined by the family of uniaxial minimizers in (13) for $A < \frac{B^2}{27C}$ and the simple quartic potential, f_B, in (7), predicts a first order isotropic-nematic transition at $A = \frac{B^2}{27C}$. It is worth noting that s_+ in (13) is less than unity if and only if

$$A \ge \frac{1}{3}\left(B - 3C\right), \tag{14}$$

so that the probabilistic bounds in (2) are not respected for low temperatures (as A decreases) [40].

The elastic energy density, f_E, is typically a quadratic function of $\nabla \mathbf{Q}$ and a commonly used version is

$$f_E(\mathbf{Q}) := \sum_{i,j,k=1}^{3} \frac{L_1}{2}\left(\frac{\partial \mathbf{Q}_{ij}}{\partial \mathbf{x}_k}\right)^2 + \frac{L_2}{2}\frac{\partial \mathbf{Q}_{ij}}{\partial \mathbf{x}_j}\frac{\partial \mathbf{Q}_{ik}}{\partial \mathbf{x}_k} + \frac{L_3}{2}\frac{\partial \mathbf{Q}_{ij}}{\partial \mathbf{x}_k}\frac{\partial \mathbf{Q}_{ik}}{\partial \mathbf{x}_j}, \quad (15)$$

where the material elastic constants, L_1, L_2, L_3 are constrained by [18]

$$L_1 > 0; \quad -L_1 < L_3 < 2L_1; \quad L_2 > -\frac{3}{5}L_1 - \frac{1}{10}L_3. \quad (16)$$

Mathematicians and modellers often work with the one-constant approximation [29, 31, 41], for which $L_2 = L_3 = 0$ and the resulting LdG energy in (6) has a nice variational structure and bears resemblance to the celebrated Ginzburg-Landau energy in superconductivity [10]:

$$I[\mathbf{Q}] := \int\int\int_{\Omega} \left[f_B(\mathbf{Q}) + \frac{L}{2}|\nabla \mathbf{Q}|^2 \right] dV, \quad (17)$$

where f_B is given by (7). As with any problem in the calculus of variations, applied mathematicians model the experimentally observed states in terms of minimizers of the energy functionals, (6) or (17). The energy minimizers are typically solutions of the associated Euler-Lagrange equations [18, 42] and the Euler-Lagrange equations associated with a three-constant elastic energy density as in (15) are not readily amenable to analytic methods. However, the Euler-Lagrange equations associated with the one-constant LdG energy in (17) have a nice mathematical structure (see below)

$$L\Delta \mathbf{Q}_{ij} = A\mathbf{Q}_{ij} - B\left(\mathbf{Q}_{ip}\mathbf{Q}_{pj} - \frac{\delta_{ij}}{3}\right) + C|\mathbf{Q}|^2\mathbf{Q}_{ij}; \quad i, j, p = 1, 2, 3, \quad (18)$$

where we have used the Einstein summation convention, and the analysis of the system (18) has generated much interest in the mathematical community in recent years. A particularly interesting observation is that all solutions of the system (18) are smooth i.e. critical points of the LdG energy (17) have no mathematical singularities and yet, experimentally observed nematic textures abound in point and line singularities [32, 41]. The mathematical characterization of defects in the LdG theory continues to attract worldwide analytical interest.

We note that the bulk potential f_B in (7) is the simplest polynomial of $\mathbf{Q}$ that allows for an isotropic-nematic transition in terms of *bulk preferred states*, as a function of the temperature. It is restrictive in the sense that it can only account for uniaxial or isotropic bulk preferred states. There are more general sixth-order bulk potentials, g_B [2],

$$g_B(\mathbf{Q}) := \frac{A}{2}\mathrm{tr}\mathbf{Q}^2 - \frac{B}{3}\mathrm{tr}\mathbf{Q}^3 + \frac{C}{4}\left(\mathrm{tr}\mathbf{Q}^2\right)^2 + \frac{D}{2}\left(\mathrm{tr}\mathbf{Q}^2\right)\left(\mathrm{tr}\mathbf{Q}^3\right) + \frac{E}{6}\left(\mathrm{tr}\mathbf{Q}^2\right)^3 + (F-E)\left(\mathrm{tr}\mathbf{Q}^3\right)^2, \tag{19}$$

where A is the re-scaled temperature as before, B, C, D, E, F are material-dependent constants and g_B admits isotropic, uniaxial and biaxial critical points [2, 34].

Nematics, in confinement, are strongly influenced by boundary effects and a natural question is—how do we incorporate boundary effects into the LdG formulation? There are multiple approaches to this question. The simplest approach is to use Dirichlet conditions for which we explicitly prescribe the LdG $\mathbf{Q}$-tensor on the domain boundary. The Dirichlet condition fixes the preferred molecular alignment on the boundary. A more realistic approach is to employ a surface anchoring energy which enforces a preferred LdG $\mathbf{Q}$-tensor on the boundary i.e. it is a weaker version of Dirichlet conditions. There are several choices for surface anchoring, such as the popular Durand-Nobili surface energy [19, 39, 46]

$$E_S[\mathbf{Q}] := \int\int_{\partial\Omega} W \mathrm{tr}\left(\mathbf{Q} - \mathbf{Q}_B\right)^2 \, dA, \tag{20}$$

where W is the surface anchoring strength and $\mathbf{Q}_B$ is the preferred LdG $\mathbf{Q}$-tensor on the boundary, $\partial\Omega$. As $W \to \infty$, we recover the Dirichlet condition, $\mathbf{Q} = \mathbf{Q}_B$ on $\partial\Omega$. The interested reader is referred to [39] for detailed numerical comparisons between different surface energies.

A related question is—how do we include external electric fields into the LdG framework? The interaction between the external field and the LdG $\mathbf{Q}$-tensor is typically described by an additional field energy density of the form

$$F[\mathbf{E}, \mathbf{Q}] := \frac{1}{2}\varepsilon_a \mathbf{E}_i \mathbf{Q}_{ij} \mathbf{E}_j, \tag{21}$$

where ε_a is the dielectric anisotropy, $\mathbf{E}$ is the electric field vector in $\mathbb{R}^3$ and we have used the Einstein summation convention [18, 22, 42, 46].

In all cases, the mathematical approach is the same: we consider the LdG energy functional, compute the associated Euler-Lagrange equations and boundary conditions and study critical points of the LdG energy functional, with special emphasis on local and global minimizers. In the next two sub-sections, we briefly review two other widely used continuum theories for uniaxial nematic phases: the Ericksen theory and Oseen-Frank theory. Both of these theories are less general than the LdG theory but have a rich mathematical theory in their own right and have been successfully used for modelling nematic systems in different contexts.

2.2 The Ericksen Theory

The Ericksen theory is often viewed as a bridge between the LdG theory and the Oseen-Frank theory [20]. The Ericksen theory is restricted to uniaxial phases, characterized by a scalar order parameter S and a director $\mathbf{n}$ as in (5). The Ericksen theory, like the LdG theory, is a variational theory and the simplest form of the Ericksen energy is

$$I_E[S, \mathbf{n}] := \int\int\int_{\Omega} \left[K\left(|\nabla S|^2 + S^2|\nabla \mathbf{n}|^2\right) + f_B(s)\right] dV, \tag{22}$$

where K is an elastic constant, the bracketed energy density penalizes any variations in S or $\mathbf{n}$ and $f_B(s)$ is a bulk potential such that $f_B \to \infty$ as $S \to -1/2$ or $S \to 1$, in accordance with the bounds established in (2). The boundary conditions can be imposed by either a fixed condition for $\mathbf{n}$ and S on $\partial\Omega$ or by employing a surface potential. As with any variational theory, modellers study energy minimizers to explain experimentally observed nematic textures. The Ericksen theory has received attention in the mathematical literature e.g. see [38], but is somewhat under-represented compared to the LdG and Oseen-Frank theories. This is simply because an analysis of the Ericksen theory is at least as demanding as the LdG theory and yet the theory contains less information by definition, for example, biaxiality is outside the scope of Ericksen theory. The Oseen-Frank theory is mathematically simpler than the LdG and Ericksen theories and yet has been adequate for a range of modelling scenarios.

2.3 The Oseen-Frank Theory

The Oseen-Frank (OF) theory is the simplest continuum theory for nematic liquid crystals, developed by Oseen and Frank [21]. The Oseen-Frank theory is restricted to uniaxial nematics with a constant scalar order parameter, and hence, the nematic state is purely characterized in terms of the director field, $\mathbf{n} \in S^2$.

The OF energy can be viewed as a Taylor expansion, in terms of $\nabla\mathbf{n}$, of the elastic Helmholtz free energy, subject to the conditions of frame indifference and equivalence under the $\mathbf{n} \to -\mathbf{n}$ relation. The second constraint, $\mathbf{n} \equiv -\mathbf{n}$, is a consequence of the *achirality* of the nematic molecules i.e. the rod-like molecules do not have a direction but have an orientation [22, 46]. These conditions naturally eliminate any linear terms in $(\nabla\mathbf{n})$ and the OF energy functional is usually taken to be

$$\begin{aligned} I_{OF}[\mathbf{n}] := &\int\int\int_{\Omega} \frac{1}{2}\left(K_1 (\nabla\cdot\mathbf{n})^2 + K_2 (\mathbf{n}\cdot\nabla\times\mathbf{n})^2 + K_3 (\mathbf{n}\times\nabla\times\mathbf{n})^2\right) dV \\ &+ \int\int\int_{\Omega} \frac{1}{2} (K_2 + K_4)\left(\nabla\cdot[(\mathbf{n}\cdot\nabla)\mathbf{n} - (\nabla\cdot\mathbf{n})\mathbf{n}]\right) dV. \end{aligned} \tag{23}$$

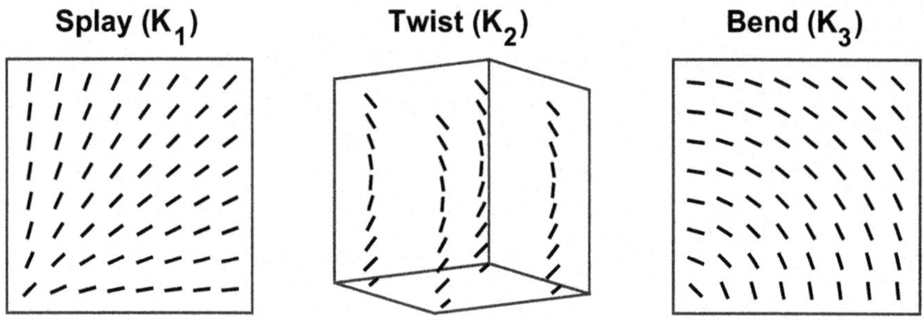

Fig. 3 The three types of director deformation (*dashed lines*) for a nematic sample

where the first three terms are the energetic penalties associated with *splay, twist and bend* deformations respectively, shown in Fig. 3, and the positive elastic constants, K_1, K_2, K_3 are a measure of the corresponding energetic penalty [22]. The elastic constants have the units of force. The last term is a surface energy, as can be seen from an immediate application of the Divergence Theorem.

The director field, **n**, is necessarily a unit-vector field that can be written in terms of polar angles as

$$\mathbf{n} = (\sin\phi\cos\theta, \sin\phi\sin\theta, \cos\phi), \tag{24}$$

and one can re-write the OF functional in (23) in terms of (ϕ, θ). As with the LdG theory and Ericksen theory, we model the experimentally observed states in terms of OF energy minimizers. Questions related to the existence and partial regularity of OF energy minimizers have been addressed in [24]. Additional effects like electric or magnetic fields can be modelled by extra terms in the OF energy functional, as with the LdG energy functional in (21). A common analytic approximation is the one-constant approximation for which $K_1 = K_2 = K_3 = K$ and $K_4 = 0$ and the OF energy reduces to the Dirichlet energy

$$I_{OF}[\mathbf{n}] := \frac{K}{2}\int\int\int_{\Omega} |\nabla\mathbf{n}|^2 \, dV, \tag{25}$$

with any additional contributions from external fields or boundary effects. The corresponding Euler Lagrange equations are

$$\Delta\mathbf{n} + |\nabla\mathbf{n}|^2\mathbf{n} = 0, \tag{26}$$

which are simply the harmonic map equations. The one-constant OF energy is mathematically appealing since it establishes a direct connection between the mathematics of liquid crystals and the theory of harmonic maps, the latter being well-studied in the mathematical community. Indeed, this connection was beautifully elucidated in the celebrated paper [11] where the authors use the analogies between the one-constant OF energy minimizers and harmonic maps to yield a complete classification of point defects in the one-constant OF energy minimizers.

As in the LdG framework, boundary conditions play a crucial role in the minimizer profiles. In the presence of strong surface forces, one might assume Dirichlet conditions for $\mathbf{n}$ on the boundary i.e. work with a fixed boundary condition $\mathbf{n} = \mathbf{n}_B$ on $\partial\Omega$, where $\mathbf{n}_B \in S^2$ is arbitrary. There are two common choices for $\mathbf{n}_B$: (i) $\mathbf{n}_B = \nu$ where ν is the outward unit normal to $\partial\Omega$, referred to as *homeotropic anchoring*, (ii) $\mathbf{n}_B \cdot \nu = 0$ which requires $\mathbf{n}_B$ to be in the plane of $\partial\Omega$, referred to as *planar anchoring*. For case (ii), $\mathbf{n}_B$ may not be prescribed so that $\mathbf{n}$ can take any value in the plane of $\partial\Omega$, referred to as *planar degenerate anchoring*. For model problems with planar degenerate anchoring, one could employ a surface energy which penalizes any out-of-plane orientation, a common choice being the Rapini-Papoular surface energy below [8]:

$$E_S[\mathbf{n}] := \int\int_{\partial\Omega} \frac{W}{2} (\mathbf{n} \cdot \boldsymbol{\nu})^2 \, dA, \tag{27}$$

where ν is the outward normal to the surface, $W > 0$ is the anchoring strength and the surface energy is a minimum when $\mathbf{n} \cdot \boldsymbol{\nu} = 0$ on $\partial\Omega$ or equivalently when $\mathbf{n}$ is tangent to $\partial\Omega$. As $W \to \infty$, we recover the Dirichlet regime.

3 The Planar Bistable Device

This section focuses on the modelling of the planar bistable nematic device reported in [45]. The planar bistable nematic device has a simple geometry comprising a periodic array of three-dimensional square or rectangular wells filled with nematic liquid crystalline material, say the conventional MBBA material [45]. We denote the edge lengths of the wells by L_x, L_y, L_z in the x, y and z-directions respectively and L_z is typically much smaller than L_x or L_y. The well dimensions are typically in the range $L_x, L_y \in (20, 80)$ microns and $L_z < \min\{L_x/5, L_y/5\}$, so that we treat these geometries as shallow wells; see Fig. 4. The well surfaces are treated to induce tangential or planar degenerate boundary conditions so that the nematic molecules on the well surfaces are constrained to lie in the plane of the surfaces [45]. In Fig. 6, we consider the bottom cross-section of a generic well in the periodic array; the tangent boundary conditions mean that the nematic molecules are free to take any orientation in the plane of the well surface or equivalently, in the xy-plane. However, they are constrained to be tangent to the edges so that they are aligned in the x-direction on the edges parallel to the x-axis and aligned in the y-direction along the pair of edges parallel to the y-axis. There is a natural mismatch in the molecular alignments at

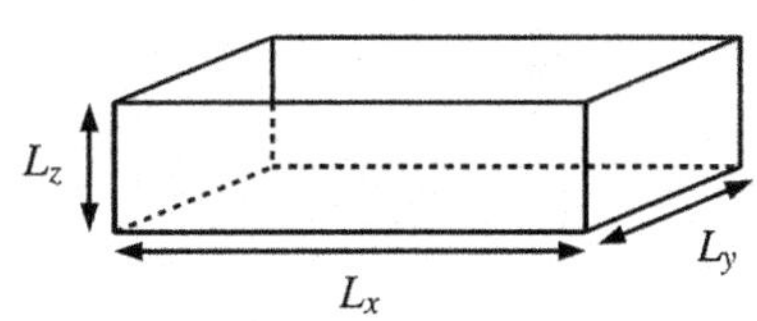

Fig. 4 Shallow rectangular well

the square vertices and hence, one expects to see some singular behaviour near the corners although the singular profile may be highly localized with negligible impact on macroscopic experimentally relevant measurements.

In [45], the authors experimentally find at least two different stable states for this simple geometry which are labelled as *diagonal* and *rotated* states respectively. The authors numerically model this device within a two-dimensional LdG theoretical framework; we omit the details for brevity but the qualitative conclusions can be summarized as follows: (i) the nematic molecules roughly align along a diagonal for the diagonal state and there are two equivalent diagonal states; (ii) the nematic molecules approximately rotate by π radians as we move between a pair of parallel edges for a rotated solution and there are four rotationally equivalent rotated states. Both of these states have long-term stability and have optically contrasting properties when viewed between crossed polarizers, as suggested by the experiments and simulations in [45].

In [15], the authors perform detailed experiments on suspensions of *fd*-viruses in rectangular microchambers (also see [34]). Suspensions of *fd*-viruses are known to exhibit liquid crystalline behaviour [23] and since these viruses are typically larger than nematic molecules, they can be readily visualized using confocal microscopy techniques. In [15], the authors use two related viruses: the semi-flexible wild-type *fd*-virus with approximately equal elastic constants [16] and the Y21M *fd*-virus with elastic constant ratio $\frac{K_3}{K_1} \approx 20$ [9]. In both cases, the authors allow the virus suspensions to settle into equilibrium and observe at least five different types of states (see Fig. 5), two of which are strongly reminiscent of the rotated and diagonal solutions and the others are variants of the rotated or diagonal solutions with internal defects. This example serves to illustrate that the diagonal and rotated solutions may be observed for a larger class of systems than conventional confined nematic systems and are hence, an interesting mathematical case study.

In the next two sub-sections, we summarize two recent papers on the modelling of this planar device. In Sect. 3.1, we summarize the results in [35] to demonstrate the strengths and predictive prowess of simple mathematical models and in Sect. 3.2, we summarize the LdG numerical simulations of this device reported in [29] to reveal a new well order reconstruction solution for nano-scale square wells.

3.1 A Two-Dimensional Oseen-Frank Model

The planar bistable nematic device is a periodic array of shallow square/rectangular wells, with L_z typically much smaller than the cross-sectional parameters. We work in the OF framework so that the nematic state is described by a unit-vector field, $\mathbf{n}$, where $\mathbf{n}$ represents the single distinguished direction of molecular alignment. We model the nematic profile within a single well and given that these are shallow wells, we take our computational domain to be a rectangle:

$$\Omega := \left\{(x, y) \in \mathbb{R}^2; 0 \le x \le 1, 0 \le y \le \lambda\right\}, \tag{28}$$

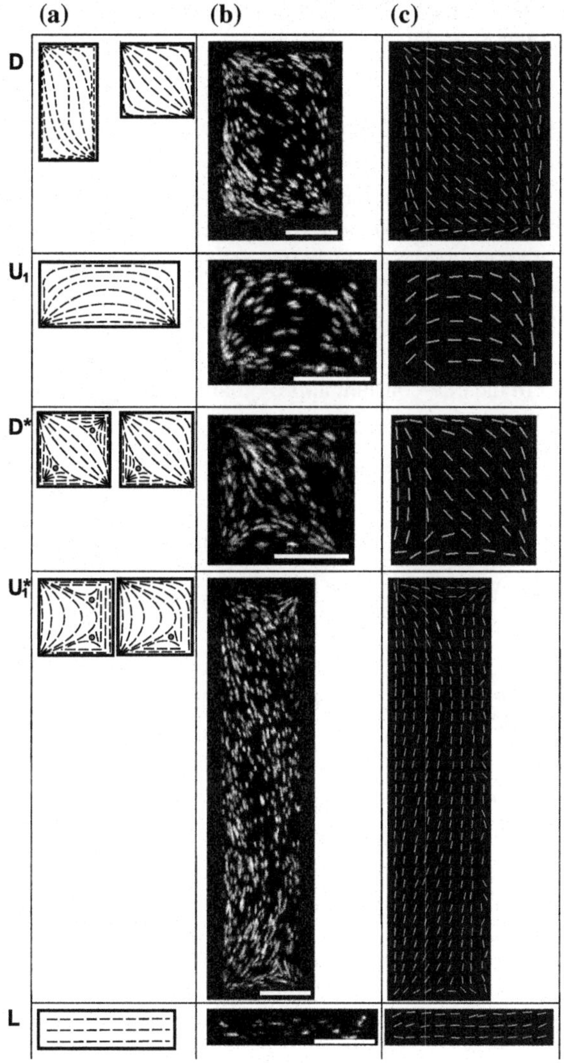

Fig. 5 Sample director fields observed in shallow rectangular wells for WT *fd* and Y21M. **a** Sketch of the director field, with internal defects indicated by *red circles*. **b** Four overlaid confocal images and scale bar = 10 μm. **c** Final experimental director field, calculated from a series of 1000 confocal images. The D and L states are wild-type *fd* and U_1, U_1^* and D^* states are the Y21M strain. Reproduced from Ref. [35] with permission from the Royal Society of Chemistry

where $\lambda = \frac{L_y}{L_x}$ is the rectangular aspect ratio and take $\mathbf{n}$ to be a two-dimensional unit-vector

$$\mathbf{n} = (\cos\theta(x, y), \sin\theta(x, y), 0)\,, \tag{29}$$

independent of z-the vertical coordinate. The OF energy in (23) then reduces to

$$E[\theta] := \frac{K_3}{2}\int\int_{\Omega}\left[|\nabla\theta|^2 - \delta\left(\theta_y\cos\theta - \theta_x\sin\theta\right)^2\right]\,d\Omega, \tag{30}$$

where $\delta := 1 - \frac{K_1}{K_3}$ is a measure of the elastic anisotropy. The corresponding Euler-Lagrange equations are

$$\Delta\theta + \delta\left(\frac{1}{2}\sin 2\theta\left(\theta_y^2 - \theta_x^2 + 2\theta_{xy}\right) + \theta_x\theta_y\cos 2\theta - \theta_{xx}\sin^2\theta - \theta_{yy}\cos^2\theta\right) = 0. \tag{31}$$

In [35], the authors study the model (28)–(31) within the one-constant approximation with $\delta = 0$, prescribe boundary-value problems for the diagonal and rotated solutions, derive analytic expressions for the corresponding **n**'s and the OF energies in terms of λ, accompanied by some asymptotics in the $\lambda \to 0$ limit. We reproduce the key mathematical results in [35] as an example of mathematical modelling for a prototype bistable device.

In the one-constant approximation, the OF energy in (30) reduces to

$$E[\theta] := \frac{K}{2}\int\int_{\Omega} |\nabla\theta|^2 \, d\Omega, \tag{32}$$

and the energy minimizers are solutions of the Laplace's equation

$$\Delta\theta \equiv \theta_{xx} + \theta_{yy} = 0, \tag{33}$$

subject to the appropriate boundary conditions on $\partial\Omega$.

Following [34, 35], we impose Dirichlet conditions for θ on the four rectangular edges. By virtue of the imposed tangent conditions, $\theta \in \{0, \pi\}$ on the edges, $y = 0$ and $y = \lambda$ and $\theta \in \left\{-\frac{\pi}{2}, \frac{\pi}{2}\right\}$ on the edges, $x = 0$ and $x = 1$. We assume that the director profile defined by $\theta : \Omega \to \mathbb{R}$ has no internal defects and any arbitrary solution of (33) subject to Dirichlet boundary conditions can be written as:

$$\theta(x, y, \lambda) := a_1 f_1(x, y; \lambda) + a_2 f_2(x, y; \lambda) + a_3 f_3(x, y; \lambda) + a_4 f_4(x, y; \lambda). \tag{34}$$

The constants a_i's are determined by the Dirichlet conditions and the function f_1 is a solution of $\Delta f_1 = 0$ subject to $f_1(x, 0; \lambda) = 1$ and $f_1(0, y; \lambda) = f_1(1, y; \lambda) = f_1(x, \lambda; \lambda) = 0$. We use the method of separation of variables i.e. assume an ansatz of the form $f_1(x, y; \lambda) = X(x; \lambda)Y(y, ; \lambda)$ with

$$\begin{aligned} &X''(x; \lambda) + \omega^2 X(x; \lambda) = 0; \quad X(0; \lambda) = X(1; \lambda) = 0, \\ &Y''(y; \lambda) - \omega^2 Y(y; \lambda) = 0; \quad Y(\lambda; \lambda) = 0, \end{aligned} \tag{35}$$

yielding a general Fourier series

$$f_1(x, y; \lambda) := \sum_{n=1}^{\infty} A_n \sin(n\pi x)\{\cosh(n\pi y) - \coth(n\pi\lambda)\sinh(n\pi y)\}, \tag{36}$$

Table 1 Values of the coefficients a_i for three equilibria

State	a_1	a_2	a_3	a_4
D	0	$\pi/2$	0	$\pi/2$
U_1	0	$-\pi/2$	0	$\pi/2$
U_2	0	$\pi/2$	π	$\pi/2$

and we use the condition $f_1(x, 0; \lambda) = 1$ to compute the Fourier coefficients A_n. A straightforward computation shows that

$$f_1(x, y; \lambda) := \sum_{n=0}^{\infty} \frac{4\sin((2n+1)\pi x)}{(2n+1)\pi} \{\cosh((2n+1)\pi y) - \coth((2n+1)\pi\lambda)\sinh((2n+1)\pi y)\}. \tag{37}$$

The remaining functions are defined by re-scaling and rotation:

$$\begin{aligned} f_2(x, y; \lambda) &:= f_1\left(\frac{y}{\lambda}, \frac{1-x}{\lambda}; \frac{1}{\lambda}\right), \\ f_3(x, y; \lambda) &:= f_1(x, \lambda - y; \lambda), \\ f_4(x, y; \lambda) &:= f_1\left(\frac{y}{\lambda}, \frac{x}{\lambda}; \frac{1}{\lambda}\right). \end{aligned} \tag{38}$$

The aim is to model the experimentally observed diagonal and rotated solutions and we assume that there are no internal defects. We model three states in terms of three distinct boundary-value problems for θ: (i) the D state where $\theta = 0$ on $\{y = 0, \lambda\}$ and $\theta = \frac{\pi}{2}$ on $x = \{0, 1\}$, rendering an average alignment along the diagonal connecting the vertices $(0, 0)$ and $(1, \lambda)$, (ii) the rotated U_1 state for which $\theta = 0$ on $\{y = 0, \lambda\}$, $\theta = \frac{\pi}{2}$ on $x = 0$ and $\theta = -\frac{\pi}{2}$ on $x = 1$, and (iii) the rotated U_2 state for which $\theta = 0$ on $y = 0$, $\theta = \pi$ on $y = \lambda$ and $\theta = \frac{\pi}{2}$ on $\{x = 0, 1\}$. The corresponding choices for $a_1, \ldots, a_4$ are enumerated in Table 1. Given the values of a_i's for each of the three competing states and the expressions, (34), (37)–(38), we obtain analytic expressions for the director fields of the D, U_1, U_2 states respectively shown in Fig. 6.

The next step is to compute the one-constant OF energies of these states as a function of λ. The Dirichlet conditions naturally imply that θ is not defined at the vertices i.e. there is a point defect at every vertex. The D, U_1, U_2 states have 4 defects each by construction: two *splay* defects where **n** has a local radial profile near the defect and two *bend* defects for which **n** has a local bend profile near the defect (see Fig. 7), and point defects have infinite energy on a two-dimensional domain [22, 46]. Hence, we cannot compute the one-constant OF energy of these 3 states on the rectangle, Ω. Since the total number and types of defects are identical for all 3 states, we argue that the energetic differences between the three states is a consequence of the bulk distortion or interior profiles. We excise a quarter-disc of

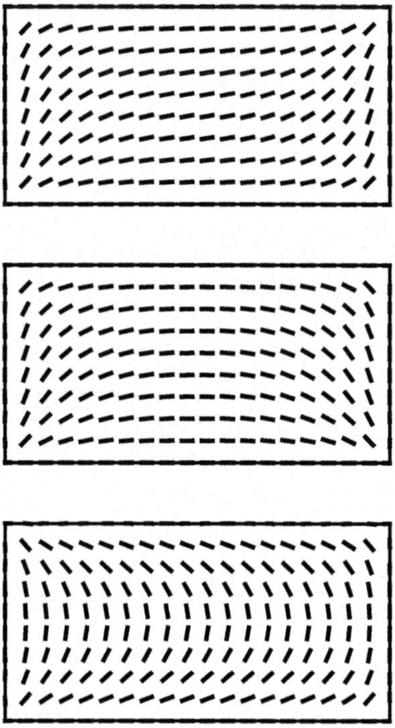

Fig. 6 The director fields for the three equilibrium states within a rectangle with rectangular aspect ratio $\lambda = 0.5$. From *top*; diagonal (D), rotated (U_1), rotated (U_2)

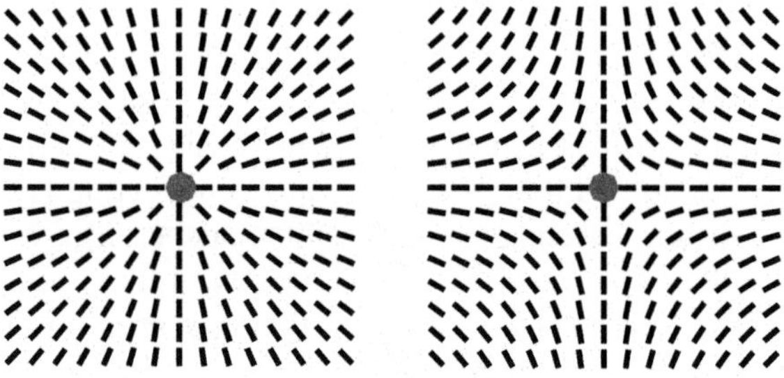

Fig. 7 Local behaviour of the director about a splay defect (*left*) and a bend defect (*right*). Defect location indicated by the *red point*

radius ε ($0 < \varepsilon \ll \lambda$) centered at each vertex and derive an asymptotic expression for the one-constant OF energy on the excised domain, denoted by Ω_ε (see Fig. 8), in the $\varepsilon \to 0$ limit [35]. The following calculations have been reproduced from [34].

The one-constant OF energy on Ω_ε can be converted to a line integral on the boundary, $\partial\Omega_\varepsilon$, by an immediate application of Green's first identity [17]:

$$E[\theta] := \frac{K}{2}\int\int_{\Omega_\varepsilon} |\nabla\theta|^2\, dx\, dy = \frac{K}{2}\int_{\partial\Omega_\varepsilon} \theta\frac{\partial\theta}{\partial\boldsymbol{\nu}}\, ds$$
$$= \frac{K}{2}\sum_{i=1}^{4}\left[\int_{C_i} \theta\frac{\partial\theta}{\partial\boldsymbol{\nu}}\, ds + \int_{\gamma_i} \theta\frac{\partial\theta}{\partial\boldsymbol{\nu}}\, ds\right], \tag{39}$$

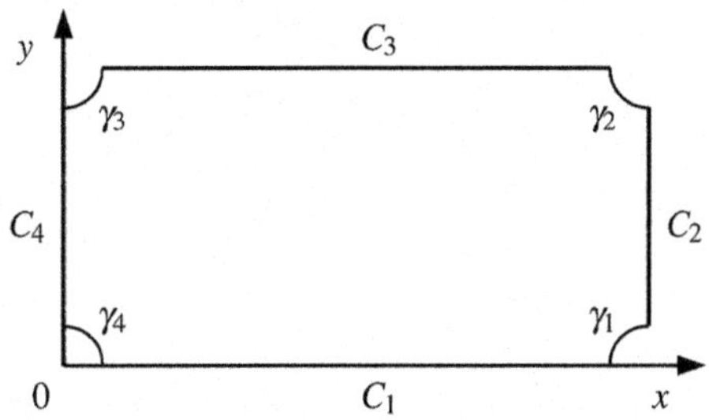

Fig. 8 The boundary $\partial\Omega_\varepsilon$ expressed in terms of straight edges C_i and arcs γ_i

where $\frac{\partial\theta}{\partial\boldsymbol{\nu}} = \nabla\theta \cdot \boldsymbol{\nu}$, $\boldsymbol{\nu}$ is the outward pointing unit normal to $\partial\Omega_\varepsilon$, $\partial\Omega_\varepsilon$ is oriented in the anti-clockwise sense and $\partial\Omega_\varepsilon$ comprises 4 straight segments, C_i, and the four quarter circles around each vertex, denoted by γ_i. The contribution from each boundary segment needs to be considered separately.

The line integral about the arc, γ_i, is approximated by considering the local solution, θ_c, near the corresponding vertex. Let us consider the arc γ_4 about the vertex $(x, y) = (0, 0)$ as an example (see Fig. 8). The function, θ_c is defined by a solution of the Laplace's equation, $\Delta\theta_c = 0$, in polar coordinates, (r, ϕ), about the origin, subject to $\theta_c(r, 0) = a_1$, $\theta_c\left(r, \frac{\pi}{2}\right) = a_4$ and the condition, $\theta_c(\varepsilon, \phi) = \theta(\varepsilon\cos(\phi), \varepsilon\sin(\phi))$ where θ is defined by (34). In this case, θ_c is given by

$$\theta_c = a_1 + \frac{2}{\pi}(a_4 - a_1)\phi + \sum_{n=1}^{\infty} b_n r^{2n}\sin(2n\phi), \tag{40}$$

where the coefficients, b_n, can in principle be determined by matching with the outer solution for θ. Then we can compute the line integral in (39) to be

$$-\int_0^{\frac{\pi}{2}} \theta_c \frac{\partial\theta_c}{\partial r}\bigg|_{r=\varepsilon} \varepsilon\, d\phi \sim -(a_1 + a_4)\, b_1\varepsilon^2 + O\left(\varepsilon^4\right), \tag{41}$$

and the remaining three line integrals about γ_1, γ_2, γ_3 can be dealt with similarly, each making an $O\left(\varepsilon^2\right)$-contribution to the energy. These energetic contributions vanish in the $\varepsilon \to 0$ limit.

It remains to consider the line integrals about the straight lines, C_i. From the definition of the functions f_i in (37)–(38), $f_i = \delta_{ij}$ on the edge C_j and hence, the line integrals about the straight edges, C_i, simplify to

$$\sum_{i=1}^{4} \int_{C_i} \theta \frac{\partial\theta}{\partial\nu}\, ds = \sum_{i}^{4} a_i \sum_{j=1}^{4} a_j \int_{C_i} \frac{\partial f_j}{\partial\nu}\, ds. \tag{42}$$

There are a total of 16 integrals and we use the symmetries of the functions f_i to reduce the 16 integrals to six integrals as shown below:

$$
\begin{aligned}
I_1 &:= \int_{C_1} \frac{\partial f_1}{\partial \nu}\, ds = \int_{C_3} \frac{\partial f_3}{\partial \nu}\, ds, \\
I_2 &:= \int_{C_1} \frac{\partial f_3}{\partial \nu}\, ds = \int_{C_3} \frac{\partial f_1}{\partial \nu}\, ds, \\
I_3 &:= \int_{C_2} \frac{\partial f_1}{\partial \nu}\, ds = \int_{C_4} \frac{\partial f_1}{\partial \nu}\, ds = \int_{C_2} \frac{\partial f_3}{\partial \nu}\, ds = \int_{C_4} \frac{\partial f_3}{\partial \nu}\, ds, \\
I_4 &:= \int_{C_2} \frac{\partial f_2}{\partial \nu}\, ds = \int_{C_4} \frac{\partial f_4}{\partial \nu}\, ds, \\
I_5 &:= \int_{C_2} \frac{\partial f_4}{\partial \nu}\, ds = \int_{C_4} \frac{\partial f_2}{\partial \nu}\, ds, \\
I_6 &:= \int_{C_1} \frac{\partial f_2}{\partial \nu}\, ds = \int_{C_3} \frac{\partial f_2}{\partial \nu}\, ds = \int_{C_1} \frac{\partial f_4}{\partial \nu}\, ds = \int_{C_3} \frac{\partial f_4}{\partial \nu}\, ds.
\end{aligned} \tag{43}
$$

These relations can be further simplified by noting that

$$
\int_{\partial \Omega_\varepsilon} \frac{\partial f_i}{\partial \nu}\, ds = 0 \quad i = 1, \dots 4, \tag{44}
$$

since $\Delta f_i = 0$. Hence, it follows that

$$
I_1 + I_2 + 2I_3 \sim O(\varepsilon^2), \quad I_4 + I_5 + 2I_6 \sim O(\varepsilon^2), \tag{45}
$$

where the ε^2 contribution originates from the curved arcs, γ_i. We use (45) to simplify the one-constant OF energy in (39) to

$$
\begin{aligned}
E[\theta] \sim &\frac{K}{2}\left(\left(a_1^2 + a_3^2\right) I_1 + 2a_1 a_3 I_2 + \left(a_2^2 + a_4^2\right) I_4 + 2a_2 a_4 I_5\right) \\
&- \frac{K}{2}\frac{(a_1 + a_3)(a_2 + a_4)}{2}(I_1 + I_2 + I_4 + I_5) + O\left(\varepsilon^2\right).
\end{aligned} \tag{46}
$$

The remaining task is to evaluate the integrals I_1, I_2, I_3, I_4 in the $\varepsilon \to 0$ limit.

Let us consider the integral I_1:

$$
\begin{aligned}
I_1 &= -\int_\varepsilon^{1-\varepsilon} \left.\frac{\partial f_1}{\partial y}\right|_{y=0} dx = \sum_{n=0}^{\infty} \frac{8\cos((2n+1)\pi\varepsilon)\coth((2n+1)\pi\lambda)}{(2n+1)\pi} \\
&= \sum_{n=0}^{\infty} \frac{8\cos((2n+1)\pi\varepsilon)}{(2n+1)\pi}(\coth((2n+1)\pi\lambda) - 1) \\
&\quad + \sum_{n=0}^{\infty} \frac{8\cos((2n+1)\pi\varepsilon)}{(2n+1)\pi}.
\end{aligned} \tag{47}
$$

The first sum is a convergent series which we denote by

$$\sum_{n=0}^{\infty} \frac{8\cos\left((2n+1)\pi\varepsilon\right)\left(\coth\left((2n+1)\pi\lambda\right)-1\right)}{(2n+1)\pi} \sim \frac{4}{\pi}s_1(\lambda) + O\left(\varepsilon^2\right), \quad (48)$$

where

$$s_1(\lambda) := 2\sum_{n=0}^{\infty} \frac{\coth\left((2n+1)\pi\lambda\right)-1}{(2n+1)}. \quad (49)$$

The second sum in (47) is a Fourier series given in [25]:

$$\sum_{n=0}^{\infty} \frac{8\cos\left((2n+1)\pi\varepsilon\right)}{(2n+1)\pi} = \frac{4}{\pi}\ln\left(\cot\left(\frac{\pi\varepsilon}{2}\right)\right) \sim \frac{4}{\pi}\ln\left(\frac{2}{\pi\varepsilon}\right) + O\left(\varepsilon^2\right), \quad (50)$$

and combining (48)–(50), we obtain

$$I_1 \sim \frac{4}{\pi}\ln\left(\frac{2}{\pi\varepsilon}\right) + \frac{4}{\pi}s_1(\lambda) + O\left(\varepsilon^2\right). \quad (51)$$

We repeat analogous arguments to compute

$$\begin{aligned} I_2 &\sim -\frac{4}{\pi}s_2(\lambda) + O(\varepsilon^2), \\ I_4 &\sim \frac{4}{\pi}\ln\left(\frac{2\lambda}{\pi\varepsilon}\right) + \frac{4}{\pi}s_1\left(\frac{1}{\lambda}\right) + O\left(\varepsilon^2\right), \\ I_5 &\sim -\frac{4}{\pi}s_2\left(\frac{1}{\lambda}\right) + O\left(\varepsilon^2\right), \end{aligned} \quad (52)$$

where

$$s_2(\lambda) := 2\sum_{n=0}^{\infty} \frac{\operatorname{csch}\left((2n+1)\pi\lambda\right)}{(2n+1)}. \quad (53)$$

Combining (51)–(53), the energies of all three competing states can be expressed as

$$E[\theta] \sim K\pi\left(\ln\left(\frac{1}{\varepsilon}\right) + \tilde{E}[\theta] + O\left(\varepsilon^2\right)\right), \quad (54)$$

where the logarithmic contribution is the *defect energy* and the defect energy is identical for all states, since they have the same number of *splay* and *bend* defects. The normalized energy, $\tilde{E}$, represents the bulk distortion energy and is a good measure of the relative energies of the three competing states as shown below:

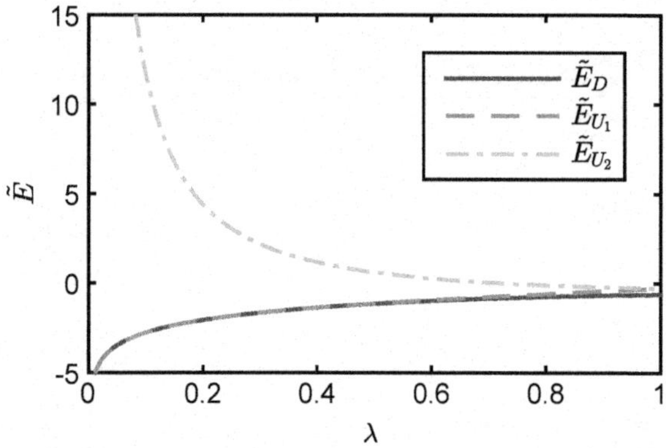

Fig. 9 The normalized energies of the three equilibrium states plotted against the rectangular aspect ratio λ

$$\tilde{E}_D = \ln\left(\frac{2\lambda}{\pi}\right) + s_1\left(\frac{1}{\lambda}\right) - s_2\left(\frac{1}{\lambda}\right), \tag{55}$$

$$\tilde{E}_{U_1} = \ln\left(\frac{2\lambda}{\pi}\right) + s_1\left(\frac{1}{\lambda}\right) + s_2\left(\frac{1}{\lambda}\right), \tag{56}$$

$$\tilde{E}_{U_2} = \ln\left(\frac{2}{\pi}\right) + s_1\left(\lambda\right) + s_2\left(\lambda\right). \tag{57}$$

One can immediately see that $\tilde{E}_D < \tilde{E}_{U_1} < \tilde{E}_{U_2}$ for $\lambda < 1$. In Fig. 9, we plot the normalized energies of the three states versus λ. These energy expressions in (55)–(57) provide quantitative information about how λ or equivalently the geometrical aspect ratio manifests in energetic trends. These expressions are consistent with the experiments on *fd*-viruses in rectangular microchambers if one hypothesises that the observational frequency is determined by the bulk distortion energy and lower energy states are more frequently observed in the sense that D state is the most frequently observed state in experiments and the U_2 state is seldom observed [35].

3.2 A Landau-de Gennes Approach

The OF approach in the previous section is a useful first approach to modelling the planar bistable wells but suffers from several limitations. The OF approach does not give us any information about the defect structures near the square vertices, does not contain any information about biaxiality and largely complements the experiments in [45] without providing any new insight that may potentially guide future experiments.

In this section, we reproduce the LdG modelling of this bistable device reported in [29] with emphasis on how the stable equilibria depend on the well cross-sectional size, denoted by R. We work with shallow square wells and measure R in units of a material-dependent length scale, the *biaxial correlation length* $\xi_b^{(0)}$, at different temperatures. The bare biaxial correlation length is typically tens of nanometers [12].

We find the conventional diagonal and rotated solutions when R is large compared to $\xi_b^{(0)}$ i.e. when R is on the micron scale, but as R decreases, there is a critical well size, R_c, such that there is a stable well order reconstruction solution (WORS) for all $R < R_c$. The WORS is featured by a star-shaped rim of maximal biaxiality connecting the four square vertices, and is hence, outside the scope of conventional OF modelling. In the following paragraphs, we discuss the modelling details and reproduce some of the numerical results from [29].

In the LdG framework, the nematic state is described by a LdG $\mathbf{Q}$-tensor order parameter or equivalently by a symmetric, traceless 3×3 matrix that contains information about the orientational ordering within its eigenvalues and eigenvectors respectively. We adopt the following three-dimensional parameterization for the $\mathbf{Q}$-tensor:

$$\mathbf{Q} = (q_3 + q_1)\,\mathbf{e}_x \otimes \mathbf{e}_x + (q_3 - q_1)\,\mathbf{e}_y \otimes \mathbf{e}_y + q_2\left(\mathbf{e}_x \otimes \mathbf{e}_y + \mathbf{e}_y \otimes \mathbf{e}_x\right) - 2q_3\mathbf{e}_z \otimes \mathbf{e}_z, \tag{58}$$

where q_1, q_2, q_3 only depend on the planar variables, x and y, and are independent of z and $\mathbf{e}_x$, $\mathbf{e}_y$, $\mathbf{e}_z$ are the unit coordinate vectors in the x, y and z directions respectively. One can prove that there exist LdG critical points of the form (58) or equivalently, the LdG Euler-Lagrange equations in (18) admits solutions of the form (58). The parameterization (58) has one fixed eigenvector, $\mathbf{e}_z$, and the remaining two eigenvectors can be arbitrary orthogonal unit-vectors in the xy-plane. When $q_2 = 0$, the three eigenvectors are the Cartesian unit-vectors, $\mathbf{e}_x$, $\mathbf{e}_y$, $\mathbf{e}_z$, respectively. The eigenvalues are given by, $\lambda_1 = q_3 + \sqrt{q_1^2 + q_2^2}$, $\lambda_2 = q_3 - \sqrt{q_1^2 + q_2^2}$ and $\lambda_3 = -2q_3$. The biaxiality parameter, β^2, is a common measure of how biaxial a $\mathbf{Q}$-tensor is and we adopt the usual definition [46]:

$$\beta^2 = 1 - \frac{6\left(\mathrm{tr}\mathbf{Q}^3\right)^2}{|\mathbf{Q}|^6}, \tag{59}$$

so that $\beta^2 \in [0, 1]$ by definition. In particular, $\beta^2 = 0$ if and only if $6\left(\mathrm{tr}\mathbf{Q}^3\right)^2 = |\mathbf{Q}|^6$ which is a sufficient condition for uniaxiality and $\beta^2 = 1$ if there is a null eigenvalue, corresponding to maximal biaxiality. If $\mathbf{Q} = 0$, then $\beta^2 = 0$ by definition.

The next step is to define the boundary conditions. We work with weak anchoring on the four lateral surfaces, defined by $x = \{0, R\}$ and $y \in \{0, R\}$, i.e. impose a preferred nematic state, described by a preferred $\mathbf{Q}_B$, on these surfaces and free boundary conditions on the top and bottom surfaces, $z = 0$ and $z = h$, respectively. We take $h < \frac{R}{10}$ by assumption since these are shallow wells and work with low temperatures that favour nematic ordering. On the surfaces, $x = 0$ and $x = R$, we have

$$\mathbf{Q}_B = \frac{S_{eq}}{3}\left(2\mathbf{e}_y \otimes \mathbf{e}_y - \mathbf{e}_x \otimes \mathbf{e}_x - \mathbf{e}_z \otimes \mathbf{e}_z\right), \tag{60}$$

where $S_{eq} = \frac{B+\sqrt{B-24AC}}{4C}$ and $\mathbf{Q}_B \in \mathbf{Q}_{\min}$ where $\mathbf{Q}_{\min}$ has been defined in (13). Similarly, on the surfaces, $y = 0$ and $y = R$, we have

$$\mathbf{Q}_B = \frac{S_{eq}}{3}\left(2\mathbf{e}_x \otimes \mathbf{e}_x - \mathbf{e}_y \otimes \mathbf{e}_y - \mathbf{e}_z \otimes \mathbf{e}_z\right), \tag{61}$$

and

$$\frac{\partial \mathbf{Q}}{\partial z} = 0 \quad \text{on } z = 0 \text{ and } z = h. \tag{62}$$

We take the LdG energy to be

$$I[\mathbf{Q}] := \int_{\mathscr{B}} \left[\frac{L}{2}|\nabla \mathbf{Q}|^2 + f_B(\mathbf{Q})\right] dV + \sum_{x\in\{0,R\};\, y\in\{0,R\}} E_S[\mathbf{Q}], \tag{63}$$

where $\mathscr{B} = \left\{(x, y, z) \in \mathbb{R}^3; 0 \leq x, y \leq R; 0 \leq z \leq h\right\}$, f_B is the bulk potential defined in (7) and the surface energies are defined in (20) over the lateral surfaces, using $\mathbf{Q}_B$ as defined in (60) and (61). In particular, $A < 0$ in (7) since we are working with low temperatures.

The nematic profiles are dictated by a complex interplay between material properties, temperature, geometry and boundary effects. In order to better quantify these effects, we measure the length R in units of the bare biaxial correlation length, $\xi_b^{(0)}$, defined to be [30]

$$\xi_b^{(0)} = \frac{2\sqrt{LC}}{B}, \tag{64}$$

which is a purely material-dependent length scale. Let t be a dimensionless temperature defined to be $t = \frac{T-T_*}{T_{**}-T_*}$, T_* is defined in (7) and T_{**} is the nematic supercooling temperature such that $\mathbf{Q} = 0$ is the unique critical point of f_B for $T > T_{**}$. We define

$$\tau := 1 + \sqrt{1-t}, \tag{65}$$

then the key material and temperature-dependent length scale is [27–30]

$$\xi_b = \frac{\xi_b^{(0)}}{\sqrt{\tau}}, \tag{66}$$

and in what follows, we work with fixed $\xi_b^{(0)}$ and vary τ. The strength of the surface interactions is typically described in terms of the surface extrapolation length [30]:

$$d = \frac{L}{W}, \tag{67}$$

where we assume that the anchoring is equally strong on all four lateral surfaces and W is the corresponding surface anchoring strength. Typical values of L are around 10^{-12}– 10^{-11} Newtons and $W \in \left(10^{-8}, 10^{-3}\right)$ N/m [42].

We non-dimensionalize the energy (63) using the following scalings: $\tilde{\mathbf{Q}} = \frac{\mathbf{Q}}{S_{**}}$, $\tilde{x} = \frac{x}{R}$, $\tilde{y} = \frac{y}{R}$, $\tilde{z} = \frac{z}{R}$, $S_{**} = \frac{B}{4C}$ and $\tilde{I} = \frac{I}{I_0}$ where $I_0 = \frac{LS_{**}^2 R^3}{\left(2\xi_b^{(0)}\right)^2}$ and reproduce the formalism in [29] verbatim. In what follows, we drop the tildes from the dimensionless variables and the corresponding dimensionless free energy densities are defined to be:

$$f_B = \frac{\tau}{6}\mathrm{tr}\mathbf{Q}^2 - \frac{2}{3}\mathrm{tr}\mathbf{Q}^3 + \frac{1}{8}\left(\mathrm{tr}\mathbf{Q}^2\right)^2 \tag{68}$$

$$= \frac{\tau}{3}\left(q_1^2 + q_2^2 + 3q_3^2\right) - 4q_3\left(q_1^2 + q_2^2 - q_3^2\right) + \frac{1}{2}\left(q_1^2 + q_2^2 + 3q_3^2\right)^2, \tag{69}$$

$$f_e = \left(\frac{\xi_b^{(0)}}{R}\right)^2 |\nabla\mathbf{Q}|^2 = 2\left(\frac{\xi_b^{(0)}}{R}\right)^2 \left(3\,|\nabla q_3|^2 + |\nabla q_1|^2 + |\nabla q_2|^2\right), \tag{70}$$

$$f_S = \frac{\xi_b^{(0)2}}{dR}\mathrm{tr}\left(\mathbf{Q} - \mathbf{Q}_s(\varphi_s)\right)^2 \tag{71}$$

$$= \frac{2}{3}\left(\frac{\xi_b^{(0)}}{R}\right)\left(\frac{\xi_b^{(0)}}{d}\right)\big(9q_3^2 + 3q_1^2 + 3q_2^2 - 3q_3\tau + \tau^2 - 3q_1\tau\cos(2\varphi_s) - 3q_2\tau\sin(2\varphi_s)\big). \tag{72}$$

where f_e is the one-constant elastic energy density, f_S is the surface energy density, the angle φ_s is either $\varphi_s = 0$ (plates at $y = 0$ and $y = R$) or $\varphi_s = \pi/2$ (plates at $x = 0$ and $x = R$). The strong anchoring limit, $W \to \infty$, corresponds to vanishing surface extrapolation length, $d \to 0$. The corresponding Euler-Lagrange equations are

$$\left(\frac{\xi_b^{(0)}}{R}\right)^2 \Delta_\perp q_1 - \frac{\tau}{6}q_1 + 2q_3q_1 - \frac{q_1}{2}(3q_3^2 + q_1^2 + q_2^2) = 0, \tag{73}$$

$$\left(\frac{\xi_b^{(0)}}{R}\right)^2 \Delta_\perp q_2 - \frac{\tau}{6}q_2 + 2q_2q_3 - \frac{q_2}{2}(3q_3^2 + q_1^2 + q_2^2) = 0, \tag{74}$$

$$\left(\frac{\xi_b^{(0)}}{R}\right)^2 \Delta_\perp q_3 - \frac{\tau}{6}q_3 + \frac{1}{3}(q_1^2 + q_2^2 - 3q_3^2) - \frac{q_3}{2}(3q_3^2 + q_1^2 + q_2^2) = 0, \tag{75}$$

where $\Delta_\perp \equiv \frac{\partial^2}{\partial x^2} + \frac{\partial^2}{\partial y^2}$.

The boundary conditions on the plates $x = 0$ and $x = R$ are

$$\frac{\partial q_1}{\partial x} = \mp\frac{R}{d}\left(q_1 - \frac{\tau}{2}\right), \tag{76}$$

$$\frac{\partial q_2}{\partial x} = \mp\frac{R}{d}q_2, \tag{77}$$

$$\frac{\partial q_3}{\partial x} = \mp \frac{R}{d}\left(q_3 - \frac{\tau}{6}\right), \tag{78}$$

$-(+)$ in $\mp$ refers to the right (left) plate and d is the surface extrapolation length defined in (67). Similarly, the boundary conditions on the plates $y = 0$ and $y = R$ ares

$$\frac{\partial q_1}{\partial y} = \mp \frac{R}{d}\left(q_1 + \frac{\tau}{2}\right), \tag{79}$$

$$\frac{\partial q_2}{\partial y} = \mp \frac{R}{d} q_2, \tag{80}$$

$$\frac{\partial q_3}{\partial y} = \mp \frac{R}{d}\left(q_3 - \frac{\tau}{6}\right), \tag{81}$$

where $-(+)$ in $\mp$ refers to the top (bottom) plate and d is the surface extrapolation length as before. The corresponding strong anchoring conditions are $\{q_1 = -\frac{\tau}{2}, q_2 = 0, q_3 = \frac{\tau}{6}\}$ on $y = \{0, R\}$ and $\left\{q_1 = \frac{\tau}{2}, q_2 = 0, q_3 = \frac{\tau}{6}\right\}$ on $x = \{0, R\}$, as can be recovered from (76)–(78) and (79)–(81) in the $d \to 0$ limit.

The Euler-Lagrange equations (73)–(75) and the boundary constraints (76)–(81) are solved using relaxation methods in [29]. These methods compute the static solutions, (q_1, q_2, q_3), by mimicking a dynamic gradient-flow procedure along which the free energy continuously decreases till the system reaches equilibrium, for an explicitly prescribed initial condition. The authors use three different kinds of initial conditions: (i) bulk uniaxial alignment along $\overrightarrow{\mathbf{e}_x}$, (ii) bulk uniaxial alignment along $\overrightarrow{\mathbf{e}_x} + \overrightarrow{\mathbf{e}_y}$, and (iii) the isotropic phase with $\mathbf{Q} = 0$. The $\mathbf{Q}$-tensor follows from the solution, (q_1, q_2, q_3), by using the parametrization (58). The converged solutions are robust with respect to different choices of initial conditions and are thus numerically stable.

We now offer some heuristic insight on nematic pattern formation as a function of the critical ratios, $\frac{\xi_b^{(0)}}{R}$ and $\frac{\xi_b^{(0)}}{d}$. On the one hand, if τ is large compared to the re-scaled elastic constant, $\left(\frac{\xi_b^{(0)}}{R}\right)^2$, then the solution predominantly minimizes the bulk potential i.e. is largely uniaxial with constant order parameter, $S = S_{eq}(T)$ (at least away from defects). On the other hand, if τ and $\left(\frac{\xi_b^{(0)}}{R}\right)^2$ are of comparable magnitude, then elastic distortions and deviations from the bulk energy minima (uniaxial phases with $S = S_{eq}(T)$) *e.g.* biaxiality, have comparable energetic costs and hence, biaxial patterns are energetically viable. Similarly, if $\frac{\xi_b^{(0)}}{R} \gg \frac{\xi_b^{(0)}}{d}$, then the uniaxial boundary conditions are relatively weakly implemented and the system has greater freedom to adopt almost "spatially homogeneous" or uniform states. This, in turn, allows the system to avoid complex structural transitions for sufficiently weak anchoring.

We reproduce Figs. 10 and 11 from [29]. The well size is $R = 4.5\xi_b^{(0)}$ at $\tau = 4$ with strong anchoring on the lateral surfaces. In dimensional terms, this corresponds to a well with size length between 120 and 150 nanometers. The authors compute solutions of the Euler-Lagrange equations (73–75) subject to different initial condi-

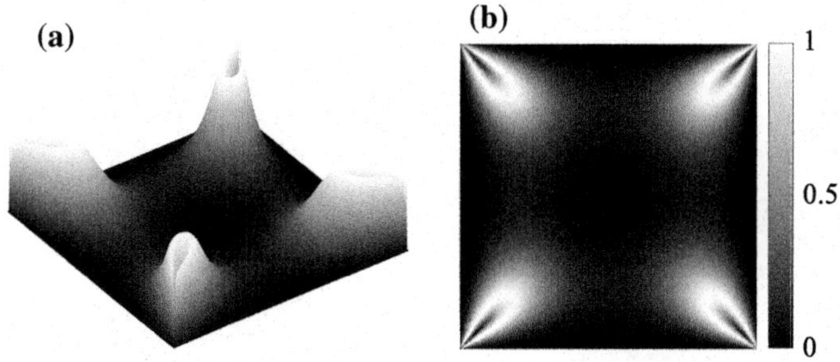

Fig. 10 The degree of biaxiality, $\beta^2(x, y)$, on the *bottom* cross-section of a square well with $R/\xi_b^{(0)} = 4.5$, $\tau = 4$ and strong anchoring conditions. The *shading code* for $\beta^2 \in [0, 1]$ is on the *right side*. Reproduced from Ref. [29] with permission from the Royal Society

Fig. 11 The maximal eigenvector (eigenvector with the largest positive eigenvalue) of the **Q**-solution with $R/\xi_b^{(0)} = 4.5$, $\tau = 4$ and strong anchoring conditions. In the uniaxial limit, this eigenvector is simply the nematic director field. The corresponding $\beta^2(x, y)$ dependence is plotted in Fig. 10. Reproduced from Ref. [29] with permission from the Royal Society

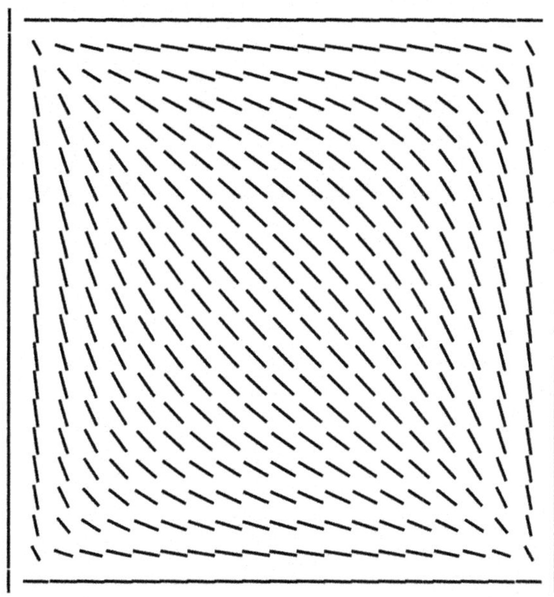

tions and in all cases, recover a diagonal structure with defects, referred to as the *DSD* structure in [29]. In Fig. 10, the authors plot β^2 on the bottom cross-section, $z = 0$ and the leading eigenvector (with the largest eigenvalue) in Fig. 11. It is evident that the leading eigenvector has a diagonal profile, strongly reminiscent of the diagonal solution in the OF framework. The Dirichlet conditions on the lateral surfaces create discontinuities at the square corners which persist as line defects along the vertical edges. This is somewhat artificial but the qualitative trends are unchanged with the more realistic weak anchoring model, e.g. for $\xi_b^{(0)}/d > 20$ i.e. for sufficiently large values of the anchoring coefficient W. The β^2 profile on $z = 0$ reveals interest-

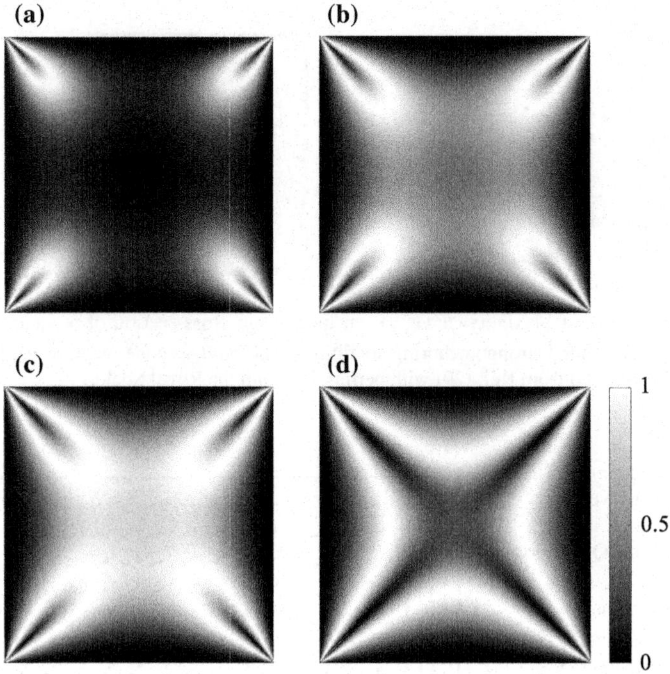

Fig. 12 Plots of $\beta^2(x, y)$ with decreasing $\eta = R/\xi_b^{(0)}$, at a fixed temperature $\tau = 4$ and strong anchoring conditions. **a** $\eta = 4.5$, **b** $\eta = 3.8$, **c** $\eta = 3.6$ and **d** $\eta = 3$. We observe the WORS below the critical value $\eta_c = 3.28 \mp 0.01$. Reproduced from Ref. [29] with permission from the Royal Society

ing information about the vertex profiles that is outside the scope of traditional OF modelling. Let us look at a small neighbourhood of the bottom left vertex, which is the point of intersection of the edges, $x = 0$ and $y = 0$. The Dirichlet conditions are uniaxial and conflicting on the two intersecting edges; consequently $\beta^2 = 0$ on both edges but the nematic profile mediates between the two prescribed uniaxial "edge" states via a biaxial profile, as is evident by a pear-shaped lobe with non-zero β^2 around the vertex. These simulations suggest that the *DSD* structure is a more refined description of the diagonal solution described in Sect. 2.3 with additional information about the biaxial defect cores near the square vertices. However, these biaxial lobes are relatively small and localized and hence, probably not significant for experimental measurements.

In Fig. 12, the authors study structural transitions induced by gradually decreasing the ratio $\eta = R/\xi_b^{(0)}$ at $\tau = 4$, with strong anchoring on four lateral surfaces. They illustrate the structural transitions by plotting the β^2-profiles on $z = 0$ for different values of η. As η decreases, the biaxial lobes around the vertices become larger, start to overlap and at a critical value $\eta = \eta_c = 3.38 \pm 0.01$, the four biaxial lobes connect yielding the *Well Order Reconstruction Solution* (WORS). The WORS is the only stable solution for $\eta < \eta_c$. The WORS is distinctive for two reasons: (i) it is featured by an uniaxial cross (with $\beta^2 = 0$) connecting the four square vertices along the square diagonals and (ii) the uniaxial cross is surrounded by a star-shaped rim of maximal biaxiality, with $\beta^2 = 1$. The uniaxial cross along the square diagonals has negative scalar order parameter. Further, in contrast to the DSD, biaxiality is not localized near the vertices but is a prominent feature of the entire spatial pattern i.e. there is a prominent biaxial ring that separates the negatively ordered uniaxial cross from the positively ordered uniaxial alignments on the square edges.

A potential concern is that the WORS is only observable (at least for $\tau = 4$) for wells with size comparable to the biaxial correlation length and the validity of the LdG theory is not clear for such small length scales. However, recent experiments show that the LdG theory can work well in the nano-regime [27, 28]. Equally importantly, the numerical results in [29] show that η_c is an increasing function of temperature and the WORS may be observable for wells with size between 240–420 nm, close to $\tau = 0$. The LdG theory is certainly valid close to $\tau = 0$, for such length scales. The LdG simulations in [29] and the WORS solution are an interesting example of new biaxial solutions in the LdG framework, outside the scope of OF methodologies. These biaxial solutions offer the possibility of new electro-optical and mechanical properties and hence, hold the promise of a new world of *biaxiality induced phenomena* with fresh perspectives.

4 Conclusions

This chapter is a self-contained introduction to the basic mathematical modelling framework for nematic liquid crystals, at a continuum level. This chapter largely focuses on the Oseen-Frank and Landau-de Gennes theories and the key ingredients are—definition of a nematic order parameter, a continuum energy functional, the governing Euler-Lagrange equations (which are typically a system of elliptic partial differential equations for a range of choices of the elastic constants) and an analysis of the corresponding solutions or equivalently the critical points of the energy. The static experimentally observed equilibria are modelled by the energy minimizers. There are of course, more refined approaches to modelling nematic liquid crystals e.g. the Gay-Berne molecular model, the Lebwohl-Lasher molecular model or the Ball-Majumdar potential (see [7, 47]). However, they present significant analytical and computational difficulties and the Oseen-Frank and Landau-de Gennes approaches are both analytically and computationally tractable, yielding a good match with experiments for a range of physically relevant scenarios. For example, in Sect. 3.1, we use separation of variables and Fourier series methods, both of which are well established in the literature, to derive analytic expressions for the nematic director fields and their energies as a function of the well aspect ratios. Further, the mathematical landscape of these continuum approaches is very rich: the analysis of continuum energies intersects many different branches of mathematics, namely calculus of variations, theory of partial differential equations and algebraic topology. A study of the multiplicity of solutions as a function of model parameters raises highly non-trivial questions in bifurcation theory and dynamical systems and the numerical simulations, particularly multiscale simulations, generate cutting-edge questions in scientific computation and numerical analysis. The natural next step is to study the dynamics of such confined nematic systems in the presence of flow and external electric fields. There are well-established theories to this effect e.g. Leslie-Ericksen [36, 37] and Beris Edwards theories of nematodynamics. The study of nematodynamics is confronted with fresh theoretical and computational challenges, originating from the complex coupling between the nematic director and fluid flow, interfacial phenomena, motion of defects and anisotropic stresses and a self-contained description of nematodynamics warrants a separate dedicated account in future work.

Acknowledgements A.M. is supported by an EPSRC Career Acceleration Fellowship EP/J001686/1 and EP/J001686/2, an OCIAM Visiting Fellowship, support from the Bath Internationalization Grant schemes and the Bath Institute for Mathematical Innovation. A. L. is supported by an Engineering Physical Sciences Research Council studentship. The authors are grateful to Peter Howell, Dirk Aarts and Samo Kralj for fruitful discussions and suggestions.

References

1. Timeline: The Early History of the Liquid Crystal Display. Available via Spectrum. http://spectrum.ieee.org/static/timeline-the-early-history-of-the-liquid-crystal-display. Cited 29 Apr 2016
2. D. Allender, L. Longa, Landau-de Gennes theory of biaxial nematics reexamined. Phys. Rev. E **78**(1), 011–704 (2008)
3. M. Ambrožič, F. Bisi, E.G. Virga, Director reorientation and order reconstruction: competing mechanisms in a nematic cell. Contin. Mech. Thermodyn. **20**(4), 193–218 (2008)
4. D. Andrienko, *Introduction to Liquid Crystals* (International Max Planck Research School, Bad Marienberg, 2006)
5. B. Bahadur, *Liquid Crystals: Applications and Uses* (World Scientific, 1991)
6. J.M. Ball, Function spaces for liquid crystals (2015). https://people.maths.ox.ac.uk/ball/Teaching/lyon2015.pdf. (Winter school, Nonlinear Function Spaces in Mathematics and Physical Sciences, Lyon)
7. J.M. Ball, A. Majumdar, Nematic liquid crystals: from Maier-Saupe to a continuum theory. Mol. Cryst. Liq. Cryst. **525**(1), 1–11 (2010)
8. G. Barbero, G. Durand, On the validity of the Rapini-Papoular surface anchoring energy form in nematic liquid crystals. J. de Phys. **47**(12), 2129–2134 (1986)
9. E. Barry, D. Beller, Z. Dogic, A model liquid crystalline system based on rodlike viruses with variable chirality and persistence length. Soft Matter **5**, 2563–2570 (2009)
10. F. Bethuel, H. Brezis, F. Hélein, Asymptotics for the minimization of a Ginzburg-Landau functional. Calc. Var. Partial Differ. Equ. **1**(2), 123–148 (1993)
11. H. Brezis, J.M. Coron, E.H. Lieb, Harmonic maps with defects. Commun. Math. Phy. **107**(4), 649–705 (1986)
12. G. Carbone, G. Lombardo, R. Barberi, I. Muševič, U. Tkalec, Mechanically induced biaxial transition in a nanoconfined nematic liquid crystal with a topological defect. Phys. Rev. Lett. **103**(16), 167–801 (2009)
13. S. Chandrasekhar, *Liquid Crystals* (Cambridge University Press, 1992)

14. J. Chen, C.T. Liu, Technology advances in flexible displays and substrates. Access IEEE **1**, 150–158 (2013)
15. O.J. Dammone, Confinement of colloidal liquid crystals. Ph.D. thesis, University College, University of Oxford, 2013
16. O.J. Dammone, I. Zacharoudiou, R.P.A. Dullens, J.M. Yeomans, M.P. Lettinga, D.G.A.L. Aarts, Confinement induced splay-to-bend transition of colloidal rods. Phys. Rev. Lett. **109**(10), 108–303 (2012)
17. A.E. Danese, *Advanced Calculus*, vol. 1 (Allyn and Bacon, 1965)
18. T.A. Davis, E.C. Gartland Jr., Finite element analysis of the Landau-de Gennes minimization problem for liquid crystals. SIAM J. Numer. Anal. **35**(1), 336–362 (1998)
19. I. Dozov, M. Nobili, G. Durand, Fast bistable nematic display using monostable surface switching. Appl. Phys. Lett. **70**(9), 1179–1181 (1997)
20. J.L. Ericksen, Liquid crystals with variable degree of orientation. Arch. Ration. Mech. Anal. **113**(2), 97–120 (1991)
21. F.C. Frank, I. liquid crystals. on the theory of liquid crystals. Discuss. Faraday Soc. **25**, 19–28 (1958)
22. P.G. de Gennes, *The Physics of Liquid Crystals* (Clarendon Press, Oxford, 1974)
23. E. Grelet, Hexagonal order in crystalline and columnar phases of hard rods. Phys. Rev. Lett. **100**, 168–301 (2008)
24. R. Hardt, D. Kinderlehrer, F.H. Lin, Existence and partial regularity of static liquid crystal configurations. Commun. Math. Phys. **105**(4), 547–570 (1986)
25. A. Jeffrey, D. Zwillinger, *Table of Integrals, Series, and Products* (Elsevier Science, 2000)
26. J. Katriel, G.F. Kventsel, G.R. Luckhurst, T.J. Sluckin, Free energies in the Landau and molecular field approaches. Liq. Cryst. **1**(4), 337–355 (1986)
27. A.V. Kityk, M. Wolff, K. Knorr, D. Morineau, R. Lefort, P. Huber, Continuous paranematic-to-nematic ordering transitions of liquid crystals in tubular silica nanochannels. Phys. Rev. Lett. **101**(18), 187–801 (2008)
28. S. Kralj, G. Cordoyiannis, A. Zidanšek, G. Lahajnar, H. Amenitsch, S. Žumer, Z. Kutnjak, Presmectic wetting and supercritical-like phase behavior of octylcyanobiphenyl liquid crystal confined to controlled-pore glass matrices. J. Chem. Phys. **127**(15), 154–905 (2007)
29. S. Kralj, A. Majumdar, Order reconstruction patterns in nematic liquid crystal wells. Proc. R. Soc. A **470**(2169), 20140276 (2014)
30. S. Kralj, E.G. Virga, Universal fine structure of nematic hedgehogs. J. Phys. A: Math. Gen. **34**(4), 829 (2001)
31. S. Kralj, E.G. Virga, S. Žumer, Biaxial torus around nematic point defects. Phys. Rev. E **60**(2), 1858 (1999)
32. J.P.F. Lagerwall, *An Introduction to the Physics of Liquid Crystals*, ed. by A. Fernandez-Nieves. Soft Materials—generation, physical properties and fundamental applications (John Wiley & Sons, 2014)
33. F.M. Leslie, Continuum theory for nematic liquid crystals. Contin. Mech. Thermodyn. **4**(3), 167–175 (1992)
34. A.H. Lewis, Defects in liquid crystals: Mathematical and experimental studies. Ph.D. thesis, University of Oxford, 2016
35. A.H. Lewis, I. Garlea, J. Alvarado, O.J. Dammone, P.D. Howell, A. Majumdar, B.M. Mulder, M.P. Lettinga, G.H. Koenderink, D.G.A.L. Aarts, Colloidal liquid crystals in rectangular confinement: theory and experiment. Soft Matter **10**, 7865–7873 (2014)
36. F. Lin, C. Wang, Recent developments of analysis for hydrodynamic flow of nematic liquid crystals. Philos. Trans. R. Soc. A **372**(2029), 20130361 (2014)
37. F.H. Lin, C. Liu, Static and dynamic theories of liquid crystals. J. Partial Diff. Equ. **14**(4), 289–330 (2001)
38. F.H. Lin, C.C. Poon, On Ericksens model for liquid crystals. J. Geom. Anal. **4**(3), 379–392 (1994)
39. C. Luo, A. Majumdar, R. Erban, Multistability in planar liquid crystal wells. Phys. Rev. E **85**, 061–702 (2012)

40. A. Majumdar, Equilibrium order parameters of nematic liquid crystals in the Landau-de Gennes theory. Eur. J. Appl. Math. **21**, 181–203 (2010)
41. A. Majumdar, A. Zarnescu, Landau-de Gennes theory of nematic liquid crystals: the Oseen-Frank limit and beyond. Arch. Ration. Mech. Anal. **196**(1), 227–280 (2010)
42. N.J. Mottram, C.J.P. Newton, Introduction to Q-tensor theory. Research report (University of Strathclyde, 2014)
43. M.J. Stephen, J.P. Straley, Physics of liquid crystals. Rev. Mod. Phys. **46**, 617–704 (1974)
44. I.W. Stewart, *The Static and Dynamic Continuum Theory of Liquid Crystals: A Mathematical Introduction* (CRC Press, Oxford, 2004)
45. C. Tsakonas, A.J. Davidson, C.V. Brown, N.J. Mottram, Multistable alignment states in nematic liquid crystal filled wells. Appl. Phys. Lett. **90**(11), 111–913 (2007)
46. E.G. Virga, *Variational Theories for Liquid Crystals* (Chapman and Hall, London, 1994)
47. M.R. Wilson, Molecular simulation of liquid crystals: progress towards a better understanding of bulk structure and the prediction of material properties. Chem. Soc. Rev. **36**, 1881–1888 (2007)

Dynamical Density Functional Theory for Brownian Dynamics of Colloidal Particles

Hartmut Löwen

1 Introduction

Variational methods play a key role in physics. Density functional theory (DFT) is a special and important example of such a variational formulation. DFT relies on the fact that there is a functional of the one-particle density which gives access to the equilibrium thermodynamics when it is minimized with respect to the density. This important theory can be both applied to quantum-mechanical electrons and to classical systems.

In this book chapter we shall consider nonequilibrium situations for completely overdamped Brownian dynamics of colloids. A dynamical version of DFT, the so-called dynamical density functional theory (DDFT), is available and makes dynamical predictions which are in good agreement with computer simulations. Here we shall derive DDFT for Brownian dynamics in a tutorial way from the microscopic Smoluchowski equation and mention some applications such as crystallization on a patterned substrate. The theory will then be generalized towards hydrodynamic interactions between the particles and to orientational degrees of freedom describing e.g. rod-like colloids. Finally some recent developments will be discussed. For parts of this tutorial we follow the ideas outlined in Ref. [60, 61].

H. Löwen (✉)
Institut für Theoretische Physik II: Weiche Materie, Heinrich-Heine-Universität Düsseldorf, Universitätsstrasse 1, D-40225 Düsseldorf, Germany
e-mail: hlowen@hhu.de

J. Wu (ed.), *Variational Methods in Molecular Modeling*,
Molecular Modeling and Simulation, DOI 10.1007/978-981-10-2502-0_9

2 Density Functional Theory (DFT) in Equilibrium

2.1 *Basics*

We shall consider density functional theory (DFT) here for classical systems at finite temperature as opposed to DFT for quantum mechanical electrons. The cornerstone of classical density functional theory is an existence theorem combined with a basic variational principle [25, 55, 57, 58, 86]. In detail, there exists a unique grand canonical free energy-density-functional $\Omega(T, \mu, [\rho])$, which gets minimal for the equilibrium density $\rho_0(\mathbf{r})$ and then coincides with the real grand canonical free energy, i.e.

$$\left. \frac{\delta\Omega(T, \mu, [\rho])}{\delta\rho(\mathbf{r})} \right|_{\rho(\mathbf{r})=\rho_0(\mathbf{r})} = 0. \tag{1}$$

Here, T is the imposed temperature and μ the prescribed chemical potential of the system. In particular DFT is also valid for systems which are inhomogeneous on a microscopic scale. In principle, all fluctuations are included if an imposed external potential $V_{ext}(\mathbf{r})$ breaks all symmetries. For interacting systems in three dimensions (3d), however, the functional $\Omega(T, \mu, [\rho])$ is not known. One can split the functional $\Omega(T, \mu, [\rho])$ exactly as follows

$$\Omega(T, \mu, [\rho]) = \mathcal{F}(T, [\rho]) + \int_V d^3r \; \rho(\mathbf{r}) \left(V_{ext}(\mathbf{r}) - \mu\right) \tag{2}$$

where $\mathcal{F}(T, [\rho])$ is a Helmholtz free energy functional and V denotes the system volume.

Fortunately, there are few exceptions where the density functional is known exactly. First, for low density, the ideal-gas-limit is reached and the density functional can be constructed analytically (see below). Next leading orders for finite densities can be incorporated via a virial expansion which is quadratic in the densities. Conversely, in the high-density-limit, the mean-field approximation (see below) becomes asymptotically exact for penetrable potentials. Indeed this approximation also works surprisingly well for finite densities beyond overlap. Furthermore, the density functional is exactly known (as the so-called Percus-functional) in one spatial dimension for the Tonks gas (hard rods on a line). However, the latter system does not exhibit freezing. Please note that the knowledge of a functional is much more than a bulk equation of state since it can be applied to any external potential $V_{ext}(\mathbf{r})$.

2.2 DFT of Freezing

One important application of DFT is freezing: First one has to chose an approximation. Then the density field is parametrized with variational parameters. In the homogeneous gas and liquid bulk phase one takes

$$\rho(\mathbf{r}) = \bar{\rho} \tag{3}$$

where $\bar{\rho}$ is a variational parameter. On the other hand, for the solid, the Gaussian approximation of density peaks on the lattice positions is an excellent choice [83], where

$$\rho(\mathbf{r}) = \left(\frac{\alpha}{\pi}\right)^{-3/2} \sum_n \exp\left(-\alpha\left(\mathbf{r} - \mathbf{R}_n\right)^2\right) . \tag{4}$$

Here both the lattice structure and spacing as well as the width α are variational parameters. Finally, for a given chemical potential μ and temperature T, one has to minimize the functional $\Omega(T, \mu, [\rho])$ with respect to all variational parameters. As a result one obtains the phase diagram in the μT plane.

The procedure itself is sketched close to a solid-liquid transition in Figs. 1 and 2. A solid-liquid transition line in the μT plane is schematically shown in Fig. 1 and we consider a path with fixed μ and increasing T crossing the solid-liquid transition at $\mu = \mu_{coex}$ and $T = T_{coex}$. Coexistence implies that temperature T, chemical potential μ and pressure p are the same in both phases. Since in the bulk $p = -\Omega/V$ (V denoting the system volume) coexistence means that at given μ and T, Ω/V has two minima with equal depth. A contour plot of the density functional in the space of variational parameters is shown in Fig. 2 for three different temperatures on the path shown in Fig. 1. The liquid minimum occurs at zero α while the solid is characterized by a minimum at finite α. The global minimum is the stable phase and, at coexistence, the two minima have equal depth.

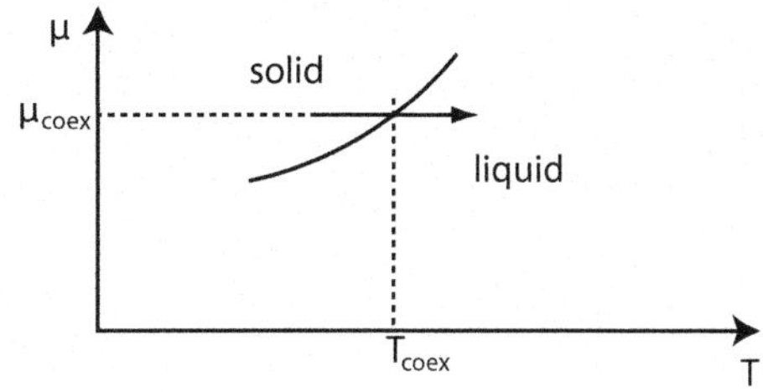

Fig. 1 Solid-liquid coexistence line in the μT plane. The path along which three state points are discussed in Fig. 2 is indicated

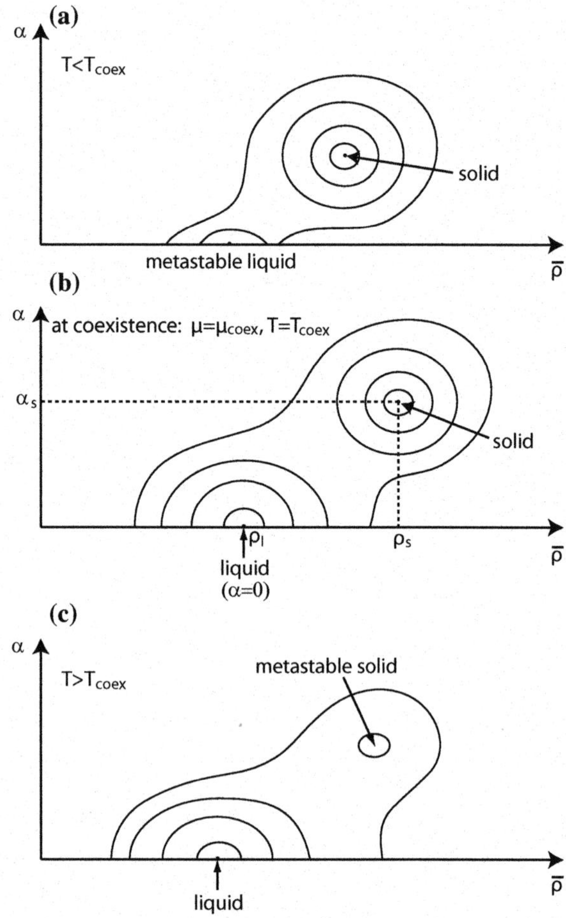

Fig. 2 Contour plot of the grand canonical free energy $\Omega(T, \mu, \bar{\rho}, \alpha)$ for fixed T and μ as a function of two variational parameters α and the averaged density $\bar{\rho}$. The latter is given by the lattice constant in the solid phase. **a** with a stable solid phase, **b** at solid-liquid coexistence, **c** with a stable liquid phase

2.3 *Approximations for the Density Functional*

Let us first recall the exact functional for the ideal gas where the pair interaction $V(r)$ between the particles vanishes, $V(r) = 0$. It reads as

$$\mathscr{F}(T, [\rho]) = \mathscr{F}_{\mathrm{id}}(T, [\rho]) = k_B T \int_V \mathrm{d}^3 r \;\; \rho(\mathbf{r}) \left[\ln(\rho(\mathbf{r})\Lambda^3) - 1\right] \tag{5}$$

where Λ is the irrelevant thermal wave length and k_B the Boltzmann constant. In this case, the minimization condition

$$\left.\frac{\delta \Omega}{\delta \rho(\mathbf{r})}\right|_0 = k_B T \ln(\rho(\mathbf{r})\Lambda^3) + V_{\mathrm{ext}}(\mathbf{r}) - \mu = 0 \tag{6}$$

leads to the generalized barometric law

$$\rho_0(\mathbf{r}) = \frac{1}{\Lambda^3} \exp\left(-\frac{V_{\mathrm{ext}}(\mathbf{r}) - \mu}{k_B T}\right) \tag{7}$$

for the inhomogeneous density. In the interacting case, $V(r) \neq 0$, one can split

$$\mathscr{F}(T, [\rho]) =: \mathscr{F}_{\mathrm{id}}(T, [\rho]) + \mathscr{F}_{\mathrm{exc}}(T, [\rho]) \tag{8}$$

which defines the excess free energy density functional $\mathscr{F}_{exc}(T, [\rho])$. Approximations work on different levels. In the *mean-field approximation*, we set

$$\mathscr{F}_{\mathrm{exc}}(T, [\rho]) \approx \frac{1}{2} \int \mathrm{d}^3 r \int \mathrm{d}^3 r' \, V(|\mathbf{r} - \mathbf{r}'|)\rho(\mathbf{r})\rho(\mathbf{r}') \tag{9}$$

In fact, the mean-field approximation (together with a correlational hole in the solid) yields freezing of a Gaussian pair potential $V(r) \sim \exp(-(r/\sigma)^2)$ [6, 50] and is the correct starting point for cluster crystals for penetrable potentials [33, 53, 54].

The Ramakrishnan-Yussouff (RY) approximation is a perturbative treatment out of the bulk liquid which needs the bulk liquid direct correlation function $c^{(2)}(r, \bar{\rho}, T)$ as an input. A functional Taylor expansion around a homogeneous reference density up to second order yields

$$\mathscr{F}_{\mathrm{exc}}(T, [\rho]) \cong -\frac{k_B T}{2} \int \mathrm{d}^3 r \int \mathrm{d}^3 r' \; c^{(2)}(|\mathbf{r} - \mathbf{r}'|, \bar{\rho}, T)(\rho(\mathbf{r}) - \bar{\rho})(\rho(\mathbf{r}') - \bar{\rho}) \tag{10}$$

The RY approximation leads to freezing for hard spheres and was historically the first demonstration that freezing can be described within DFT. The RY functional can readily be generalized to soft interactions [106] (as the one-component plasma) and gives reasonable results for freezing there (though it is better to improve the functional by including triplet correlations). The input $c^{(2)}(|\mathbf{r} - \mathbf{r}'|, \bar{\rho}, T)$ can be gained by liquid integral equation theory [11].

A non-perturbative functional is based on Rosenfeld's fundamental measure theory (FMT) [105]. This works, however, only for a hard interaction, such as hard spheres. In FMT we have

$$\frac{\mathscr{F}_{\mathrm{exc}}[\rho]}{k_B T} = \int \mathrm{d}^3 r \;\; \Phi[\{n_\alpha(\mathbf{r})\}] \tag{11}$$

with

$$n_\alpha(\mathbf{r}) = \int \mathrm{d}^3 r' \;\; w^{(\alpha)}(\mathbf{r} - \mathbf{r}')\rho(\mathbf{r}') \tag{12}$$

where the six weight function are given explicitly as

$$w^{(0)}(\mathbf{r}) = \frac{w^{(2)}(\mathbf{r})}{\pi\sigma^2} \tag{13}$$

$$w^{(1)}(\mathbf{r}) = \frac{w^{(2)}(\mathbf{r})}{2\pi\sigma} \tag{14}$$

$$w^{(2)}(\mathbf{r}) = \delta\left(\frac{\sigma}{2} - r\right) \tag{15}$$

$$w^{(3)}(\mathbf{r}) = \Theta\left(\frac{\sigma}{2} - r\right) \tag{16}$$

$$w^{(V_1)}(\mathbf{r}) = \frac{\mathbf{w}^{(V_2)}(\mathbf{r})}{2\pi\sigma} \tag{17}$$

$$w^{(V_2)}(\mathbf{r}) = \frac{\mathbf{r}}{r}\delta\left(\frac{\sigma}{2} - r\right) \tag{18}$$

with σ denoting the hard sphere diameter and

$$\Phi = \Phi_1 + \Phi_2 + \Phi_3 \tag{19}$$

$$\Phi_1 = -n_0 \ln(1 - n_3) \tag{20}$$

$$\Phi_2 = \frac{n_1 n_2 - \mathbf{n}_{v_1} \cdot \mathbf{n}_{v_2}}{1 - n_3} \tag{21}$$

$$\Phi_3 = \frac{\frac{1}{3}n_2^3 - n_2(\mathbf{n}_{v_2} \cdot \mathbf{n}_{v_2})}{8\pi(1 - n_3)^2} \tag{22}$$

This FMT functional yields the Percus-Yevick solution of the direct correlation function as an output. It furthermore survives the dimensional crossover [96]: If the three-dimensional hard sphere system is confined within a one-dimensional tube, the exact Percus functional is recovered. Moreover, in a spherical cavity which holds one or no particle at all, the exact functional is recovered. This helps to understand that the constraint packing argument of freezing is geometrically included in the FMT. In fact (also with a tensor modification in Φ_3 [103]), the FMT gives excellent data for hard-sphere freezing [96], see Table 1 and the review by Roth [97].

Last but not least we mention perturbation theories which can be used for attractive tails. The total potential $V(r)$ is then split into a purely repulsive short-ranged part $V_{rep}(r)$ and a longer-ranged attractive part $V_{attr}(r)$ such that $V(r) = V_{rep}(r) + V_{attr}(r)$. The repulsive part is treated as an effective hard core with an effective (temperature-dependent) Barker-Henderson diameter

$$\sigma(T) = \int_0^\infty \mathrm{d}r \ \left(1 - e^{-\beta V_{\mathrm{rep}}(r)}\right) \tag{23}$$

Table 1 Coexisting number densities and solid Lindemann parameter at coexistence for the hard sphere system. "Exact" computer simulation data are shown as well as DFT data using the Ramakrishnan-Yussouff (RY) or Rosenfeld's fundamental measure theory

	$\rho_l\,\sigma^3$	$\rho_s\,\sigma^3$	L (: Lindemann)
Computer simulations	0.94	1.04	0.129
RY	0.97	1.15	0.06
Rosenfeld	0.94	1.03	0.101

where $\beta = 1/k_B T$ and the attractive part is treated within mean-field approximation. Accordingly, the total excess free energy functional reads as

$$\mathscr{F}_{\text{exc}}(T, [\rho]) \cong \mathscr{F}_{\text{exc}}^{\text{HS}}(T, [\rho])\big|_{\sigma=\sigma(T)} + \frac{1}{2}\int d^3r \int d^3r' \; \rho(\mathbf{r})\rho(\mathbf{r}')V_{\text{attr}}(|\mathbf{r}-\mathbf{r}'|) \tag{24}$$

This procedure yields good phase diagrams for both Lennard-Jones potentials and sticky-hard-sphere systems including the isostructural solid-solid transition [52].

To summarize:

1. Rosenfeld's FMT yields excellent data for hard sphere freezing.
2. The much less justified RY perturbative approach works in principle for softer repulsions.
3. The mean-field density functional approximation works for penetrable potentials.
4. Hard sphere perturbation theory yields stability of liquids and solid-solid isostructural transitions.

3 Brownian Dynamics (BD)

We now proceed to consider dynamics described by time-dependent DFT. Time-dependent DFT is well-known in the quantum-mechanical context of electrons where the time-dependence is described by Schrödinger's equation [99]. Here we consider classical systems with a completely overdamped dynamics which experiences thermal fluctuations, i.e. we consider dynamical density functional theory (DDFT) for Brownian dynamics as appropriate for colloidal particles. Colloidal particles are embedded in a molecular solvent and are therefore randomly kicked by the solvent molecules on timescales much smaller than the drift of the colloidal motion [17, 88].

3.1 Noninteracting Brownian Particles

Let us first discuss the Smoluchowski picture. Here the time-dependent density field is the central quantity. It should follow a simple deterministic diffusion equation. For noninteracting particles with an inhomogeneous time-dependent particle density $\rho(\mathbf{r}, t)$, Fick's law states that the current density $\mathbf{j}(\mathbf{r}, t)$ is

$$\mathbf{j}(\mathbf{r}, t) = -D_0 \nabla \rho(\mathbf{r}, t) \tag{25}$$

where D_0 is a phenomenological diffusion coefficient.

The continuity equation of particle number conservation

$$\frac{\partial \rho(\mathbf{r},t)}{\partial t} + \nabla \cdot \mathbf{j}(\mathbf{r},t) = 0 \tag{26}$$

then leads to the well-known diffusion equation for $\rho(\mathbf{r},t)$:

$$\frac{\partial \rho(\mathbf{r},t)}{\partial t} = D_0 \Delta \rho(\mathbf{r},t) \tag{27}$$

In the presence of an external potential $V_{\text{ext}}(\mathbf{r})$, the force $\mathbf{F} = -\nabla V_{\text{ext}}(\mathbf{r})$ acts on the particles and will induce a drift velocity $\mathbf{v}_\text{D}$ giving rise to an additional current density

$$\mathbf{j}_\text{D} = \rho \mathbf{v}_\text{D} \,. \tag{28}$$

We now assume totally overdamped motion since inertia effects are small as the colloids are much bigger than the solvent molecules. This results in

$$\mathbf{v}_\text{D} = \frac{\mathbf{F}}{\xi} = -\frac{1}{\xi}\nabla V_{\text{ext}}(\mathbf{r}) \tag{29}$$

with ξ denoting a friction coefficient. For a sphere of radius R in a viscous solvent, $\xi = 6\pi\eta_\text{s} R$, with η_s denoting the shear viscosity of the solvent (Stokes law). Now the total current density is

$$\mathbf{j} = -D_0 \nabla \rho(\mathbf{r},t) - \rho(\mathbf{r},t)\frac{1}{\xi}\nabla V_{\text{ext}}(\mathbf{r}) \tag{30}$$

In equilibrium, the one-particle density is a Boltzmann distribution, see Eq. (7).

$$\rho(\mathbf{r},t) \equiv \rho^{(1)}(\mathbf{r}) = \rho^{(0)}(\mathbf{r}) = A \exp(-\beta V_{\text{ext}}(\mathbf{r})) \tag{31}$$

Furthermore, in equilibrium, the total current has to vanish. Therefore, necessarily

$$D_0 = \frac{k_B T}{\xi} \tag{32}$$

which is the so-called Stokes-Einstein relation. Hence $\mathbf{j} = -\frac{1}{\xi}(k_B T \nabla \rho + \rho \nabla V_{\text{ext}})$ and the continuity equation yields

$$\frac{\partial \rho(\mathbf{r},t)}{\partial t} = \frac{1}{\xi}(k_B T \Delta \rho(\mathbf{r},t) + \nabla(\rho(\mathbf{r},t)\nabla V_{\text{ext}}(\mathbf{r}))) \tag{33}$$

which is called Smoluchowski equation (for non-interacting particles).

The same equation holds for the probability density $w(\mathbf{r},t)$ to find a particle at position $\mathbf{r}$ for time t. For N non-interacting particles,

$$w(\mathbf{r}, t) = \frac{1}{N}\rho(\mathbf{r}, t)\,, \tag{34}$$

and the Smoluchowski equation reads:

$$\frac{\partial w}{\partial t} = \frac{1}{\xi}(k_B T \Delta w - \nabla(w \nabla V_{\text{ext}})) \tag{35}$$

3.2 Interacting Brownian Particles

Now we consider N **interacting** particles. Using a compact notation for the particle positions

$$\{x_i\} = \{\mathbf{r}_i\} = \{\underbrace{x_1, x_2, x_3}_{\mathbf{r}_1}, \underbrace{x_4, x_5, x_6}_{\mathbf{r}_2}, \cdots, \underbrace{x_{3N-2}, x_{3N-1}, x_{3N}}_{\mathbf{r}_N}\} \tag{36}$$

we assume a linear relation between acting forces on the particles and the resulting drift velocities. (The same compact notation is used for other multiple vectors.) The details of this relation embody the so-called hydrodynamic interactions mediated between the colloidal particles by the solvent flow field induced by the moving colloidal particles. This linear relation is in general

$$v_i = \sum_{j=1}^{3N} L_{ij}(\{x_n\})\mathbf{F}_j \tag{37}$$

where $\mathbf{F}_j = -\frac{\partial}{\partial x_j} U_{\text{tot}}$ where U_{tot} involves both the internal and the interaction potential energy and v is the drift velocity. The underlying assumption in (37) is that the hydrodynamic interactions act quasi-instantaneously. This is justified by the fact that the timescale upon which a shear perturbation is traveling through the suspension within an interparticle distance is much smaller than that of Brownian motion. The coefficients L_{ij} constitute the so-called $3N \times 3N$ mobility matrix and can in principle be obtained by solving the Stokes equations of N spheres with appropriate stick boundary conditions of the solvent flow field on the particle's surfaces.

In general, L_{ij} depends on $\mathbf{r}^N$, and we postulate:

- symmetry

$$L_{ij} = L_{ji} \tag{38}$$

- positivity

$$\sum_{ij} F_i F_j L_{ij} > 0 \text{ for all } F_{i,j} \neq 0 \tag{39}$$

with $w(\{\mathbf{r}_i\}, t)$ denoting the probability density for interacting particles, the suitable generalization of the continuity equation is

$$\frac{\partial w}{\partial t} = -\sum_{n=1}^{3N} \frac{\partial}{\partial x_n}(v_{\mathrm{tot},n} w) \tag{40}$$

with

$$v_{\mathrm{tot},n} = \sum_{m=1}^{3N} L_{mn} \frac{\partial}{\partial x_m}(k_B T \ln w + U_{\mathrm{tot}}) \tag{41}$$

which leads to the generalized Smoluchowski equation for interacting particles.

$$\frac{\partial w}{\partial t} = \hat{\mathcal{O}} w \tag{42}$$

with the Smoluchowski operator

$$\hat{\mathcal{O}} = \sum_{n,m=1}^{3N} \frac{\partial}{\partial x_n} L_{nm} (k_B T \frac{\partial}{\partial x_m} + \frac{\partial U_{\mathrm{tot}}}{\partial x_m}) \tag{43}$$

In many applications, hydrodynamic interactions are neglected. This means that the mobility matrix is constant and a diagonal

$$L_{nm} = \frac{1}{\xi} \delta_{nm} \tag{44}$$

This assumption, however, is only true for small volume fractions of the colloidal particles.

4 Dynamical Density Functional Theory (DDFT)

Here we derive a deterministic equation for the time-dependent one-particle density from the Smoluchowski equation [6]. We follow the idea of Archer and Evans [7]. First, we recall Smoluchowski equation for the N-particle density

$$w(\mathbf{r}_1, \ldots, \mathbf{r}_N, t) \equiv w(\mathbf{r}^N, t),\ \mathbf{r}^N = \{\mathbf{r}_1, \ldots, \mathbf{r}_N\} \text{ as}$$

$$\frac{\partial w}{\partial t} = \hat{\mathcal{O}} w = \frac{1}{\xi} \sum_{i=1}^{N} \nabla_i \cdot [k_B T \nabla_i + \nabla_i U_{\mathrm{tot}}(\mathbf{r}^N, t)] w \tag{45}$$

with

$$U_{\text{tot}}(\mathbf{r}^N, t) = \sum_{i=1}^{N} V_{\text{ext}}(\mathbf{r}_i, t) + \sum_{\substack{i,j=1 \\ i<j}}^{N} V(|\mathbf{r}_i - \mathbf{r}_j|) \tag{46}$$

involving a time-dependent external potential $V_{ext}(\mathbf{r}, t)$ and a time-independent interaction $V(r)$. Here, we assume a pairwise additivity of the interactions and neglect hydrodynamic interactions. Now the idea is to integrate out degrees of freedom. In fact, an integration yields

$$\rho(\mathbf{r}_1, t) = N \int \mathrm{d}^3 r_2 \ldots \int \mathrm{d}^3 r_N \; w(\mathbf{r}^N, t) \tag{47}$$

The 2-particle density is analogously obtained as

$$\rho^{(2)}(\mathbf{r}_1, \mathbf{r}_2, t) = N(N-1) \int \mathrm{d}^3 r_3 \ldots \int \mathrm{d}^3 r_N \; w(\mathbf{r}^N, t) \tag{48}$$

Integrating the Smoluchowski equation with $N \int \mathrm{d}^3 r_2 \ldots \int \mathrm{d}^3 r_N$ yields

$$\begin{aligned} \Rightarrow \quad \frac{\partial}{\partial t}\rho(\mathbf{r}_1, t) = N \int \mathrm{d}^3 r_2 \ldots \int \mathrm{d}^3 r_N \{ \sum_{i=1}^{N} (k_B T \Delta_i w(\mathbf{r}^N, t) \\ + \nabla_i (\nabla_i V_{\text{ext}}(\mathbf{r}_i, t) w(\mathbf{r}^N, t))) \\ + \sum_{\substack{i=1 \\ i<j}}^{N} \nabla_i (\nabla_i (V(|\mathbf{r}_i - \mathbf{r}_j|) w(\mathbf{r}^N, t)))\} \end{aligned}$$

The right-hand-side of this equation is a sum of three terms which we now discuss step by step. The first term can be rearranged as follows:

$$\begin{aligned} N \int \mathrm{d}^3 r_2 \ldots \int \mathrm{d}^3 r_N \sum_{i=1}^{N} k_B T \Delta_i w(\mathbf{r}^N, t) &= k_B T \Delta_1 \rho(\mathbf{r}_1, t) \\ &+ N \int \mathrm{d}^3 r_2 \ldots \int \mathrm{d}^3 r_N k_B T \sum_{i=2}^{N} \Delta_i w(\mathbf{r}^N, t) \\ &= k_B T \Delta_1 \rho(\mathbf{r}_1, t) \\ &+ \sum_{i=2}^{N} N k_B T \int \mathrm{d}^3 r_i \nabla_i \underbrace{(\nabla_i \int \mathrm{d}^3 r_2 \ldots \int \mathrm{d}^3 r_{i-1} \int \mathrm{d}^3 r_{i+1} \ldots \int \mathrm{d}^3 r_N \; w(\mathbf{r}^N, t))}_{f(\mathbf{r}_1, \mathbf{r}_i, t)} \end{aligned}$$

Now the theorem of Gauss tell us that $\int_S \mathrm{d}^2 r_i \nabla_i f(\mathbf{r}_1, \mathbf{r}_i, t) = 0$ such that the first term becomes equal to:

$$k_B T \Delta_1 \rho(\mathbf{r}_1, t) + \sum_{i=2}^{N} N k_B T \underbrace{\int \mathrm{d}^3 r_i \nabla_i (\nabla_i (f(\mathbf{r}_1, \mathbf{r}_i, t)))}_{=0} = k_B T \Delta_1 \rho(\mathbf{r}_1, t)$$

Using the same argument, the second term becomes

$$\begin{aligned} N \int \mathrm{d}^3 r_2 \ldots \int \mathrm{d}^3 r_N \sum_{i=1}^{N} \nabla_i (\nabla_i V_{\text{ext}}(\mathbf{r}_i, t) w(\mathbf{r}^N, t)) \\ = N \int \mathrm{d}^3 r_2 \ldots \int \mathrm{d}^3 r_N \nabla_1 (\nabla_1 V_{\text{ext}}(\mathbf{r}_1, t)) w(\mathbf{r}^N, t) + 0 \\ = \nabla_1 ((\nabla_1 V_{\text{ext}}(\mathbf{r}_1, t)) \rho(\mathbf{r}_1, t)) \end{aligned}$$

Finally the third terms which involves the interactions is:

$$\begin{aligned} N \int \mathrm{d}^3 r_2 \ldots \int \mathrm{d}^3 r_N \sum_{\substack{i,j=1 \\ i<j}}^{N} \nabla_i \cdot (\nabla_i V(|\mathbf{r}_i - \mathbf{r}_j|)) w(\mathbf{r}^N, t) \\ = N \int \mathrm{d}^3 r_2 \ldots \int \mathrm{d}^3 r_N \nabla_1 (\sum_{j=2}^{N} \nabla_1 V(|\mathbf{r}_1 - \mathbf{r}_j|) w(\mathbf{r}^N, t)) \end{aligned}$$

The integral is symmetric in the coordinates, therefore without loss of generality we can set $j = 2$ and obtain for the third term

$$\begin{aligned} N(N-1)\nabla_1 \int \mathrm{d}^3 r_2 \nabla_1 V(|\mathbf{r}_1 - \mathbf{r}_2|) \int \mathrm{d}^3 r_3 \ldots \int \mathrm{d}^3 r_N \; w(\mathbf{r}^N, t) \\ = \int \mathrm{d}^3 r_2 \nabla_1 (\nabla_1 V(|\mathbf{r}_1 - \mathbf{r}_2|)) \rho^{(2)}(\mathbf{r}_1, \mathbf{r}_2, t) \end{aligned}$$

Summarizing the previous calculations we end up with the following integrated Smoluchowski equation:

$$\begin{aligned} \xi \frac{\partial}{\partial t} \rho(\mathbf{r}_1, t) = k_B T \Delta_1 \rho(\mathbf{r}_1, t) + \nabla_1 (\rho(\mathbf{r}_1, t) \nabla_1 V_{\text{ext}}(\mathbf{r}_1, t) \\ + \nabla_1 \int \mathrm{d}^3 r_2 \; \rho^{(2)}(\mathbf{r}_1, \mathbf{r}_2, t) \nabla_1 V(|\mathbf{r}_1 - \mathbf{r}_2|) \end{aligned} \quad (49)$$

In equilibrium, necessarily $\frac{\partial \rho(\mathbf{r}_1, t)}{\partial t} = 0$ which implies

$$0 = \nabla \left(k_B T \nabla \rho(\mathbf{r}) + \rho(\mathbf{r}) \nabla V_{\text{ext}}(\mathbf{r}) + \int \mathrm{d}^3 r' \; \rho^{(2)}(\mathbf{r}, \mathbf{r}') \nabla V(|\mathbf{r} - \mathbf{r}'|) \right) \quad (50)$$

The constant must vanish for $r \to \infty$ and is thus identical to zero. Therefore

$$0 = k_B T \nabla \rho(\mathbf{r}) + \rho(\mathbf{r}) \nabla V_{\text{ext}}(\mathbf{r}) + \int \mathrm{d}^3 r' \ \rho^{(2)}(\mathbf{r}, \mathbf{r}') \nabla V(|\mathbf{r} - \mathbf{r}'|) \tag{51}$$

This is also known as Yvon-Born-Green-hierarchy (YBG). This equation also reflects a balance of fluxes generated by entropic, external and internal forces.

In equilibrium, DFT implies, see Eq. (1):

$$\frac{\delta \mathcal{F}}{\delta \rho(\mathbf{r})} = \mu - V_{\text{ext}}(\mathbf{r}) \tag{52}$$

$$= k_B T \ln(\Lambda^3 \rho(\mathbf{r})) + \frac{\delta \mathcal{F}_{\text{exc}}}{\delta \rho(\mathbf{r})} \tag{53}$$

We now apply the gradient which gives:

$$\nabla V_{\text{ext}}(\mathbf{r}) + k_B T \nabla \ln(\Lambda^3 \rho(\mathbf{r})) + \nabla \frac{\delta \mathcal{F}_{\text{exc}}}{\delta \rho(\mathbf{r})} = 0 \tag{54}$$

combined with YBG we obtain

$$\int \mathrm{d}^3 r' \ \rho^{(2)}(\mathbf{r}, \mathbf{r}') \nabla V(|\mathbf{r} - \mathbf{r}'|) = \rho(\mathbf{r}) \nabla \frac{\delta \mathcal{F}_{\text{exc}}[\rho]}{\delta \rho(\mathbf{r})} \tag{55}$$

We postulate that this argument holds also in nonequilibrium. In doing so, non-equilibrium correlations are approximated by equilibrium ones at the same $\rho(\mathbf{r}, t)$ (via a suitable $V_{\text{ext}}(\mathbf{r})$ in equilibrium). Equivalently, one can say that it is postulated that pair correlations decay much faster to their equilibrium one than the one-body density. Therefore the basic approximation of DDFT is sometimes called adiabatic approximation. (For an alternate derivation, see Marconi and Tarazona [67–70] or Español and Löwen [24].)

Hence:

$$\xi \frac{\partial \rho(\mathbf{r}, t)}{\partial t} = \nabla (k_B T \nabla \rho(\mathbf{r}, t) + \rho(\mathbf{r}, t) \nabla V_{\text{ext}}(\mathbf{r}, t) + \rho(\mathbf{r}, t) \nabla \frac{\delta \mathcal{F}_{\text{exc}}}{\delta \rho(\mathbf{r}, t)}) \tag{56}$$

or equivalently

$$\boxed{\xi \frac{\partial \rho(\mathbf{r}, t)}{\partial t} = \nabla \rho(\mathbf{r}, t) \nabla \frac{\delta \Omega[\rho]}{\delta \rho(\mathbf{r}, t)}} \tag{57}$$

which constitutes the basic equation of dynamical density functional theory (DDFT).

The applications of DDFT are numerous. The dynamics of a strongly inhomogeneous Brownian fluid has found to be in good agreement with BD computer simulations [43, 44].

5 An Example: Crystal Growth on Patterned Substrates

Dynamical density functional theory can be used for the dynamics of crystallization [20, 46]. This is demonstrated by an example in the following. A two-dimensional hard disk system is considered. The freezing behavior is reasonably well described by a fundamental-measure-like density functional [98]. The stable two-dimensional solid has a hexagonal structure corresponding to the close-packing of hard disks. Here we study the two-dimensional crystallization on a patterned substrate which has a square symmetry. This is described by an external potential of the form

$$V_{\text{ext}}(\mathbf{r}) = V_0 \left[1 - \frac{1}{4} (1 - \cos(k_x x)) \left(\cos(k_y y) \right) \right] \tag{58}$$

where $k_x = k_y = \pi / a_V$ are the components of a reciprocal lattice vector and V_0 denotes the strength of the external potential. We chose an amplitude of $V_0 = 0.5 k_B T$ and an area fraction of 0.74. Furthermore, the length scale of the substrate a_V is chosen such that on average there is always one particle per minimum. For these parameters, in equilibrium the system is in a triangular crystalline phase [82].

Now an initial crystalline nucleus with 19 density peaks is exposed to the external potential. Obviously there is a mismatch between the stable hexagonal lattice and the underlying substrate. DDFT predicts that the crystal nucleus grows in time but in an strongly anisotropic way, see Fig. 3. This can be understood by overlaying the hexagonal structure with the square structure resulting in a Moirée pattern. It is the compatibility of the two structures which dictates the solid growth. Regions of high compatibility grow the crystal earlier than others.

This example shows that dynamical density functional theory represents a microscopic approach to nonequilibrium phenomena like crystal growth in an external field. A similar behavior has been seen for growing crystallites [80, 107] and crystal fronts [107, 109] as well as for vacancy dynamics in two-dimensional solids [108]. As long as the data can be compared to Brownian dynamics simulations [107, 109] there is agreement, at least for the underlying trends in nonequilibrium. Stille one has to keep in mind that DDFT is a mean-field theory such that subtle dynamical effects caused by fluctuations may not be encaptured completely [9]. An interesting example is the flow of colloidal solids and fluids through geometric constrictions [117] where intermittent flow effects of the solid are predicted by the theory and confirmed by the simulations. However the dynamical correlations are periodic in time in the mean-field DDFT but decay as a function of time in the BD computer simulations.

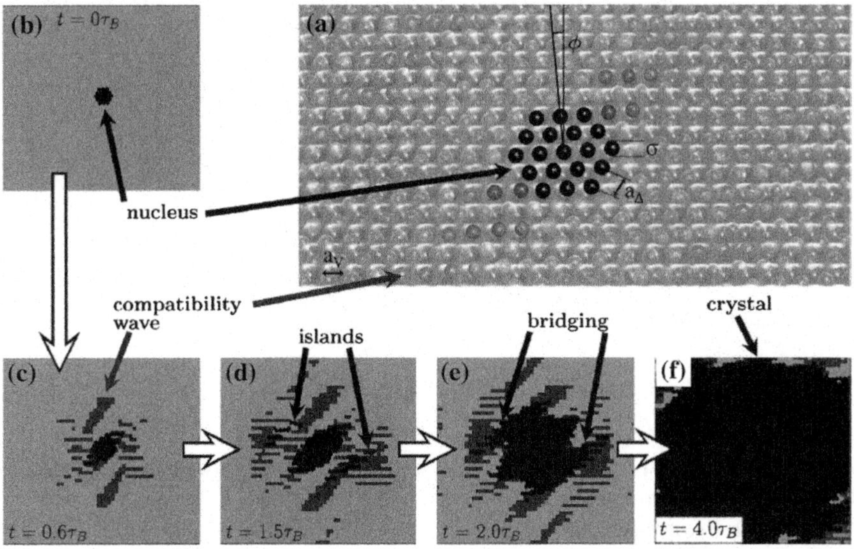

Fig. 3 **a** Schematic representation of a nucleus of hard discs with diameter σ located on a 2D substrate with square symmetry rotated counterclockwise by an angle $\phi = 5$ relative to a symmetry direction of the substrate and lattice constant a_V. The snapshots **b**–**f** are density contour plots and illustrate the growth of a spherical nucleus influenced by the substrate with amplitude $V_0 = 0.5k_BT$ at (**b**) $t = 0.0$, **c** $t = 0.6\tau_B$, **d** $t = 1.0\tau_B$, **e** $t = 2.0\tau_B$ and **f** $t = 4.0\tau_B$, where τ_B is a suitable Brownian time. *Red regions* display the compatibility wave and are defined by density peaks above a threshold value $\rho_{\text{th}}\sigma^2 = 1.5061$. *Black regions* denote crystalline areas ($\rho_{\text{th}}\sigma^2 = 2.0$), whereas blue regions remain fluid ($\rho_{\text{th}}\sigma^2 \leq 1.5060$). From Ref. [81]

6 Hydrodynamic Interactions

Now we address the question how $L_{nm}(\{x_j\})$ looks like explicitly. Solving the Stokes equations with appropriate stick boundary conditions on the particle surfaces is a difficult problem [15]. Furthermore it is problematic that

(i) $L_{nm}(\{x_j\})$ is long-ranged in terms of distances between particles
(ii) H.I. have a **many-body character**, a pair expansion is only possible at low concentrations.
(iii) H.I. have quite different near-field behavior. There are divergent lubrication terms.

The linear relationship (37) can be rewritten as

$$\mathbf{v}_n = \sum_{m=1}^{N} \bar{\bar{H}}_{nm}\mathbf{F}_m \tag{59}$$

where each quantity $\bar{\bar{H}}_{nm}$ is a 3×3 matrix. In particular, we can discriminate the following cases:

(1) **no H.I.** $H_{nm} = \mathbb{1}\frac{\delta_{nm}}{\xi}$

(2) **Oseen-tensor** In the Oseen approximation [17], $\bar{\bar{H}}_{nn} = \frac{\mathbb{1}}{\xi}$

$$\bar{\bar{H}}_{nm} = \bar{\bar{H}}(\underbrace{\mathbf{r}_n - \mathbf{r}_m}_{\mathbf{r}}) \text{ for } n \neq m \tag{60}$$

with the Oseen tensor

$$\bar{\bar{H}}(\mathbf{r}) = \frac{1}{8\pi\eta_s}(\mathbb{1} + \hat{r} \otimes \hat{r})\frac{1}{r},\ \hat{r} = \frac{\mathbf{r}}{r} \tag{61}$$

This is the leading far field term for two particles at large distance $\mathbf{r}$. The symbol $\otimes$ denotes the dyadic product or tensor product.

(3) **Rotne-Prager-tensor**

In this approximation (as for more details, see e.g. [17]), the next leading correction is included.

$$H_{nn} = \frac{\mathbb{1}}{\xi}, \quad H_{nm} = \bar{\bar{H}}_{RP}(\mathbf{r}_n - \mathbf{r}_m) \tag{62}$$

with

$$\bar{\bar{H}}_{RP}(\mathbf{r}) = \frac{D_0}{k_B T}\left(\frac{3}{4}\frac{R_H}{r}[\mathbb{1} + \hat{r} \otimes \hat{r}] + \frac{1}{2}\frac{R_H^3}{r^3}[1 - 3\hat{r} \otimes \hat{r}]\right) \tag{63}$$

Higher order expansions of higher order than $\frac{1}{r^3}$ are possible. These include also terms of sphere rotation. Finally the triplet contribution can be estimated.

The DDFT can be generalized to hydrodynamic interactions. The first study was done in Refs. [92, 94] and subsequent investigations [18, 29–32, 90] have extended and detailed the theoretical derivation. Again the starting point is the Smoluchowski equation which we now write in the form

$$\frac{\partial w(\mathbf{r}^N, t)}{\partial t} = \sum_{i,j=1}^{N} \nabla_i \cdot \bar{\bar{H}}_{ij}(\mathbf{r}^N) \cdot \left[\nabla_j + \nabla_j \frac{U_{\text{tot}}(\mathbf{r}^N, t)}{k_B T}\right] w(\mathbf{r}^N, t) \tag{64}$$

We use the two particle approximation

$$\bar{\bar{H}}_{ij}(\mathbf{r}^N) \approx \frac{D_0}{k_B T}\left(\mathbb{1}\delta_{ij} + \delta_{ij}\sum_{i \neq j}\omega_{11}(\mathbf{r}_i - \mathbf{r}_e) + (1 - \delta_{ij})\omega_{12}(\mathbf{r}_i - \mathbf{r}_e)\right) \tag{65}$$

on the level of the Rotne-Prager expression

$$\omega_{11}(\mathbf{r}) = 0 \tag{66}$$

$$\omega_{12}(\mathbf{r}) = \frac{3}{8}\frac{\sigma_H}{r}(\mathbb{1} + \hat{r} \otimes \hat{r}) + \frac{1}{16}(\frac{\sigma_H}{r})^3(1 - 3\hat{r} \otimes \hat{r}) + O((\frac{\sigma_H}{r})^7) \tag{67}$$

where σ_H is the hydrodynamic diameter.

Integrating Smoluchowski equation [6] then yields [92]

$$\begin{aligned}\frac{k_B T}{D_0}\frac{\partial \rho(\mathbf{r},t)}{\partial t} = \nabla_r \cdot \Bigg[&\rho(\mathbf{r},t)\nabla_r \frac{\delta \mathscr{F}[\rho]}{\delta \rho(\mathbf{r},t)} \\ &+ \int \mathrm{d}\mathbf{r}' \ \rho^{(2)}(\mathbf{r},\mathbf{r}',t)\omega_{11}(\mathbf{r}-\mathbf{r}') \cdot \nabla_r \frac{\delta \mathscr{F}[\rho]}{\delta \rho(\mathbf{r},t)} \\ &+ \int \mathrm{d}\mathbf{r}' \ \rho^{(2)}(\mathbf{r},\mathbf{r}',t)\omega_{12}(\mathbf{r}-\mathbf{r}') \cdot \nabla_r \frac{\delta \mathscr{F}[\rho]}{\delta \rho(\mathbf{r},t)} \Bigg]\end{aligned} \tag{68}$$

A possible closure is via the Ornstein-Zernike equation

$$\begin{aligned}\rho^{(2)}(\mathbf{r},\mathbf{r}',t) = &(1 + c^{(2)}(\mathbf{r},\mathbf{r}'))\rho(\mathbf{r},t)\rho(\mathbf{r}',t) \\ &+ \rho(\mathbf{r}',t)\int \mathrm{d}\mathbf{r}''((\rho^{(2)}(\mathbf{r},\mathbf{r}'',t) - \rho(\mathbf{r},t)\rho(\mathbf{r}'',t))c^{(2)}(\mathbf{r}'',\mathbf{r}'))\end{aligned} \tag{69}$$

with

$$c^{(2)}(\mathbf{r},\mathbf{r}') = -\beta \frac{\delta^2 \mathscr{F}_{\text{exc}}[\rho]}{\delta \rho(\mathbf{r},t)\delta \rho(\mathbf{r}',t)} \tag{70}$$

In an easier attempt, one can approximate

$$\rho^{(2)}(\mathbf{r},\mathbf{r}',t) \approx \rho(r,t)\rho(r',t)g(|\mathbf{r}-\mathbf{r}'|,\bar{\rho}) \tag{71}$$

where $\bar{\rho}$ is a suitable averaged density and $g(r,\bar{\rho})$ is a pair distribution function in the equilibrium bulk fluid. The latter function $g(r,\bar{\rho})$ can be obtained by integral equation theories (such as the hypernetted-chain theory) as derived within equilibrium statistical mechanics [39].

An example involves hard sphere colloids of diameter σ moving in a time-dependent oscillating radial-symmetric trapping potential $V_{\text{ext}}(r,t)$. The potential reads as

$$V_{\text{ext}}(r,t) = V_1 r^4 + V_2 \cos(\omega t) r^2 \tag{72}$$

and corresponds to a shape which switches between stable and unstable situations at the origin $r = 0$ but is globally stable. The amplitudes are $V_1 = 10k_B T/4096\sigma^4$, $V_2 = k_B T/\sigma^2$ and the external switching frequency is $\omega = 4\pi D_0/\sigma^2$. As a result, the density profile is picking up the external frequency ω and exhibits a breathing mode, i.e. it is periodically expanding and compressed again by the external potential (72). Time-dependent density profiles $\rho(r,t)$ are presented in Fig. 4 in the steady breathing state.

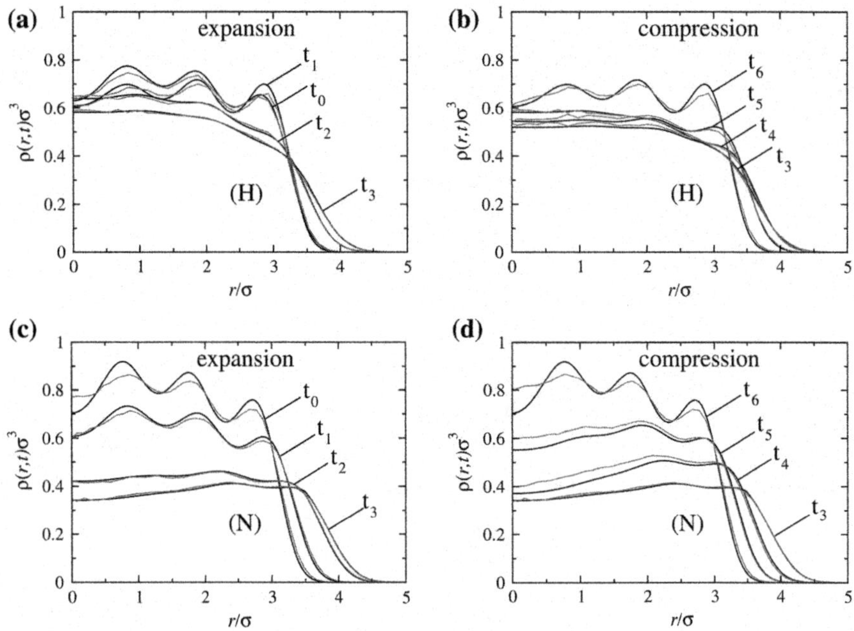

Fig. 4 *Steady-state* DDFT (*solid curves*) and BD (*noisy curves*) results for the time-dependent density profile $\rho(r, t)$. In Fig. **a** and **b** hydrodynamic interactions are taken into account while in (**c**) and (**d**) they are neglected. **a**, **c** correspond to the expanding half period and **b** and **d** to the compressing half period, respectively. The profiles correspond to the following time sequence: $t_0 = 2.5\tau_B$, $t_1 = 2.6\tau_B$, $t_2 = 2.7\tau_B$, $t_3 = 2.75\tau_B$ in (**a**) and (**c**), and $t_3 = 2.75\tau_B$, $t_4 = 2.85\tau_B$, $t_5 = 2.9\tau_B$ and $t_6 = 3.0\tau_B$, $\tau_B = \sigma^2/D_0$. From Ref. [92]

Figure 4a, b show results for hydrodynamic interactions included on the Rotne-Prager level. DDFT data are in very good agreement with Brownian dynamics computer simulations which include hydrodynamic interactions on the same Rotne-Prager level. On the other hand, in Fig. 4c, d, hydrodynamic interactions are ignored. The density profiles are qualitatively different to that shown in Fig. 4a, b but again DDFT data are in agreement with Brownian dynamics computer simulations. This demonstrates that DDFT is a reliable microscopic theory both if hydrodynamic interactions are included or ignored.

7 Rod-Like Particles

7.1 Statistical Mechanics of Rod-Like Particles

Density functional theory can readily be extended to rod-like particles which possess an additional orientational degree of freedom described by a unit vector $\hat{u}$ [3, 13, 28].

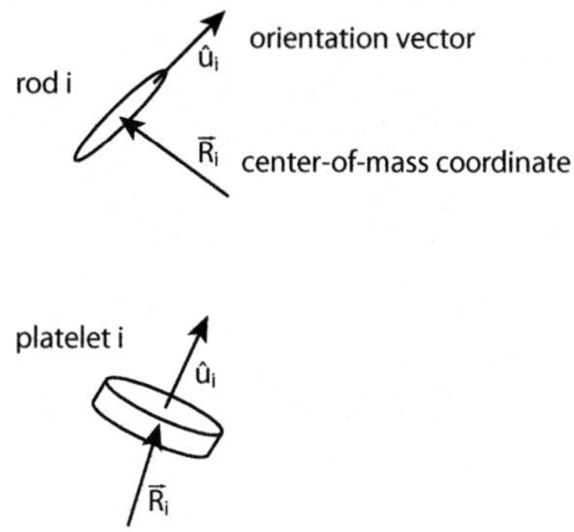

Fig. 5 Sketch of the center-of-mass position $\mathbf{R}_i$ and the orientational unit vector $\hat{u}_i$ for the ith particle both for a rod-like and plate-like particle

A configuration of N particles is now fully specified by the set of positions of the center of masses and the corresponding orientations $\{\mathbf{R}_i, \hat{u}_i, i = 1, \ldots, N\}$, see Fig. 5.

Examples for anisotropic particles include

(1) molecular dipolar fluids (e.g. H_2O molecule)
(2) rod-like colloids (e.g. tobacco-mosaic viruses)
(3) molecular fluids without dipole moment (apolar), (e.g. H_2 molecule)
(4) plate-like objects (clays)

The canonical partition function for rod-like particles now reads [27]

$$Z = \frac{1}{h^{6N} N!} \int_V d^3 R_1 \ldots \int_V d^3 R_N \int_{\mathbb{R}^3} d^3 p_1 \ldots \int_{\mathbb{R}^3} d^3 p_N \times \int_{S_2} d^2 u_1 \ldots \int_{S_2} d^2 u_N \int_{\mathbb{R}^3} d^3 L_1 \ldots \int_{\mathbb{R}^3} d^3 L_N \, e^{-\beta \mathscr{H}} \tag{73}$$

with the total Hamilton function

$$\mathscr{H} = \sum_{i=1}^{N} \left\{ \frac{\mathbf{p}_i^2}{2m} + \frac{1}{2} \mathbf{L}_i (\bar{\bar{\Theta}})^{-1} \mathbf{L}_i \right\} + \frac{1}{2} \sum_{i,j=1}^{N} v(\mathbf{R}_i - \mathbf{R}_j, \hat{u}_i, \hat{u}_j) + \sum_{i=1}^{N} V_{\text{ext}}(\mathbf{R}_i, \hat{u}_i) \tag{74}$$

which comprises the kinetic energy, the pair interaction energy and the external energy. Here $\bar{\bar{\Theta}}$ is the inertia tensor and S_2 the unit-sphere in 3d. Moreover $\mathbf{p}_i$ and $\mathbf{L}_i$ denote the translational and angular momenta.

Again the central quantity is the one-particle density $\rho_0^{(1)}(\mathbf{r}, \hat{u})$ which is defined as

$$\rho_0^{(1)}(\mathbf{r}, \hat{u}) := \left\langle \sum_{i=1}^{N} \delta(\mathbf{r} - \mathbf{R}_i)\delta(\hat{u} - \hat{u}_i) \right\rangle \tag{75}$$

with $\langle \ldots \rangle$ denoting a canonical average. Integrating the orientations over the unit sphere S_2 results in the density of the center-of-masses

$$\rho_0(\mathbf{r}) = \frac{1}{4\pi} \int_{S_2} \mathrm{d}^2 u \; \rho_0^{(1)}(\mathbf{r}, \hat{u}) \tag{76}$$

On the other hand, the globally averaged orientational order is gained by integrating over the center-of-mass coordinates and given by

$$f(\hat{u}) = \frac{1}{V} \int_{V} \mathrm{d}^3 r \; \rho_0^{(1)}(\mathbf{r}, \hat{u}) \tag{77}$$

7.2 *Density Functional Theory*

Again density functional theory tells us that there exists a unique grand canonical free energy functional $\Omega(T, \mu, [\rho^{(1)}])$ (functional of the one-particle density) which becomes minimal for the equilibrium density $\rho_0^{(1)}(\mathbf{r}, \hat{u})$ and then equals the real grand canonical free energy, i.e.

$$\boxed{\left. \frac{\delta \Omega(T, \mu, [\rho^{(1)}])}{\delta \rho^{(1)}(\mathbf{r}, \hat{u})} \right|_{\rho^{(1)} = \rho_0^{(1)}(\mathbf{r}, \hat{u})} = 0} \tag{78}$$

Here, the functional can be decomposed as follows

$$\begin{aligned} \Omega(T, \mu, [\rho^{(1)}]) = k_B T \int \mathrm{d}^3 r \int \mathrm{d}^2 u \; \rho^{(1)}(\mathbf{r}, \hat{u})[\ln(\Lambda^3 \rho^{(1)}(\mathbf{r}, \hat{u})) - 1] \\ + \int \mathrm{d}^3 r \mathrm{d}^2 u \; (V_{\mathrm{ext}}(\mathbf{r}, \hat{u}) - \mu)\rho^{(1)}(\mathbf{r}, \hat{u}) + \mathscr{F}_{\mathrm{exc}}(T, [\rho^{(1)}]) \end{aligned} \tag{79}$$

The first term on the right hand side of Eq. (79) is the functional $\mathscr{F}_{\mathrm{id}}[\rho^{(1)}]$ for ideal rotators. The excess part $\mathscr{F}_{\mathrm{exc}}(T, [\rho^{(1)}])$ is in general unknown and requires approximative treatments.

For hard spherocylinders, $\mathscr{F}_{\mathrm{exc}}(T, [\rho^{(1)}])$ can be approximated by a smoothed density approximation (SMA) [86] yielding several stable liquid-crystalline phases, namely: isotropic, nematic, smectic-A and ABC crystalline. A modified weighted density approximation (MWDA) was subsequently proposed [34] which improves upon the SMA by exhibiting stable plastic crystalline and AAA crystals as well.

An important recent progress was achieved by generalizing Rosenfeld's fundamental measure theory from hard spheres to hard objects with any shape [40]. For spherocylinders the functional was worked out explicitly and recently more data have been obtained for hard particles of different shape, such as dumbbells [74], rounded cubes [75] and polyhedra [73]. This functional could be exploited also for attractions by employing a perturbation theory for the attractive parts in the potential. Finally, a mean-field density functional for rods with soft segments was proposed and studied [93].

7.3 *Brownian Dynamics of Rod-Like Particles and DDFT*

In order to derive a dynamical density functional theory (DDFT) for rod-like particles one can start from the Smoluchowski equation for the full probability density distribution $w(\mathbf{r}_1, \dots, \mathbf{r}_N; \mathbf{u}_1, \dots, \mathbf{u}_N, t)$ of N rods with their corresponding center-of-mass positions $\mathbf{r}^N = (\mathbf{r}_1, \dots, \mathbf{r}_N)$ and orientations $\hat{u}^N = (\hat{u}_1, \dots, \hat{u}_N)$ which reads [16]

$$\frac{\partial w}{\partial t} = \hat{\mathcal{O}}_S w \tag{80}$$

where the Smoluchowski operator is now given by

$$\hat{\mathcal{O}}_S = \sum_{i=1}^{N} \Bigg[\nabla_{\mathbf{r}_i} \cdot \bar{\bar{D}}(\hat{u}_i) \cdot \left(\nabla_{\mathbf{r}_i} + \frac{1}{k_B T} \nabla_{\mathbf{r}_i} U(\mathbf{r}^N, \hat{u}^N, t) \right) \\ + D_r \hat{\mathbf{R}}_i \cdot \left(\hat{\mathbf{R}}_i + \frac{1}{k_B T} \hat{\mathbf{R}}_i U(\mathbf{r}^N, \hat{u}^N, t) \right) \Bigg] \tag{81}$$

where $U(\mathbf{r}^N, \hat{u}^N, t)$ is the total potential energy. Here the rotation operator $\hat{\mathbf{R}}_i$ is defined as $\hat{\mathbf{R}}_i = \hat{u}_i \times \nabla_{\hat{u}_i}$ and the anisotropic translational diffusion tensor is given by

$$\bar{\bar{D}}(\hat{u}_i) = D^{\parallel} \hat{u}_i \otimes \hat{u}_i + D^{\perp} (\mathbf{1} - \hat{u}_i \otimes \hat{u}_i) \tag{82}$$

The two diffusion constants $D^{\parallel}$ and $D^{\perp}$, parallel and perpendicular to the orientations reflect the fact that the translational diffusion is anisotropic. For hard spherocylinders there are valid approximations for $D^{\parallel}$ and $D^{\perp}$ [56].

Following the idea of Archer and Evans [7] one can integrate the Smoluchowski equation by now applying $N \int d^3 r_2 \cdots \int d^3 r_N \int d^2 u_1 \cdots \int_d^2 u_n$ on both sides of Eq. (80). This results in [93]:

$$\frac{\partial\rho(\mathbf{r},\hat{u},t)}{\partial t} = \nabla_\mathbf{r}\cdot\bar{\bar{D}}(\hat{u})\cdot\left[\nabla_\mathbf{r}\rho(\mathbf{r},\hat{u},t)+\frac{1}{k_BT}\rho(\mathbf{r},\hat{u},t)\cdot\nabla_\mathbf{r}V_{\text{ext}}(\mathbf{r},\hat{u},t)-\frac{\mathbf{F}(\mathbf{r},\hat{u},t)}{k_BT}\right]$$
$$+D_r\hat{R}\cdot\left[\hat{R}\rho(\mathbf{r},\hat{u},t)+\frac{1}{k_BT}\rho(\mathbf{r},\hat{u},t)\nabla_\mathbf{r}V_{\text{ext}}(\mathbf{r},\hat{u},t)-\frac{1}{k_BT}\mathbf{T}(\mathbf{r},\hat{u},t)\right] \tag{83}$$

with an average force

$$\mathbf{F}(\mathbf{r},\hat{u},t) = -\int d^3r'\int d^2u'\ \rho^{(2)}(\mathbf{r},\mathbf{r}',\hat{u},\hat{u}',t)\nabla_\mathbf{r}v(\mathbf{r}-\mathbf{r}',\hat{u},\hat{u}') \tag{84}$$

and average torque

$$\mathbf{T}(\mathbf{r},\hat{u},t) = -\int d^3r'\int d^2u'\ \rho^{(2)}(\mathbf{r},\mathbf{r}',\hat{u},\hat{u}',t)\hat{R}v(\mathbf{r}-\mathbf{r}',\hat{u},\hat{u}') \tag{85}$$

The two-particle density which is in general unknown can be approximated in equilibrium by using

$$\mathbf{F}(\mathbf{r},\hat{u},t) = \rho_0(\mathbf{r},\hat{u})\nabla_\mathbf{r}\frac{\delta\mathcal{F}_{\text{exc}}(T,[\rho_0])}{\delta\rho_0(\mathbf{r},\hat{u})} \tag{86}$$

respectively

$$\mathbf{T}(\mathbf{r},\hat{u},t) = \rho_0(\mathbf{r},\hat{u})\hat{R}\frac{\delta\mathcal{F}_{\text{exc}}[\rho]}{\delta\rho_0(\mathbf{r},\hat{u})} \tag{87}$$

Similar as in the isotropic (spherical) case we now employ the "adiabatic" approximation. We assume that the pair correlations in nonequilibrium are the same as those for an equilibrium system with the same one-body density profile (established by a suitable $V_{\text{ext}}(\mathbf{r},\hat{u},t)$). The resulting dynamical equation for the time-dependent one particle density $\rho(\mathbf{r},\hat{u},t)$ is then given by [93]:

$$k_BT\frac{\partial\rho(\mathbf{r},\hat{u},t)}{\partial t} = \nabla_\mathbf{r}\cdot\bar{\bar{D}}(\hat{u})\cdot\left[\rho(\mathbf{r},\hat{u},t)\nabla_\mathbf{r}\frac{\delta\mathcal{F}[\rho(\mathbf{r},\hat{u},t)]}{\delta\rho(\mathbf{r},\hat{u},t)}\right]$$
$$+D_r\hat{R}\left[\rho(\mathbf{r},\hat{u},t)\hat{R}\frac{\delta\mathcal{F}[\rho(\mathbf{r},\hat{u},t)]}{\delta\rho(\mathbf{r},\hat{u},t)}\right] \tag{88}$$

with the equilibrium Helmholtz free energy density functional

$$\mathcal{F}[\rho] = k_BT\int d^3r\int d\hat{u}\ \rho(\mathbf{r},\hat{u})\left[\ln(\Lambda^3\rho(\mathbf{r},\hat{u}))-1\right]$$
$$+\mathcal{F}_{\text{exc}}(T,[\rho])+\int d^3r\int d\hat{u}\ \rho(\mathbf{r},\hat{u})V_{\text{ext}}(\mathbf{r},\hat{u},t) \tag{89}$$

This sets the frame for dynamical density functional theory (DDFT) for rods.

As for a special application of DDFT to dynamics in the confined isotropic phase we refer to [93] where the mean-field approximation for the functional was employed. More recent work has used the Rosenfeld functional for hard spherocylinders [40] for driven nematic phases [37, 38].

As a final remark the DDFT can also derived for anisotropic particles of arbitrary shape which have no rotational symmetry. The corresponding Brownian dynamics of biaxial particles is much more complicated [26, 47, 63]. The most general framework to describe BD correctly and the corresponding derivation of DDFT based on the Smoluchowski equation was recently proposed in Ref. [112].

8 Recent Developments

8.1 Derivation of the Phase Field Crystal (PFC) Model from DDFT

A less detailed approach which can be derived from DDFT of freezing is the so-called phase-field-crystal (PFC) model, for a recent review, see [22]. The dynamics of the PFC model can be systematically derived by a gradient expansion [109]. Though more approximative than DDFT, it is numerically much more efficient to obtain results on larger length and time scales and is thereby a standard tool to obtain results on the dynamics of freezing and melting in material science. As an example, the diffusion of vacancies in the two-dimensional crystal has been explored both by DDFT and PFC [108].

More recently, the PFC model was also derived within similar approximations for *liquid crystals* both in two [59, 113, 114] and three [111] spatial dimensions. The resulting PFC model keeps both the translational density and the orientational ordering as conserved and non-conserved order parameters and describes different coupling terms in the free energy functional which are allowed by phenomenological symmetry considerations. As an additional benefit from the derivation based on DDFT, the corresponding prefactors (or coupling parameters) can be expressed as moments over microscopic correlation functions. The PFC model for liquid crystals was used to calculate the bulk phase diagram [2] and surface free energies between coexisting phases [87].

8.2 More Recent Applications of DDFT

There is a variety of problems to which DDFT of Brownian particles was applied recently, beyond to what was discussed previously in the context of crystallization. One important problem is colloids in an external *shear* field such they are advected with the solvent flow. A suitable generalization of DDFT is not straightforward but again the Smoluchowski equation can be used as a starting point [14, 49, 89, 91]. A similar problem is that of *microrheology* where a particle is driven through a colloidal background and its response is recorded [4]. Very good data have been achieved for *diffusion* in hard sphere fluids even at high packing fractions [102], for the dynamics of binary *mixtures* [10, 41, 64], for set-ups describing nonlinear *feedback control* of colloids [51] and for the *collapse* of a colloidal monolayer as governed by attractive interactions [12].

Furthermore we mention that the growth of *quasicrystals* has become an interesting general problem as there is a competition between local and long-range ordering. Recent experiments have revealed quite complex growth scenarios [79] which were modelled by DDFT-like approaches [1, 8].

Finally it is very challenging to construct a DDFT for *molecular dynamics* which is not overdamped as realized for example in atoms and granulates. In particular the work of Marconi and Melchionna has worked out the conditions under which DDFT can be applied to such classical systems and has proposed modifications leading towards a kinetic density functional theory [68–72].

8.3 DDFT for Active Brownian Particles

Active particles are self-propelled by their own intrinsic motor. In fact, apart from swimming bacteria, there are artificial microswimmers made by colloidal particles [19, 21, 23, 65, 95]. Ignoring hydrodynamic interactions, these swimmers can simply be modeled by rod-like particles which are driven by a constant force along their orientations; the force corresponds to an effective drift velocity and mimics the actual propulsion mechanism. On top of the intrinsic propulsion, the particles feel Brownian noise of the solvent. The corresponding motion is intrinsically a nonequilibrium one and even the dynamics of a single Brownian swimmer was solved only recently [35, 42, 105].

Starting from the Smoluchowski equation with an appropriate intrinsic drift term, a DDFT can be derived using the same adiabatic approximation (55) as in the passive case. The resulting equation of motion for the one-particle density then [110] has an extra term on the right-hand side of Eq. (88), namely

$$- F_0 \nabla \bar{\bar{D}}(\hat{u}) \rho(\mathbf{r}, \hat{u}, t) \hat{u} \tag{90}$$

where F_0 denotes the internal driving force which gives rise to the propulsion [36]. This represents a microscopic theory for concentrated "active" matter. For swimmers in a two-dimensional channel, the time-dependent density profiles were found to be in agreement with Brownian dynamics computer simulations [110] even if a crude Onsager-like density functional approximation [84] was used. Qualitatively, the transient formation of hedgehog-like clusters near the channel boundaries was reproduced by the DDFT. Recently the DDFT was generalized towards three spatial dimensions for swimmers of arbitrary shape with complicated hydrodynamic friction tensors [112]. The most general DDFT framework for microswimmers can be found in Ref. [77].

Finally the DDFT for active Brownian particles can be reduced to a PFC theory as briefly discussed in the previous paragraph. This theory has been evaluated and a new state of a traveling crystal was found [76] where the whole system moves collectively in one direction.

8.4 Modern Derivation of DDFT Using Projection Operator Techniques and Extended DDFT (EDDFT)

In this book chapter we have derived DDFT from the Smoluchowski equation. An alternate route which provides new physical insight can be performed using the Mori-Zwanzig (sometimes called Mori-Zwanzig-Forster) projection operator technique. In this approach, one splits all observables into quickly relaxing and slowly relaxing ones. All degrees of freedom which are fast are integrated out leaving only equations of motion for the slow variables. If one *assumes* that the one-particle density itself is the only slowly varying variable in the system, a consequent application of the projection operator technique [24] yields the standard DDFT equation. This highlights the role of the adiabatic approximation in both derivation routes: if one assumes adiabaticity in the traditional derivation of DDFT using the Smoluchowski equation, then by assumption the one-particle density is indeed the only slow variable, completely consistent with the alternate derivation put forward in Ref. [24].

The great advantage of the Mori-Zwanzig technique is that it provides a consistent framework to derive systematically equations of motion of arbitrary slow order parameter fields which possess a microscopic expression. Therefore it starts in principle on the full microscopic level and unravels the physical spirit of the underlying approximations by splitting the variables into slow and fast ones. This basic idea of an "extended DDFT" (EDDFT) was proposed by Wittkowski and coworkers [115] and was evaluated explicitly for the case where the concentration fields and the internal energy density are slow variables. If the entropy density itself is considered to be a relevant slow variable, the situation is more subtle [116] since the entropy density does not have a corresponding microscopic expression. This problem can be solved by invoking linear irreversible thermodynamics which yields a microscopic expression for the entropy density in terms of the other microscopic variables [116]. As a side remark, in the original EDDFT approach [115, 116] one considers functionals for the free energy. Alternatively one can consider and motivate dynamical equations for the particle number density and internal energy density or entropy density from functionals for the internal energy [100] or entropy [5].

Let us now discuss briefly the basic idea of EDDFT introduced by Wittkowski and coworkers [115, 116], which allows for a compact notation of the equations of motion. We assume a set of slow conserved order parameters fields summarized in the vector field $\mathbf{a}(\mathbf{r}, t)$ and only dissipative currents. In equilibrium, there exists a generalized free energy functional $\mathscr{F}$ of these order parameter fields such that one can define the *thermodynamic conjugates* as

$$a_i^\natural(\mathbf{r}, t) = \frac{\delta \mathscr{F}[a_1, \ldots, a_n]}{\delta a_i(\mathbf{r}, t)} \tag{91}$$

We can then define generalized *thermodynamic forces* as

$$\mathbf{a}_i^\sharp = -\nabla_\mathbf{r} a_i^\natural . \tag{92}$$

Within linear irreversible thermodynamics there exists a dissipation functional $\mathscr{R}$ of the thermodynamic forces such that one can define the generalized currents as

$$\mathbf{J}^{(i)} = \frac{\delta \mathfrak{R}}{\delta \mathbf{a}_i^\sharp} . \tag{93}$$

The final equation of motion for the slow order parameters is then given by the continuity equation

$$\dot{a}_i(\mathbf{r}, t) + \nabla_\mathbf{r} \cdot \mathbf{J}^{(i)}(\mathbf{r}, t) = 0 . \tag{94}$$

In this EDDFT framework, for any choice of the slow variables the dissipation functional yields consistent equations of motion. Obviously, the traditional DDFT is a special case where the particle number density is the only slow variable and the dissipation functional is given by [116]

$$\mathfrak{R} = \int_{\mathbb{R}^3} \mathrm{d}^3 r \; \mathfrak{r}(\mathbf{r}, t) \tag{95}$$

with

$$\mathfrak{r}(\mathbf{r}, t) = \frac{1}{2}\frac{1}{\xi}\rho(\mathbf{r}, t)\big(\nabla_\mathbf{r}\rho^\natural(\mathbf{r}, t)\big)^2 . \tag{96}$$

It is important to note that the equilibrium situation is characterized by a vanishing current such that the equilibrium condition corresponds to the **variational problem**

$$\frac{\delta \mathfrak{R}}{\delta \mathbf{a}_i^\sharp} = \mathbf{J}^{(i)} = 0 . \tag{97}$$

A nonequilibrium steady state is characterized by a constant (time and space independent) current as given by the equation

$$\frac{\delta \mathfrak{R}}{\delta \mathbf{a}_i^\sharp} = \mathbf{J}^{(i)} = \mathrm{const} . \tag{98}$$

Also this steady-state condition can be viewed as a variational problem where the constant is a suitable Lagrange multiplier for a constraint. This shows the relevance of the variational calculus even in the nonequilibrium steady state.

Another approach towards a generalization of DDFT which is more special than the EDDFT framework outlined above, was proposed by Schmidt and Brader in Ref. [101]. Here the density alone was considered as relevant variable and a phenomenologically augmented dissipation functional was considered which is different to that of Eq. (96) and was called a "power functional". The resulting additional terms in the dynamical equations as compared to standard DDFT imply that the adiabaticity condition of standard DDFT is broken within the power functional theory. We are therefore now at the beginning to understand the relevance of nonadiabatic terms in DDFT. Still a full microscopic justification and a precise and systematic test of nonadiabatic terms against exact Brownian dynamics computer simulations needs to be performed.

9 Conclusions

In conclusion, density functional theory can be extended towards dynamics, so-called dynamical density functional theory, to tackle various nonequilibrium phenomena for Brownian dynamics. This was demonstrated for colloidal crystal growth on patterned substrates, for colloids in an oscillating trap and for the collective behavior of "active" Brownian particles. In general, dynamical density functional theory is in good agreement with the simulations but despite of recent progress going systematically beyond the underlying adiabaticity assumption remains a challenge for the future. Another fascinating challenge is to include inertia effects systematically arising from undamped Newtonian dynamics which occurs for example in molecular liquid crystals or in a complex plasma [45]. This would require to generalize the EDDFT framework towards non-dissipative currents. Therefore the future will see more applications of DDFT to a bunch of different problems. To mention just a few, these could involve the dynamics of mixtures (for a recent DFT application see [78]), the description of the dynamics of injection of particles where corresponding source terms have to be added to the equations of motion [62], and the dynamics of colloids driven over energy barriers [48].

Acknowledgements I thank R. Wittkowski, M. Marechal and B. ten Hagen for many helpful suggestions. This work was supported by the DFG within project LO 418/20-1.

References

1. C.V. Achim, M. Schmiedeberg, H. Löwen, Phys. Rev. Lett. **112**, 255501 (1–5) (2014)
2. C.V. Achim, R. Wittkowski, H. Löwen, Phys. Rev. E **83**, 061712 (1–8) (2011)

3. M.P. Allen, D.J. Tildesley, *Computer Simulation of Liquids* (Oxford Science Publications, Clarendon Press, Oxford, 1987)
4. L. Almenar, M. Rauscher, J. Phys.: Condens. Matter **23**, 184115 (1–9) (2011)
5. J.G. Anero, P. Español, P. Tarazona, J. Chem. Phys. **139**, 034106 (1–14) (2013)
6. A.J. Archer, Phys. Rev. E **72**, 051501 (1–7) (2005)
7. A.J. Archer, R. Evans, J. Chem. Phys. **121**, 4246–4254 (2004)
8. A.J. Archer, A.M. Rucklidge, E. Knobloch, Phys. Rev. E **92**, 012324 (1–14) (2015)
9. A.J. Archer, M. Rauscher, J. Phys. A **37**, 9325–9333 (2004)
10. L. Assoud, F. Ebert, P. Keim, R. Messina, G. Maret, H. Löwen, Phys. Rev. Lett. **102**, 238301 (1–4) (2009)
11. J.-L. Barrat, J.-P. Hansen, *Basic Concepts for Simple and Complex Liquids* (Cambridge University Press, 2003)
12. J. Bleibel, A. Dominguez, M. Oettel, S. Dietrich, Soft Matter **10**, 4091–4109 (2014)
13. P. Bolhuis, D. Frenkel, J. Chem. Phys **106**, 666–687 (1997)
14. J.M. Brader, M. Krüger, Mol. Phys. **109**, 1029–1041 (2011)
15. J.F. Brady, G. Bossis, Ann. Rev. Fluid Mech. **20**, 111–157 (1988)
16. J.K.G. Dhont, *An Introduction to Dynamics of Colloids* (Elsevier, Amsterdam, 1996)
17. M. Doi, S.F. Edwards, *The Theory of Polymer Dynamics* (Oxford Science Publications, Clarendon Press, Oxford, 1986)
18. A. Donev, E. Vanden-Eijnden, J. Chem. Phys. **140**, 234115 (1–18) (2014)
19. R. Dreyfus, J. Baudry, M.L. Roper, M. Fermigier, H.A. Stone, J. Bibette, Nature **437**, 862–865 (2005)
20. J. Dzubiella, H. Löwen, C.N. Likos, Phys. Rev. Lett. **91**, 248301 (1–4) (2003)
21. J. Elgeti, R.G. Winkler, G. Gompper, Rep. Prog. Phys. **78**, 056601 (1–50) (2015)
22. H. Emmerich, H. Löwen, R. Wittkowski, T. Gruhn, G.I. Tóth, G. Tegze, L. Gránásy, Adv. Phys. **61**, 665–743 (2012)
23. A. Erbe, M. Zientara, L. Baraban, C. Kreidler, P. Leiderer, J. Phys.: Condens. Matter **20**, 404215 (1–5) (2008)
24. P. Español, H. Löwen, J. Chem. Phys. **131**, 244101 (1–7) (2009)
25. R. Evans, Adv. Phys. **28**, 143–200 (1979)
26. M.X. Fernandes, J.G. de la Torre, Biophys. J. **83**, 3039–3045 (2002)
27. D. Frenkel, in *Liquids, Freezing and Glass Transition*, ed. by J.P. Hansen, D. Levesque, J. Zinn-Justin (North Holland, Amsterdam, 1991), pp. 689–756
28. D. Frenkel, B.M. Mulder, J.P. McTague, Phys. Rev. Lett. **52**, 287–290 (1984)
29. B.D. Goddard, A. Nold, S. Kalliadasis, J. Chem. Phys **138**, 144904 (1–9) (2013)
30. B.D. Goddard, A. Nold, N. Savva, P. Yatsyshin, S. Kalliadasis, J. Phys.: Condens. Matter **25**, 035101 (1–14) (2013)
31. B.D. Goddard, A. Nold, N. Savva, G.A. Pavliotis, S. Kalliadasis, Phys. Rev. Lett. **109**, 120603 (1–4) (2012)
32. B.D. Goddard, G.A. Pavliotis, S. Kalliadasis, Multiscale Model. Simul. **10**, 633–663 (2012)
33. D. Gottwald, C.N. Likos, G. Kahl, H. Löwen, Phys. Rev. Lett. **92**, 068301 (1–4) (2004)
34. H. Graf, H. Löwen, J. Phys.: Condens. Matter **11**, 1435–1452 (1999)
35. B. ten Hagen, S. van Teeffelen, H. Löwen, J. Phys.: Condens. Matter **23**, 194119 (1–16) (2011)
36. B. ten Hagen, R. Wittkowski, D. Takagi, F. Kümmel, C. Bechinger, H. Löwen, J. Phys.: Condens. Matter **27**,194110 (1–10) (2015)
37. A. Härtel, H. Löwen, J. Phys.: Condens. Matter **22**, 104112 (1–11) (2010)
38. A. Härtel, R. Blaak, H. Löwen, Phys. Rev. E **81**, 051703 (1–5) (2010)
39. J.P. Hansen, I. McDonald, *Theory of Simple Liquids*, 3rd edn. (Elsevier, Amsterdam, Academic Press, 2005)
40. H. Hansen-Goos, K. Mecke, Phys. Rev. Lett. **102**, 018302 (1–4) (2009)
41. N. Hoffmann, F. Ebert, C.N. Likos, H. Löwen, G. Maret, Phys. Rev. Lett. **97**, 078301 (1–4) (2006)

42. J.R. Howse, R.A.L. Jones, A.J. Ryan, T. Gough, R. Vafabakhsh, R. Golestanian, Phys. Rev. Lett. **99**, 048102 (1–4) (2007)
43. J.E. Hug, F. van Swol, C.F. Zukoski, Langmuir **11**, 111–118 (1995)
44. R.J. Hunter, *Foundations of Colloid Science*, 2nd edn. (Oxford University Press, Oxford, 1989)
45. A.V. Ivlev, H. Löwen, G.E. Morfill, C.P. Royall, *Complex Plasmas and Colloidal Dispersions: Particle-Resolved Studies of Classical Liquids and Solids* (World Scientific, Singapore, 2012)
46. G. Kahl, H. Löwen, J. Phys.: Condens. Matter **21**, 464101 (1–7) (2009)
47. D.J. Kraft, R. Wittkowski, B. ten Hagen, K.V. Edmond, D.J. Pine, H. Löwen, Phys. Rev. E **88**, 050301(R) (1–4) (2013)
48. C. Kreuter, U. Siems, P. Nielaba, P. Leiderer, A. Erbe, Eur. Phys. J. - Special Topics **222**, 2923–2939 (2013)
49. M. Krüger, J.M. Brader, EPL **96** 68006 (1–6) (2011)
50. A. Lang, C.N. Likos, M. Watzlawek, H. Löwen, J. Phys.: Condens. Matter **12**, 5087–5108 (2000)
51. K. Lichtner, S.H.L. Klapp, Phys. Rev. E **86**, 051405 (1–10) (2012)
52. C.N. Likos, Z.T. Németh, H. Löwen, J. Phys.: Condens. Matter **6**, 10965–10976 (1994)
53. C.N. Likos, H. Löwen, M. Watzlawek, B. Abbas, O. Jucknischke, J. Allgaier, D. Richter, Phys. Rev. Lett. **80**, 4450–4453 (1998)
54. C.N. Likos, A. Lang, M. Watzlawek, H. Löwen, Phys. Rev. E **63**, 031206 (1–9) (2001)
55. H. Löwen, Phys. Rep. **237**, 249–324 (1994)
56. H. Löwen, Phys. Rev. E **50**, 1232–1242 (1994)
57. H. Löwen, J. Phys.: Condens. Matter **13**, R415–R432 (2001)
58. H. Löwen, J. Phys.: Condens. Matter **14**, 11897–11905 (2002)
59. H. Löwen, J. Phys.: Condens. Matter **22**, 364105 (1–6) (2010)
60. H. Löwen, in 3rd Warsaw School of Statistical Physics, (Warsaw University Press, 2010), pp. 87–121
61. H. Löwen, in *Understanding Soft Condensed Matter via Modeling and Computation*, ed. by W. Hu, A.-C. Shi. Applications of Density Functional Theory in Soft Condensed Matter, Series in Soft Condensed Matter (World Scientific, 2011), vol. 3, pp. 9–45
62. H. Löwen, M. Heinen, Eur. Phys. J. - Special Topics **223**, 3113–3127 (2014)
63. M. Makino, M. Doi, J. Phys. Soc. Jpn. **73**, 2739–2745 (2004)
64. A. Malijevsky, A.J. Archer, J. Chem. Phys. **139** 144901 (1–13) (2013)
65. M.C. Marchetti, J.F. Joanny, S. Ramaswamy, T.B. Liverpool, J. Prost, M. Rao, R.A. Simha, Rev. Mod. Phys. **85**, 1143–1189 (2013)
66. U.M.B. Marconi, P. Tarazona, J. Chem. Phys. **110**, 8032–8044 (1999)
67. U.M.B. Marconi, P. Tarazona, J. Phys.: Condens. Matter **12**, A413–A418 (2000)
68. U.M.B. Marconi, S. Melchionna, J. Chem. Phys. **126** 184109 (1–9) (2007)
69. U.M.B. Marconi, P. Tarazona, F. Cecconi, S. Melchionna, J. Phys.: Condens. Matter **20**, 494233 (1–6) (2008)
70. U.M.B. Marconi, S. Melchionna, J. Chem. Phys. **131** 014105 (1–14) (2009)
71. U.M.B. Marconi, S. Melchionna, J. Phys.: Condens. Matter **22**, 364110 (1–8) (2010)
72. U.M.B. Marconi, S. Melchionna, Comm. in Theor. Phys. **62**, 596–606 (2014)
73. M. Marechal, H. Löwen, Phys. Rev. Lett. **110**, 137801 (1–5) (2013)
74. M. Marechal, H.-H. Goetzke, A. Härtel, H. Löwen, J. Chem. Phys. **135**, 234510 (1–13) (2011)
75. M. Marechal, U. Zimmermann, H. Löwen, J. Chem. Phys. **136**, 144506 (1–14) (2012)
76. A.M. Menzel, H. Löwen, Phys. Rev. Lett. **110**, 055702 (1–5) (2013)
77. A.M. Menzel, A. Saha, C. Hoell, H. Löwen, J. Chem. Phys. **144**, 024115 (1–13) (2016)
78. M. Mukherjee, P. Mishra, H. Löwen, J. Phys.: Condens. Matter **26**, 465101 (1–9) (2014)
79. K. Nagao, T. Inuzuka, K. Nishimoto, K. Edagawa, Phys. Rev. Lett. **115**, 075501 (1–5) (2015)
80. T. Neuhaus, M. Schmiedeberg, H. Löwen, Phys. Rev. E **88**, 062316 (1–6) (2013)
81. T. Neuhaus, M. Schmiedeberg, H. Löwen, New J. Phys. **15**, 073013 (1–11) (2013)
82. T. Neuhaus, M. Marechal, M. Schmiedeberg, H. Löwen, Phys. Rev. Lett. **110**, 118301 (1–5) (2013)

83. R. Ohnesorge, H. Löwen, H. Wagner, Europhys. Lett. **22**, 245–249 (1993)
84. L. Onsager, Proc. N. Y. Acad. Sci. **51**, 627–659 (1949)
85. D.W. Oxtoby, in *Liquids, Freezing and Glass Transition*, ed. by J.P. Hansen, D. Levesque, J. Zinn-Justin (North Holland, Amsterdam, 1991), pp. 145–189
86. A. Poniewierski, R. Holyst, Phys. Rev. Lett. **61**, 2461 (1988)
87. S. Praetorius, A. Voigt, R. Wittkowski, H. Löwen, Phys. Rev. E **87**, 052406 (1–8) (2013)
88. P.N. Pusey, in *Liquids, Freezing and Glass Transition*, ed. by J.P. Hansen, D. Levesque, J. Zinn-Justin (North Holland, Amsterdam, 1991), pp. 145–189
89. M. Rauscher, A. Dominguez, M. Krüger, F. Penna, J. Chem. Phys **127**, 244906 (1–8) (2007)
90. M. Rauscher, J. Phys.: Condens. Matter **22**, 364109 (1–6) (2010)
91. J. Reinhardt, F. Weysser, J. M. Brader, EPL **102**, 28011 (1–6) (2013)
92. M. Rex, H. Löwen, Phys. Rev. Lett. **101**, 148302 (1–4) (2008)
93. M. Rex, H.H. Wensink, H. Löwen, Phys. Rev. E **76**, 021403 (1–10) (2007)
94. M. Rex, H. Löwen, Eur. Phys. J. E **28**, 139–146 (2009)
95. P. Romanczuk, M. Bär, W. Ebeling, B. Linder, L. Schimansky-Geier, Eur. Phys. J. - Special Topics **202**, 1–162 (2012)
96. Y. Rosenfeld, M. Schmidt, H. Löwen, P. Tarazona, Phys. Rev. E **55**, 4245–4263 (1997)
97. R. Roth, J. Phys.: Condens. Matter **22**, 063102 (1–18) (2010)
98. R. Roth, K. Mecke, M. Oettel, J. Chem. Phys. **136** 081101 (1–4) (2012)
99. E. Runge, E.K.U. Gross, Phys. Rev. Lett. **52**(12), 997–1000 (1984)
100. M. Schmidt, Phys. Rev. E **84**, 051203 (1–8) (2011)
101. M. Schmidt, J. Brader, J. Chem. Phys. **138**, 214101 (1–8) (2013)
102. D. Stopper, K. Marolt, R. Roth, H. Hansen-Goos, Phys. Rev. E **92**, 022151 (1–11) (2015)
103. P. Tarazona, Phys. Rev. Lett. **84**, 694–697 (2000)
104. P. Tarazona, J.A. Cuesta, Y. Martinez-Raton, *Density Functional Theories of Hard Particle Systems*, Lecture Notes in Physics, vol. 753 (Springer, Berlin, 2008), pp. 247–341
105. S. van Teeffelen, H. Löwen, Phys. Rev. E **78**, 020101(R) (1–4) (2008)
106. S. van Teeffelen, N. Hoffmann, C.N. Likos, H. Löwen, Europhys. Lett. **75**, 583–589 (2006)
107. S. van Teeffelen, C.N. Likos, H. Löwen, Phys. Rev. Lett. **100**, 108302 (1–4) (2008)
108. S. van Teeffelen, C.V. Achim, H. Löwen, Phys. Rev. E **87**, 022306 (1–6) (2013)
109. S. van Teeffelen, R. Backofen, H. Löwen, A. Voigt, Phys. Rev. E **79**, 051404 (1–10) (2009)
110. H.H. Wensink, H. Löwen, Phys. Rev. E **78**, 031409 (1–4) (2008)
111. R. Wittkowski, H. Löwen, H.R. Brand, Phys. Rev. E **82**, 031708 (1–7) (2010)
112. R. Wittkowski, H. Löwen, Mol. Phys. **109**, 2935–2943 (2011)
113. R. Wittkowski, H. Löwen, H.R. Brand, Phys. Rev. E **83**, 061706 (1–10) (2011)
114. R. Wittkowski, H. Löwen, H.R. Brand, Phys. Rev. E **84**, 041708 (1–9) (2011)
115. R. Wittkowski, H. Löwen, H.R. Brand, J. Chem. Phys. **137**, 224904 (1–14) (2012)
116. R. Wittkowski, H. Löwen, H.R. Brand, J. Phys. A: Math. Theor. **46**, 355003 (1–17) (2013)
117. U. Zimmermann, F. Smallenburg, H. Löwen, J. Phys.: Condens. Matter **28**, 244019 (1–10) (2016)

Introduction to the Variational Monte Carlo Method in Quantum Chemistry and Physics

Brenda Rubenstein

1 Overview of Quantum Monte Carlo Methods

Variational Monte Carlo is but one member of a much broader class of algorithms known as quantum Monte Carlo methods. Over the past fifty or so years, an ever growing list of QMC algorithms have been developed to solve a wide variety of problems in physics and chemistry. This list includes (among many others):

1. **Diffusion Monte Carlo** for ground state, continuum systems of bosons or fermions [1–5], which was famously used to study the ground state homogenous electron gas, later employed to parameterize density functionals [6];
2. **Path Integral Monte Carlo** for finite temperature, continuum systems of bosons or fermions in equilibrium [7–9], which is commonly used to study the properties of superfluids and plasmas [7, 10];
3. **Auxiliary Field Quantum Monte Carlo** for ground state or finite temperature, continuum and lattice systems of bosons or fermions in equilibrium [11–14], which was originally developed to explore lattice models in nuclear and condensed matter physics, but has more recently been generalized to chemical and solid state systems [15–17];
4. **Diagrammatic Quantum Monte Carlo** for either equilibrium or non-equilibrium fermion lattice models [18–21], which has most recently been applied to the Kondo problem [22, 23];

B. Rubenstein (✉)
Lawrence Livermore National Laboratory, Quantum Simulations Group, Livermore, CA 94551, USA
e-mail: brenda_rubenstein@brown.edu

B. Rubenstein
Department of Chemistry, Brown University, Providence, RI 02912, USA

J. Wu (ed.), *Variational Methods in Molecular Modeling*,
Molecular Modeling and Simulation, DOI 10.1007/978-981-10-2502-0_10

5. **Full Configuration Interaction Quantum Monte Carlo** for ground state and finite temperature continuum systems of fermions in equilibrium [24–26], which has been applied to a variety of chemical, and most recently, solid state systems [27–29].

As this brief list suggests, there are nearly as many QMC methods as problems to solve and workers in the field, which serves as a testament to QMC's utility and overall impact in the physical sciences.

What unites QMC methods is that they all harness the power of the Monte Carlo algorithm used throughout classical physics [30] to solve quantum many body problems. In general, solving quantum many body problems involves being able to accurately predict the properties of large quantum systems consisting of many interacting particles. Highly accurate deterministic approaches for solving these problems commonly scale exponentially (as in Full Configuration Interaction/Exact Diagonalization [31, 32]) or with a large power (as in Coupled Cluster or higher order perturbation theories [33]) as a function of system size, severely restricting their applicability to minimal models consisting of only a few particles. Less accurate mean field methods such as Density Functional Theory [34, 35] or Dynamical Mean Field Theory [36] come with a significantly reduced cost, but are far less accurate, particularly when particles are strongly interacting. Thus, current deterministic approaches tend to either be accurate, but too expensive or approximate, but cheap. QMC methods are particularly attractive because they lie in between these two extremes: by making careful use of stochasticity, they are capable of solving large quantum many body problems with a high level of accuracy while scaling less than $O(N^4)$. Thus, QMC algorithms are often the methods of choice when reliable simulations of practical systems are desired.

To harness QMC's benefits, however, physicists make a Faustian bargain: for QMC's speed and accuracy, one must tolerate the randomness that lies at the heart of all MC methods. Because of its stochasticity, all properties within QMC are measured by taking averages over samples, often called "walkers" in the Monte Carlo community. These averages are inherently accompanied by a statistical error bar that scales as $1/\sqrt{N}$, where N denotes the number of samples. In contrast, the error bar that accompanies most deterministic methods scales either as an exponential or as $1/N^p$, where p is typically some small integer power. Consider, for example, Simpson's rule, used to evaluate integrals over a region using a set of N grid points. For smooth integrands, the Simpson's rule error bar scales as $N^{-4/d}$, where d is the dimensionality of the region of interest. For $d < 8$, Simpson's rule is thus much more efficient than MC methods [37, 38]. This example illustrates a thread that runs through all MC simulations: the power of MC simulations is only fully harnessed when studying high dimensional spaces, for which MC errors bars outcompete those from deterministic methods. Nevertheless, even when MC methods are most efficient, $1/\sqrt{N}$ scaling can be quite costly and one must always deal with some level of uncertainty. This uncertainty can be disquieting for those accustomed to performing electronic structure calculations that typically go unaccompanied by error bars. However, these error bars often provide information about the quality and reproducibility of the simulations being performed that cannot be readily provided by other computational techniques. The $1/\sqrt{N}$ scaling of MC's error bars is also becoming less of an issue given the trend toward increasingly parallel computers. Unlike other techniques, Monte Carlo methods are trivially parallelizable: in MC, samples can be generated completely independently of one another, which makes it possible to simply assign different samples to different processors. MC techniques are thus also primed to make optimal use of the computing architectures of the future.

Variational Monte Carlo fits into this larger QMC picture by serving as another means by which to reduce statistical errors in QMC. As will be discussed in detail below, while VMC can itself provide highly accurate estimates of quantum observables such as a system's ground state energy, VMC is typically used in tandem with other QMC methods (the most popular tandem is VMC with Diffusion Monte Carlo [4, 39]). Many, so-called projector QMC methods, including Diffusion Monte Carlo (DMC) and Auxiliary Field Quantum Monte Carlo (AFQMC), determine a system's ground state energy by projecting out a ground state wave function from some initial guess and then using that initial guess to direct the random sampling toward the most important regions of parameter space [4, 13]. Ultimately, the better the initial guess is, the smaller the prefactor in front of the unavoidable $1/\sqrt{N}$ error scaling will be. VMC is a method capable of sampling ground state[1] wave functions and can thus be used to substantially reduce error bars by providing reliable trial wave functions which other methods can then refine. In the remainder of this chapter, I will first describe how VMC works in its simplest incarnations. I will then discuss the improvements that have been made to this basic algorithm over the last twenty or so years which have made it a robust tool in quantum chemistry and physics. In Sect. 5, I will outline several example applications of VMC from the chemical physics literature. Lastly, I will comment on fruitful directions future VMC research can take.

2 Variational Monte Carlo: A Basic Introduction

The Variational Monte Carlo algorithm is a combination of two key ingredients: the variational principle from quantum mechanics and Monte Carlo sampling.

2.1 *The Variational Principle*

According to the variational principle, the energy computed from any normalized wave function that satisfies the appropriate boundary conditions is always greater than or equal to the ground state energy of the system, where the equality only comes

[1]Recent work has shown that VMC, with some modifications, can also be applied to excited state wave functions (see Sect. 6.2 and [40–50]).

into play when the normalized wave function is the ground state wave function [32]. More specifically, if one considers the normalized wave function $|\Psi_N\rangle$, for which

$$\langle \Psi_N | \Psi_N \rangle = 1, \tag{1}$$

then

$$\langle \Psi_N | \hat{H} | \Psi_N \rangle \geq E_0, \tag{2}$$

where $\hat{H}$ is the system Hamiltonian and E_0 is the ground state energy. The proof of this principle is straightforward. If E_α denotes the eigenvalue corresponding to the eigenfunction $|\Psi_\alpha\rangle$ of $\hat{H}$, then, using the resolution of the identity,

$$\begin{aligned} \langle \Psi_N | \hat{H} | \Psi_N \rangle &= \sum_{\alpha\beta} \langle \Psi_N | \Psi_\alpha \rangle \langle \Psi_\alpha | \hat{H} | \Psi_\beta \rangle \langle \Psi_\beta | \Psi_N \rangle \\ &= \sum_{\alpha} E_\alpha |\langle \Psi_\alpha | \Psi_N \rangle|^2. \end{aligned} \tag{3}$$

Since $E_\alpha \geq E_0$, it follows that

$$\begin{aligned} \langle \Psi_N | \hat{H} | \Psi_N \rangle &= \sum_{\alpha} E_\alpha |\langle \Psi_\alpha | \Psi_N \rangle|^2 \\ &\geq \sum_{\alpha} E_0 |\langle \Psi_\alpha | \Psi_N \rangle|^2 \\ &= E_0 \sum_{\alpha} |\langle \Psi_\alpha | \Psi_N \rangle|^2 = E_0. \end{aligned} \tag{4}$$

What the variational theorem implies is that a strategy for finding an approximation to the ground state energy of a system is to start with a normalized wave function and vary its parameters until it yields the minimum possible energy. In general, the wave function will be a function of a large number of parameters and its expectation values will be evaluated over a high-dimensional space. Finding the minimum possible energy will therefore require an extremely efficient way to vary the parameters while also continually re-evaluating the energy. This is where the Monte Carlo aspect of VMC comes into play.

2.2 *The Metropolis Monte Carlo Method*

In order to find an approximation to the ground state energy using the variational principle, Eq. 2 will have to be re-evaluated a large number of times for a variety of wave functions. For a general wave function in the first quantized framework, this equation can be rewritten as

$$E_v = \frac{\langle \Psi | \hat{H} | \Psi \rangle}{\langle \Psi | \Psi \rangle} = \frac{\int d\mathbf{R} \Psi^*(\mathbf{R}) \hat{H} \Psi(\mathbf{R})}{\int d\mathbf{R} \Psi^*(\mathbf{R}) \Psi(\mathbf{R})}. \tag{5}$$

Here, **R** denotes the $3N$-dimensional vector of electron coordinates and E_v represents the current approximation to the variational energy. As discussed in Sect. 1, if N is small, the above integral can readily be evaluated using deterministic quadrature methods. However, for practical applications, the magnitude of N very quickly exhausts such methods. The primary alternative is to use the Monte Carlo method.

The goal of all Monte Carlo algorithms is to efficiently evaluate integrals of the form[2]

$$I = \int_a^b dx f(x) \tag{6}$$

using random numbers. The simplest Monte Carlo method that can accomplish this is the direct sampling technique. This technique exploits the fact that any integral can be rewritten as

$$I = (b - a)\langle f(x) \rangle, \tag{7}$$

where $\langle f(x) \rangle$ is the unweighted average of $f(x)$ over the interval $[a, b]$. As such, I can be evaluated by randomly sampling points over the interval $[a, b]$, evaluating $f(x)$ at those points, and averaging over all $f(x)$ values to obtain $\langle f(x) \rangle$. For a small number of samples, this procedure yields only a rough estimate of I, but as the number of points sampled approaches infinity, direct sampling becomes exact. How many samples it takes for direct sampling to converge is highly dependent on the form of $f(x)$. However, for most functions, sampling a uniform distribution over the interval is highly inefficient: because the uniform distribution contains no information about $f(x)$, at many of the points sampled, the value of $f(x)$ will be negligible and thus minimally contribute to the average. This is particularly true as the dimensionality of the system increases. What is therefore needed is a way to prioritize sampling toward the values of x that will significantly contribute to the average.

The primary way of achieving this is through importance sampling. Note that Eq. 6 can always be rewritten as

$$I = \int_a^b dx w(x) \frac{f(x)}{w(x)}, \tag{8}$$

[2]Here, I have chosen a simple one-dimensional integral to illustrate the basic ideas behind MC. Nevertheless, MC is most advantageous in many dimensions, as previously discussed.

where $w(x)$ is some probability distribution. If $w(x)$ is uniform, Eq. 8 maps onto Eq. 7. However, in importance sampling, $w(x)$ is purposefully constructed so as to favor points that will maximally contribute to the integral and minimize the variance [38, 51]. Once a form for $w(x)$ is chosen, Eq. 8 can then be evaluated by choosing points distributed according to $w(x)$ and then averaging over all values of $\frac{f(x)}{w(x)}$.

What Metropolis et al. showed in their landmark 1953 paper was that any probability distribution, $w(x)$, may be sampled by constructing a Markov chain that obeys the detailed balance condition [52]. A Markov chain is a trajectory of successive states, $x_1, x_2, x_3, \ldots, x_N$, in which the probability of transitioning from state x_{k-1} to x_k is given by a probability, $P(x_k|x_{k-1})$, that only depends on the previous state, x_{k-1} [53, 54]. $P(x_k|x_{k-1})$ is termed the transition probability (or matrix) and must satisfy the basic requirements that

$$P(x_k|x_{k-1}) \geq 0 \tag{9}$$

and

$$\sum_{x_{k-1}} P(x_k|x_{k-1}) = 1, \tag{10}$$

meaning that the system must evolve and that the total probability of transitioning to a given final state from any initial state must be 1. In order to calculate an integral like that in Eq. 8, a Markov chain of states x that can be used to evaluate $\frac{f(x)}{w(x)}$ must be constructed so that its equilibrium distribution is $w(x)$. To construct such a Markov chain, an initial state, x_i, is typically selected and then the transition matrix is repeatedly sampled to obtain the next states in the chain. In order to sample the distribution $w(x)$, the transition probability must satisfy the stationarity condition

$$\sum_i P(x_f|x_i)w(x_i) = \sum_i P(x_i|x_f)w(x_f) = w(x_f) \tag{11}$$

for all final states, x_f. This condition requires that, if we start with some distribution $w(x)$, after repeated sampling of the transition probability, we continue to sample this distribution. Another practical requirement for efficient sampling is ergodicity: given any initial distribution, the Markov chain should evolve to the final, desired distribution in a finite number of steps. Even with these conditions, a wide range of transition probabilities may be selected. In the Metropolis method, the transition probability is constructed so that it satisfies the detailed balance condition

$$P(x_f|x_i)w(x_i) = P(x_i|x_f)w(x_f) \tag{12}$$

for any states i, f. This condition requires that the rate of flow from x_i is the same as the rate of flow from x_f. If the transition probability is defined as the product of a trial probability, α, and an acceptance probability, acc,

$$P(x_f|x_i) = \alpha(x_f|x_i)acc(x_f|x_i), \tag{13}$$

then Eq. 12 may be rewritten as

$$\alpha(x_f|x_i)acc(x_f|x_i)w(x_i) = \alpha(x_i|x_f)acc(x_i|x_f)w(x_f). \tag{14}$$

This can be rearranged to yield

$$\frac{acc(x_f|x_i)}{acc(x_i|x_f)} = \frac{\alpha(x_i|x_f)w(x_f)}{\alpha(x_f|x_i)w(x_i)}. \tag{15}$$

Metropolis et al. found that the optimal choice for the acceptance probability is

$$acc(x_f|x_i) = min\{1, \frac{\alpha(x_i|x_f)w(x_f)}{\alpha(x_f|x_i)w(x_i)}\}. \tag{16}$$

In other words, when a transition from x_i to x_f is attempted, if

$$\frac{\alpha(x_i|x_f)w(x_f)}{\alpha(x_f|x_i)w(x_i)} < 1, \tag{17}$$

it will be accepted with probability

$$\frac{\alpha(x_i|x_f)w(x_f)}{\alpha(x_f|x_i)w(x_i)}; \tag{18}$$

otherwise, it will be accepted with a probability of 1. As will be discussed in the next section, $w(x)$ often has a normalization constant that cannot be explicitly evaluated. The beauty of Eq. 16 is that it does not depend on this normalization constant since it cancels out in the above expression. This enables the sampling of complex probability distribution functions, such as partition functions in statistical mechanics, that would not otherwise be able to be sampled. In Eq. 16, the probability distribution to be sampled is determined by the problem. The trial probability, however, can assume a wide variety of forms, some of which are more efficient than others as measured by simulation autocorrelation times. Even though Metropolis et al. employed a symmetric[3] α in their seminal paper, it has since been shown by Hastings that using a nonsymmetric α can significantly improve algorithmic efficiency [55]. In the context of VMC calculations of electronic systems, a very efficient choice of α was presented by Umrigar [56]. Optimal selection can help minimize the number of samples that must be taken to evaluate the integral of interest and thus substantially reduce simulation time.

[3]If α is symmetric, $\alpha(x_f|x_i) = \alpha(x_i|x_f)$ and these quantities cancel in Eq. 16. If α is nonsymmetric, $\alpha(x_f|x_i) \neq \alpha(x_i|x_f)$.

2.3 Putting the Two Together: Variational Monte Carlo

In the simplest version of Variational Monte Carlo, the Metropolis algorithm is used to evaluate the variational energy of a predetermined wave function whose variational energy is too difficult or time-consuming to evaluate using alternative means based upon Eq. 5. How to accomplish this becomes clearer after rewriting Eq. 5

$$\begin{aligned} E_V &= \frac{\int d\mathbf{R}\Psi^*(\mathbf{R})\hat{H}\Psi(\mathbf{R})}{\int d\mathbf{R}\Psi^*(\mathbf{R})\Psi(\mathbf{R})} \\ &= \frac{\int d\mathbf{R}|\Psi(\mathbf{R})|^2\Psi^{-1}(\mathbf{R})\hat{H}\Psi(\mathbf{R})}{\int d\mathbf{R}|\Psi(\mathbf{R})|^2} \\ &= \int d\mathbf{R}w(\mathbf{R})E_L(\mathbf{R}), \end{aligned} \tag{19}$$

where E_L is the local energy

$$E_L = \Psi^{-1}(\mathbf{R})\hat{H}\Psi(\mathbf{R}) \tag{20}$$

and $w(\mathbf{R})$ is the probability distribution that must be sampled

$$w(\mathbf{R}) = \frac{|\Psi(\mathbf{R})|^2}{\int d\mathbf{R}'|\Psi(\mathbf{R}')|^2}. \tag{21}$$

$E_L(\mathbf{R})$ is the multidimensional equivalent of $\frac{f(x)}{w(x)}$ in Eq. 8 and $w(\mathbf{R})$ is akin to $w(x)$. At this point, the astute reader may recognize that Eq. 21 cannot be evaluated as it is written. Doing so would require knowing the value of $\int d\mathbf{R}'|\Psi(\mathbf{R}')|^2$, a quantity just as, if not more, difficult to determine than the variational energy itself in most situations. Fortunately, this quantity cancels out of the expression for the acceptance probability, just like the partition function does in classical Monte Carlo simulations. Substituting Eq. 21 into Eq. 16, one obtains

$$acc(\mathbf{R_f}|\mathbf{R_i}) = min\{1, \frac{\alpha(\mathbf{R_i}|\mathbf{R_f})\frac{|\Psi(\mathbf{R_f})|^2}{\int d\mathbf{R}'|\Psi(\mathbf{R}')|^2}}{\alpha(\mathbf{R_f}|\mathbf{R_i})\frac{|\Psi(\mathbf{R_i})|^2}{\int d\mathbf{R}'|\Psi(\mathbf{R}')|^2}}\} \tag{22}$$

$$= min\{1, \frac{\alpha(\mathbf{R_i}|\mathbf{R_f})|\Psi(\mathbf{R_f})|^2}{\alpha(\mathbf{R_f}|\mathbf{R_i})|\Psi(\mathbf{R_i})|^2}\} \tag{23}$$

for the acceptance probability. Thus, in VMC, the variational energy can be evaluated by starting with some random configuration of particles at positions, $\mathbf{R_i}$, displacing these particles to a new set of positions, $\mathbf{R_f}$, according to the transition probability, $\alpha(\mathbf{R_i}|\mathbf{R_f})$, and determining whether to accept or reject these new coordinates based upon Eq. 23. If the new coordinates are accepted, the particles are moved to $\mathbf{R_f}$; if they are rejected, they remain at $\mathbf{R_i}$. The process is then continued iteratively. As long as the transition probability is ergodic (meaning that it makes it possible for the particles to visit any set of coordinates in their configuration space in a finite number of Monte Carlo steps), this Markov process will converge to the equilibrium distribution, $w(\mathbf{R})$. The simplest type of trial particle move that samples the transition probability distribution is the "box" move, in which a maximum box dimension, Δ, is selected at the beginning of the simulation, and during each iteration, the new particle positions, $\mathbf{R_f}$, are generated as

$$\mathbf{R_f} = \boldsymbol{\xi}\Delta + \mathbf{R_i}, \tag{24}$$

where $\boldsymbol{\xi}$ is a $3N$-dimensional vector (N is the number of particles) of random numbers uniformly distributed between -1 and 1. Typically, the box size is varied until roughly 50 % of the trial moves are accepted to optimize efficiency. As the particles are evolved, the local energy (Eq. 20) is computed based upon the current particle positions. An estimate of the variational energy of the wave function is obtained by averaging over these local energy values

$$E_{var} = \frac{1}{M}\sum_{i=1}^{M} E_L(\mathbf{R_i}). \tag{25}$$

In the above, M denotes the number of samples taken. It should be noted that the consecutive samples generated using the above algorithm are generally serially correlated, meaning that they are not statistically independent. One must therefore use a blocking algorithm [54, 57] or more sophisticated sampling schemes [58] to reduce correlation.

Aside from being used to sample real-space wave functions, VMC has proven to be an invaluable tool for condensed matter physicists interested in sampling the second-quantized wave functions of particular relevance in the study of lattice models [59–63]. In this context, VMC is used to sample spin configurations as opposed to particle positions. As such, instead of moving particles from place to place, at each iteration, one or more spins are sampled (in the case of discrete spin models, the spins are flipped; in the case of continuous spins, new spin values are sampled from a spin probability distribution). A recent application of VMC methods to second-quantized wave functions will be described in Sect. 6.3.

Up to this point in the discussion, it has been implied that researchers typically use VMC to calculate the variational energy of wave functions they have already identified as being able to approximate the true ground state. While VMC is useful for this purpose, in practice, researchers typically have limited knowledge of the ground state wave function for a given system and therefore equally limited knowledge of wave function forms that may approximate it. Fortunately, VMC is capable of not only evaluating the variational energy of a given wave function, but of identifying and parameterizing wave functions that may yield the overall lowest variational energies.

3 VMC Wave Functions

Before discussing how VMC can be used to parameterize wave functions, a few words should be said about the trial wave function forms most often used in variational Monte Carlo calculations. In real-space VMC calculations, N-electron wave functions are typically represented as Slater-Jastrow wave functions [4, 39, 64–66]

$$|\Psi_0(\mathbf{R})\rangle = e^{\hat{J}(\alpha_0,\mathbf{R})}|\Phi_0(\mathbf{R})\rangle, \tag{26}$$

where $\hat{J}(\alpha_0, \mathbf{R})$ denotes the Jastrow factor and $|\Phi_0(\mathbf{R})\rangle$ is either a single- or multi-determinant wave function. In its most general form, $|\Phi_0(\mathbf{R})\rangle$ may be expressed as a sum over orthogonal configuration state functions (CSFs)

$$|\Phi_0(\mathbf{R})\rangle = \sum_i^{N_{CSF}} c_i|C_I(\mathbf{R})\rangle. \tag{27}$$

The CSFs, $|C_I(\mathbf{R})\rangle$, may in turn be rewritten as short sums over products of spin-up and spin-down Slater determinants

$$|C_I(\mathbf{R})\rangle = \sum_k d_{I,k}|D_k^{\uparrow}\rangle|D_k^{\downarrow}\rangle. \tag{28}$$

Per usual in quantum chemistry, the spin-up and spin-down Slater determinants are composed of a set of $N_{\uparrow}$ and $N_{\downarrow}$ orthonormal orbitals

$$|D_k^{\uparrow}\rangle = \hat{a}_{k_1\uparrow}\hat{a}_{k_2\uparrow}...\hat{a}_{k_N\uparrow}|0\rangle \tag{29}$$

and

$$|D_k^{\downarrow}\rangle = \hat{a}_{k_1\downarrow}\hat{a}_{k_2\downarrow}...\hat{a}_{k_N\downarrow}|0\rangle. \tag{30}$$

In the above, $|0\rangle$ denotes the vacuum state and $\hat{a}_{k\sigma}$ is the fermion operator that creates an electron with spin σ in spatial orbital $|\phi_k^0\rangle$. The spatial orbitals are linear combinations of N_{bas} Slater or Gaussian basis functions $|\Xi_\mu\rangle$

$$|\phi_k^0\rangle = \sum_\mu^{N_{bas}} \lambda_{k,\mu}^0|\Xi_\mu\rangle, \tag{31}$$

where $\lambda_{k,\mu}^0$ are the linear coefficients. The orbitals used in the Slater determinants are typically taken from Hartree-Fock, Density Functional Theory, or, most commonly, Multiconfigurational Self Consistent Field calculations. The simplest Jastrow factors may be written as a sum over electron-electron interactions

$$J(\alpha) = \sum_{\sigma_i, \sigma_j} \sum_{ij} U_{\sigma_i, \sigma_j}(r_{ij}, \alpha). \tag{32}$$

Here, U represents a spin-dependent functional form for the electron-electron interactions and r_{ij} denotes the distance between electron i and electron j.

Slater-Jastrow wave functions of the form proposed above are highly compact approximations to the true ground state wave function that can be rapidly manipulated within quantum Monte Carlo calculations. It is well-known that the multideterminant expansions that make up the Slater part of the wave function can successfully capture the non-dynamical correlation arising from the degeneracy of different system states [32]. Nevertheless, the Slater part is highly inefficient at capturing analytical features of the wave function such as electron-electron cusps. Indeed, it can sometimes take millions of determinants to recover these features if Gaussian or other basis functions that do not explicitly take cusp conditions into account are employed.[4] Jastrow factors make up for what multideterminant expansions lack by containing terms that explicitly enforce electron-electron cusp conditions [67]. Multiplying Slater determinants by Jastrow factors can therefore dramatically reduce the size of wave function expansions from millions to thousands of determinants, significantly decreasing the cost of QMC simulations. Using Slater-Jastrow wave functions to represent VMC trial wave functions has likewise become the norm within the VMC community.[5]

Based on Eqs. 26–32, Slater-Jastrow wave functions may be viewed as functions of the Jastrow parameters, $\boldsymbol{\alpha}$, the configuration state function coefficients, c_i, and the orbital coefficients, $\lambda_{k,\mu}$. It is these parameters that are optimized when VMC is used to optimize wave functions.

4 Wave Function Optimization

In modern-day simulation, VMC is most often employed to optimize wave functions for later use as trial wave functions in projector quantum Monte Carlo methods, such as the Diffusion and Auxiliary Field Quantum Monte Carlo techniques [4, 13]. In this flavor of VMC, a specific form of the wave function is proposed (see Sect. 3), but the constants, $\mathbf{p}$, used to parameterize that wave function form are left undetermined. VMC is then used to determine the set of parameters that optimize a selected cost function.

[4]This issue is eliminated and only a single determinant is necessary for capturing electron-electron cusps if Slater basis functions, which are tailored to satisfy cusp conditions, are used instead.

[5]Other, more powerful wave function ansatzes that include more explicit forms of correlation, including Antisymmetric Geminal Power (AGP) wave functions [68–70], Pfaffians [71, 72], and Matrix Product States [73], are currently being explored by a number of groups. Because these wave functions take on significantly different forms than the wave functions discussed in this chapter, I will not discuss their variational optimization here.

4.1 Choosing the Cost Function

The most obvious cost function is the variational energy. Nevertheless, most researchers have historically chosen to minimize the variance of the energy [74–78]

$$\sigma_E^2(\mathbf{p}) = \frac{\int \Psi_T^2(\mathbf{p})[E_L(\mathbf{p}) - E_V(\mathbf{p})]^2 d\mathbf{R}}{\int \Psi_T^2(\mathbf{p}) d\mathbf{R}}. \tag{33}$$

There are two key reasons for this choice. Firstly, the variance of the energy has a strict lower bound of zero [74, 79]. This is in stark contrast with the variational energy, whose lower bound is unknown before calculations are performed. Perhaps, even more importantly, early attempts to minimize the variational energy were plagued by numerical instabilities [80]. When a finite number of walkers were used to perform the minimization, one configuration tended to dominate the rest, causing the minimization to converge to a poor set of parameter values. While these considerations are still pertinent, with the advent of the linear method—a highly efficient, highly robust minimization technique—by Umrigar et al. (see Sect. 5 for details), attention has returned to minimizing the variational energy itself. This is because

- Minimizing the variational energy typically yields wave functions that minimize the energy produced by projector quantum Monte Carlo methods.
- Energy minimized wave functions yield more accurate values for other observables than variance minimized wave functions [81, 82].

In what follows, I therefore describe the simplest (although not necessarily the most practical) way to minimize the variational energy.

4.2 Minimizing the Variational Energy

Upon first thought, one might imagine that the easiest way to minimize the variational energy would be to start with a set of constants, $\mathbf{p}$, that parameterizes the wave function and then to continually vary those constants and recalculate the variational energy until a set of constants that minimize the energy is obtained. The fundamental problem with this idea is that the variational energies estimated by VMC always come with a finite statistical error. It is therefore often practically difficult to determine which VMC energy is truly the lowest without resorting to infinite sample sizes.

The way to overcome this deficit is to use correlated sampling [58, 83]. In this approach, energy differences are calculated as a function of each set of parameters instead of the total energy. Energy differences are, in general, much less noisy than total energy evaluations because they enable statistical fluctuations to cancel. The way this noise cancellation is achieved in VMC is by generating a set of configurations distributed according to some initial guess at the wave function and then evaluating the variational energy of each new set of parameters as a function of these same configurations. More specifically, Eq. 19 may be rewritten as

$$E_V = \frac{\int d\mathbf{R}|\Psi(\mathbf{R})|^2 E_L(\mathbf{R})}{\int d\mathbf{R}|\Psi(\mathbf{R})|^2}$$
$$= \frac{\int d\mathbf{R}|\Psi_T(\mathbf{R}, \mathbf{p}_0)|^2 w(\mathbf{R}, \mathbf{p}) E_L(\mathbf{p})}{\int d\mathbf{R}|\Psi_T(\mathbf{R}, \mathbf{p}_0)|^2 w(\mathbf{R}, \mathbf{p})}, \tag{34}$$

where

$$w(\mathbf{R}, \mathbf{p}) = \frac{|\Psi_T(\mathbf{R}, \mathbf{p})|^2}{|\Psi_T(\mathbf{R}, \mathbf{p}_0)|^2} \tag{35}$$

is a reweighting factor. As such, one can compute the variational energy for a given set of parameters, $\mathbf{p}$, by starting with an initial set of parameters, $\mathbf{p}_0$, sampling a set of configurations from $|\Psi_T(\mathbf{R}, \mathbf{p}_0)|^2$, and then accumulating $w(\mathbf{R}, \mathbf{p})E_L(\mathbf{p})$ and $w(\mathbf{R}, \mathbf{p})$ as a function of those configurations. Thus, one can use the same set of configurations to determine the energy for many different sets of parameters and the statistical fluctuations that would have arisen from using different configurations are eliminated. The one caveat to using Eq. 34 is that $w(\mathbf{R}, \mathbf{p})$ must remain of order unity. If it becomes either too large or too small, only a few points will dominate the sums that must be accumulated, leading to large errors. Thus, one must ensure that the functions $\Psi_T(\mathbf{R}, \mathbf{p})$ and $\Psi_T(\mathbf{R}, \mathbf{p}_0)$ are not too different from one another. In modern day simulations, this is typically achieved by starting with some $\Psi_T(\mathbf{R}, \mathbf{p}_0)$, finding a set of parameters, $\mathbf{p}_1$, that minimizes the energy as a function of the original Ψ_T, replacing $\Psi_T(\mathbf{R}, \mathbf{p}_0)$ with $\Psi_T(\mathbf{R}, \mathbf{p}_1)$, and then starting over again. Usually three or four iterations of sampling parameters and replacing leads to convergence.

While the conceptually simplest way to determine the optimal set of parameters would be to start with $\mathbf{p}_0$ and vary each of the parameters sequentially until the energy minimum is found, this process becomes prohibitively expensive as the number of dimensions grows. Researchers therefore avoid this bottleneck by combining correlated sampling with standard minimization techniques, such as the steepest descent, conjugate gradient, or Broyden-Fletcher-Goldfarb-Shanno (BFGS) algorithms [84]. To illustrate how this synthesis works, consider combining correlated sampling with the method of steepest descent. According to the method of steepest descent, if there exists a multivariable function, $F(\mathbf{x})$, defined and differentiable at the point $\mathbf{a}$, then the function decreases fastest away from the point $\mathbf{a}$ along the negative of its gradient at $\mathbf{a}$, $-\nabla F(\mathbf{a})$. Thus, if one starts at a point $\mathbf{x}_n$, one can arrive at a lower point, $\mathbf{x}_{n+1}$, by subtracting a multiple of the gradient from $\mathbf{x}_n$

$$\mathbf{x}_{n+1} = \mathbf{x}_n - \gamma_n \nabla F(\mathbf{x}_n), \tag{36}$$

where γ_n denotes the size of the step taken along the gradient. This process may be iterated until $\mathbf{x}_n$ hopefully converges (under *certain* conditions, the steepest descent algorithm is guaranteed to converge, but may not in the most general case) [84]. In order to combine the method of steepest descent with correlated sampling [85], one must start with a set of parameters, $\mathbf{p}_0$, and perturb it by γ_k along each dimension, k, of K total dimensions, such that

$$\mathbf{p}_k = \{p_1, p_2, \ldots, p_k + \gamma_k, \ldots, p_K\}. \tag{37}$$

The variational energy from each set of parameters, $E_{var,k}$, may then be determined using Eq. 34 and substituted into the finite difference formula

$$\nabla E(\mathbf{p}_0) \equiv \frac{\partial E_{\mathbf{p}}}{\partial \gamma_k} \cong \frac{E_k - E_0}{\gamma_k} \tag{38}$$

to estimate the gradient. The gradient may in turn be inserted into Eq. 36 to obtain the next set of parameters

$$\mathbf{p}_1 = \mathbf{p}_0 - \gamma \nabla E(\mathbf{p}_0), \tag{39}$$

and the process is iterated until the lowest energy is attained. A more accurate way of obtaining the gradient is to explicitly differentiate the energy with respect to the parameters [58]

$$\frac{\partial \langle \hat{H}(\mathbf{p}) \rangle}{\partial \mathbf{p}} = \langle \frac{\partial}{\partial \mathbf{p}} ln|\Psi_T(\mathbf{p})|^2 E_L(\mathbf{p}) \rangle - \langle \frac{\partial}{\partial \mathbf{p}} ln|\Psi_T(\mathbf{p})|^2 \rangle \langle E_L(\mathbf{p}) \rangle \tag{40}$$

and to use this in place of the finite difference value. Needless to say, this estimate of the gradient depends upon being able to obtain the average of a number of different quantities with small error bars, which can be costly, depending upon the number of parameters that need to be varied. Using more robust and efficient minimization techniques such as the Newton or BFGS methods may accelerate this process substantially [76, 78, 86–90].

5 A Selection of State-of-the-Art VMC Algorithms

5.1 The Linear Method

While it is possible to use the steepest descent minimization technique presented in Sect. 4.2 to optimize wave functions, this technique requires many more minimization steps, and therefore, significantly more computation than competing minimization techniques. After years of debate, the technique that is now viewed as both the most efficient and robust method for optimizing wave functions is the linear method of Umrigar and coworkers [64, 65, 91]. The basic idea behind this method is that, if a wave function can be expanded linearly in terms of its parameters, it is possible to determine those parameters after each minimization step by solving a simple linear eigenvalue equation. More specifically, consider a normalized wave function

$$|\bar{\Psi}(\mathbf{p})\rangle = \frac{|\Psi(\mathbf{p})\rangle}{\sqrt{\langle\Psi(\mathbf{p})|\Psi(\mathbf{p})\rangle}}, \tag{41}$$

where $\mathbf{p}$ denotes its parameters. For wave functions of the form discussed in Sect. 3, for example, $\mathbf{p} = \{\mathbf{c}, \boldsymbol{\alpha}, \boldsymbol{\lambda}\}$, the set of all configuration state function coefficients, nonlinear Jastrow parameters, and orbital coefficients. Such a wave function can always be expanded to first order in its parameters $\mathbf{p}$ around its current parameters $\mathbf{p}_0$

$$|\bar{\Psi}_{lin}(\mathbf{p})\rangle = |\Psi_0\rangle + \sum_{i}^{N_{opt}} \Delta p_i |\bar{\Psi}_i\rangle. \tag{42}$$

Here, $|\Psi_0\rangle = |\bar{\Psi}_0\rangle = |\bar{\Psi}(\mathbf{p}_0)\rangle$ and $|\bar{\Psi}_i\rangle$ are the derivatives of $|\bar{\Psi}(\mathbf{p})\rangle$ with respect to the N_{opt} parameters made orthogonal to $|\Psi_0\rangle$

$$\begin{aligned} |\bar{\Psi}_i\rangle &= \left(\frac{\partial|\bar{\Psi}(\mathbf{p})\rangle}{\partial p_i}\right)_{\mathbf{p}=\mathbf{p}_0} \\ &= |\Psi_i\rangle - S_{0i}|\Psi_0\rangle, \end{aligned} \tag{43}$$

with $S_{0i} = \langle\Psi_0|\Psi_i\rangle$.[6] Minimizing the energy evaluated in terms of the linearized wave function

$$E_{lin}(\mathbf{p}) = \frac{\langle\bar{\Psi}_{lin}(\mathbf{p})|\hat{H}|\bar{\Psi}_{lin}(\mathbf{p})\rangle}{\langle\bar{\Psi}_{lin}(\mathbf{p})|\bar{\Psi}_{lin}(\mathbf{p})\rangle}. \tag{44}$$

with respect to the parameters then leads to the generalized eigenvalue equation

$$\bar{\mathbf{H}} \cdot \Delta\mathbf{p} = E_{lin}\bar{\mathbf{S}} \cdot \Delta\mathbf{p}. \tag{45}$$

In the above, $\bar{\mathbf{H}}$ represents the Hamiltonian matrix with matrix elements $\bar{H}_{ij} = \langle\bar{\Psi}_i|\hat{H}|\bar{\Psi}_j\rangle$ evaluated in the $N_{opt} + 1$-dimensional basis of the normalized wave function and its derivatives. Similarly, $\bar{\mathbf{S}}$ denotes the overlap matrix with matrix elements $\bar{S}_{ij} = \langle\bar{\Psi}_i|\bar{\Psi}_j\rangle$ evaluated in the same $N_{opt} + 1$-dimensional space ($\bar{S}_{00} = 1$ and $\bar{S}_{i0} = \bar{S}_{0i} = 0$). $\Delta\mathbf{p}$ is the $N_{opt} + 1$-dimensional vector of parameter variations

[6] If the parameters are linear, the generalized eigenvalue equation can be solved without orthogonalizing the derivatives with respect to $|\Psi_0\rangle$. Orthogonalization as in Eq. 43 is only necessary if the parameters are nonlinear.

with $\Delta p_0 = 1$. The parameter variations that correspond to the lowest eigenvalue solution of Eq. 45 are the parameter variations that lead to the minimum energy.

What these equations thus imply is that, if a wave function is linear in its parameters or otherwise can be well approximated by a linear expansion, the parameterization that minimizes the energy can be determined by iteratively computing the normalized wave function and its derivatives, forming the matrices $\bar{\mathbf{H}}$ and $\bar{\mathbf{S}}$, solving Eq. 45 for the lowest eigenvalue $\Delta\mathbf{p}$s, and updating the starting parameters $\mathbf{p}_0' = \mathbf{p}_0 + \Delta\mathbf{p}$. Following the correlated sampling approach described in Sect. 4.2, the $\bar{\mathbf{H}}$ and $\bar{\mathbf{S}}$ matrices may be evaluated in practice by averaging over configurations sampled from the $|\Psi_0|^2$ distribution such that

$$\bar{H}_{ij} = \langle \frac{\bar{\Psi}_i \hat{H} \bar{\Psi}_j}{\Psi_0 \Psi_0} \rangle_{\Psi_0^2}, \tag{46}$$

and

$$\bar{S}_{ij} = \langle \frac{\bar{\Psi}_i \bar{\Psi}_j}{\Psi_0 \Psi_0} \rangle_{\Psi_0^2}. \tag{47}$$

For a discussion of improved versions of these estimators, please see [64–66, 91].

Thus far, this discussion has focused on wave functions that may readily be linearized with respect to their parameters. Wave functions containing Jastrow factors and orbital coefficients, however, are typically highly *nonlinear* with respect to their parameters. The breakthrough made by Umrigar and coworkers that has made the linear method widely applicable is the realization that, by carefully exploiting the normalization of the wave function, one can still apply the linear method to highly nonlinear wave functions [64, 65]. Consider the wave function normalized such that

$$|\tilde{\Psi}(\mathbf{p})\rangle = N(\mathbf{p})|\Psi(\mathbf{p})\rangle, \tag{48}$$

where $N(\mathbf{p})$ depends only on nonlinear parameters, and as before, $|\tilde{\Psi}(\mathbf{p}_0)\rangle = |\Psi(\mathbf{p}_0)\rangle = |\Psi_0\rangle$. The derivatives of $|\tilde{\Psi}'(\mathbf{p})\rangle$ with respect to the parameters are then

$$|\tilde{\Psi}_i\rangle = |\Psi_i\rangle + N_i|\Psi_0\rangle, \tag{49}$$

where

$$N_i = \left(\frac{\partial N(\mathbf{p})}{\partial p_i}\right)_{\mathbf{p}=\mathbf{p}_0}. \tag{50}$$

$N_i = 0$ for linear parameters. $|\tilde{\Psi}\rangle$ may linearized just like $|\bar{\Psi}\rangle$ yielding

$$|\tilde{\Psi}\rangle_{lin} = |\Psi_0\rangle + \sum_i^{N_{opt}} \Delta\tilde{p}_i|\tilde{\Psi}_i\rangle. \tag{51}$$

Umrigar and coworkers then cleverly argue that, since $|\tilde{\Psi}\rangle$ and $|\bar{\Psi}\rangle$ are optimized in the same variational space, they must be proportional to each other, implying that

$$\Delta\tilde{\mathbf{p}} = \frac{\Delta\bar{\mathbf{p}}}{1 - \sum_i^{N_{opt}} N_i \Delta\bar{p}_i}. \tag{52}$$

Thus, if one solves the generalized eigenvalue equation for the $\Delta\bar{\mathbf{p}}$, one can obtain the $\Delta\tilde{\mathbf{p}}$ by rescaling. It has been found that the fastest, most stable optimization is achieved if the N_i are determined by imposing the condition that all of the derivatives $|\tilde{\Psi}_i\rangle$ are orthogonal to a linear combination of $|\Psi_0\rangle$ and $|\Psi_{lin}\rangle$ such that

$$\langle \xi \frac{\Psi_0}{||\Psi_0||} + (1-\xi)\frac{\Psi_{lin}}{||\Psi_{lin}||}|\tilde{\Psi}_i\rangle = 0. \tag{53}$$

If this condition is enforced, the N_i may be expressed as

$$N_i = \frac{-\xi D S_{0i} + (1-\xi)(S_{0i} + \sum_j^{nonlin} S_{ij}\Delta p_j)}{\xi D + (1-\xi)(1 + \sum_j^{nonlin} S_{0j}\Delta p_j}, \tag{54}$$

where ξ can vary between 0 and 1 and

$$D = \sqrt{1 + 2\sum_j^{nonlin} S_{0j}\Delta p_j + \sum_{j,k}^{nonlin} S_{jk}\Delta p_j \Delta p_k}. \tag{55}$$

The sums are only over nonlinear parameters. Umrigar and coworkers typically suggest setting $\xi = 1/2$ so that finite parameter changes are made until the energy minimum is reached. In order to ensure that the algorithm is stable when the MC sample size is small or the original wave function parameterization is poor, a positive constant, a_{diag}, is often added to all of the diagonal entries of H_{ij} except for the first:

$$H'_{ij} = H_{ij} + a_{diag}\delta_{ij}(1 - \delta_{i0}), \tag{56}$$

where δ denotes the Kroenecker delta. The optimal value of a_{diag} is typically computed on-the-fly by guessing several values of a_{diag}, each an order of magnitude larger than the last, using those a_{diag}s to determine the optimal parameterization of the wave function and the energies that stem from those wave functions, and then interpolating which value of a_{diag} yields the optimal energy.

The linear algorithm as just described is now widely adopted to optimize wave functions to be later used as trial wave functions in Diffusion Monte Carlo calculations [92, 93]. The highly accurate variational wave functions it yields have been crucial for reducing the fixed node errors in DMC calculations and have paved the way toward achieving chemical accuracy using quantum Monte Carlo techniques (see Sect. 6.1).

5.2 Stochastic Reconfiguration

A second modern VMC algorithm that has been widely used to study lattice models [94–97] and more recently applied to chemical problems [98–100] is the stochastic reconfiguration with Hessian acceleration algorithm of Sorella and coworkers [96, 97, 101]. Like the linear method, the stochastic reconfiguration with Hessian acceleration method expands the wave function with respect to its variational parameters and then minimizes the energy based upon this expansion. What differs between these methods is the exact form of the expansion and how many orders of the expansion are kept. In the stochastic reconfiguration method, an operator

$$O_k(\mathbf{R}) = \frac{\partial_{p_k} \Psi_p(\mathbf{R})}{\Psi_p(\mathbf{R})} \tag{57}$$

is defined for each of the k variational parameters p_k in the wave function $\Psi(\mathbf{R}, \mathbf{p})$ [96, 101]. These operators are motivated by operators, such as density-density terms, which often appear in the Jastrow factor as simple exponentials, $\exp(\sum_k p_k(O_k - \langle O_k \rangle))$, where $\langle O_k \rangle$ denotes the average of the operator commonly subtracted from the operator itself in order to reduce fluctuations [14, 15]. Of course, operators corresponding to the Slater portion of the wave function do not enter into the wave function as exponentials. Nevertheless, even for non-exponential operators, the approximation can be made that

$$|\Psi(\mathbf{p}+\gamma)\rangle \approx e^{\sum_k \gamma_k (O_k - \langle O_k \rangle)} |\Psi(\mathbf{p})\rangle \tag{58}$$

if the changes in the variational parameters, γ, are sufficiently small. Expanding this expression for the wave function to second order in γ, one obtains

$$|\Psi(\mathbf{p}+\gamma)\rangle \approx \left[1 + \sum_k \gamma_k (O_k - \langle O_k \rangle) + \frac{\beta}{2} \sum_{k,k'} \gamma_k \gamma_{k'} (O_k - \langle O_k \rangle)(O_{k'} - \langle O_{k'} \rangle) \right] |\Psi(\mathbf{p})\rangle. \tag{59}$$

Here, β is a constant. In the case of Jastrow terms, this expansion is valid to second order if $\beta = 1$. However, for general operators, this expansion is only valid to first order for small parameters. In the general case, the quadratic terms are therefore highly approximate, but become negligible, and consequently, irrelevant, as the minimum is approached. Based upon this idea, the stochastic reconfiguration method uses the value of β as a knob that can be used to improve the efficiency of the algorithm. Substituting Eq. 59 into the formula for the expectation value of the energy, the energy can be systematically expanded in powers of the parameters

$$\Delta E = -\sum_k p_k f_k + \frac{1}{2} \sum_{k,k'} \gamma_k \gamma_{k'} [S_h + (1+\beta) G]^{k,k'}, \tag{60}$$

where

$$S_h^{k,k'} = \langle [O_k, [H, O_{k'}]] \rangle, \tag{61}$$

$$G^{k,k'} = 2\langle (H - E_p)(O_k - \langle O_k \rangle)(O_{k'} - \langle O_{k'} \rangle) \rangle, \tag{62}$$

and

$$f_k = -\partial_{p_k} E_p = -2\langle (H - E_p) O_k \rangle. \tag{63}$$

The optimal wave function parameters may then be obtained by iteratively minimizing Eq. 60 with respect to γ and then updating the parameters such that $p_k{}' = p_k + \gamma_k$, where the prime denotes the new set of parameters. Minimizing Eq. 60 yields

$$\gamma = B^{-1}\mathbf{f} \tag{64}$$

with

$$B = S_h + (1 + \beta)G. \tag{65}$$

It can only be guaranteed that each iteration is reducing the energy when B is positive definite. To ensure that this is the case, a multiple of the overlap matrix, S, is typically added to B

$$B' = B + \mu S, \tag{66}$$

with

$$S^{k,k'} = \langle (O_k - \langle O_k \rangle)(O_{k'} - \langle O_{k'} \rangle) \rangle. \tag{67}$$

In practice, this minimization is performed under the constraint that

$$\Delta\Psi = (|\Psi(\mathbf{p} + \gamma)\rangle - |\Psi(\mathbf{p})\rangle)/|\Psi(\mathbf{p})\rangle \leq r, \tag{68}$$

where r is a sufficiently small control parameter to ensure that the energy is reduced at each step during the minimization. The quantities in Eqs. 61–63 may be computed using correlated sampling techniques. Namely, configurations may be sampled from the square of the initial wave function and, during each step, S_h, $\mathbf{f}$, and G may be obtained by averaging over these configurations.

When $\beta = 1$ and the wave function is linear in all of its parameters, this technique may be mapped onto the standard Newton method [86, 89]. It has been shown that selecting a $\beta \neq 1$, however, results in substantially faster convergence to the optimal parameters [96, 101]. While stochastic reconfiguration *without* Hessian acceleration [97] often converges exceedingly slowly to the optimal parameters, often taking hundreds to thousands of iterations, if it even converges, the stochastic reconfiguration *with* Hessian acceleration algorithm just described can converge in just a few iterations. Moreover, unlike Newton's method, both stochastic reconfiguration and the linear method do not necessitate the cumbersome calculation of the second derivative of the wave function with respect to the variational parameters. The stochastic reconfiguration technique is thus an alternative, yet highly efficient technique for the optimization of the most general wave function forms.

6 Applications of Variational Monte Carlo Methods in Physics and Chemistry

Variational Monte Carlo algorithms are now widely used throughout physics and chemistry. In the following subsections, I highlight recent uses of VMC to determine the electronic structure of a wide variety of compounds, explore the excited states of photoactive molecules, and shed light on the ground state of the Hubbard model in different parameter regimes.

6.1 Quantum Chemistry

One of the long-standing goals of the quantum Monte Carlo community has been to determine the atomization energies of a wide variety of atoms and molecules to within chemical accuracy, defined as 1 kcal/mol. For years, achieving this goal has been stymied by the fermion sign problem—the appearance of negative probabilities that can result in insurmountable levels of noise when projector quantum Monte Carlo methods[7] are applied to fermions [4]. One approximate way to avoid the sign problem is to restrict the configurations sampled in projector quantum Monte Carlo algorithms to lie within a nodal surface determined by a trial wave function, as is done when the fixed node [4] or constrained path approximations [102] are applied. The more accurate the trial wave function that provides these constraints, the more accurate these approximate algorithms become. Indeed, these algorithms are exact in the limit that the trial wave function becomes the exact ground state wave function. One path toward surmounting the fermion sign problem has therefore been to try to improve VMC techniques to the point where the wave functions they produce are so accurate that they can eliminate any approximations associated with constraining algorithms.

---

[7] Such "projector" quantum Monte Carlo methods include Diffusion Monte Carlo, Green's Function Monte Carlo, and Auxiliary Field Quantum Monte Caro. Variational Monte Carlo methods are not projector quantum Monte Carlo methods and do not suffer from the sign problem. Nevertheless, projector quantum Monte Carlo algorithms are designed to converge to the ground state, while no such promise can be made for variational algorithms.

In recent years, this goal has nearly been achieved owing in large part to the success of the linear method. As discussed above, the linear method represents the first highly efficient, highly robust method for optimizing *all* wave function parameters. Before the advent of this method, researchers frequently attempted to break the 1 kcal/mol level of accuracy barrier, but failed because of the inadequacies of their trial wave functions [103, 104]. More specifically, Grossman was one of the first researchers to calculate the mean absolute deviation (MAD) of the Gaussian-1 (G1) test set [105, 106] atomization energies produced using Diffusion Monte Carlo from those obtained via experiment [103]. Using a Slater-Jastrow wave function with a single determinant whose orbitals were directly taken from quantum chemistry calculations that employed pseudopotentials, he was able to obtain a MAD from the experimental values of 2.9 kcal/mol with a maximum deviation from experiment of 14 kcal/mol over the 55 molecules in the test set. A MAD from the experimental values of 3.2 kcal/mol was obtained by Nemec and coworkers [104] on the G2 test set [107]. Using the linear method to optimize both Jastrow and Slater (orbital and determinantal) parameters, however, Petruzielo and coworkers were able to reduce the G2 MAD from experiment to 2.1 kcal/mol [92]. By moving from a single determinant to a complete active space trial wave function, Petruzielo reduced his MAD over the G2 set to 1.2 kcal/mol, within striking distance of chemical accuracy. These results suggest that, to achieve chemical accuracy, it is crucial that all wave function parameters are optimized at the same time. Equally important is the use of multideterminant trial wave functions. Grossman, Petruzielo, Morales, and coworkers have all shown that the more determinants included in the Slater portion of the wave function, the more accurate the trial wave function is at reproducing the nodal features needed to constrain projector quantum Monte Carlo calculations [92, 103, 108]. The most recent research has therefore turned toward finding improved ways of efficiently representing and manipulating multideterminant expansions in quantum Monte Carlo calculations [108–110] and mitigating/eliminating the remaining pseudopotential errors in such calculations.

Emboldened by this success, many researchers are now beginning to apply quantum Monte Carlo methods to complex solids, including magnetic materials and high-temperature superconductors, for which the combined speed and accuracy of quantum Monte Carlo methods may answer a number of questions left unresolved by conventional methods [111–113].

6.2 Excited States

Another recent application of quantum Monte Carlo techniques made possible by robust variational Monte Carlo methods has been to excited states [40–49]. Excited states are of physical and chemical interest because they play a central role in determining the spectra of excited state absorption experiments and the band gaps of solid state materials. Typically, projector and variational quantum Monte Carlo techniques are designed to determine the electronic ground state of a system (or the lowest energy excited state of a given symmetry). However, these methods can be targeted toward arbitrary excited states by employing variational techniques with alternative forms for their cost functions.

One of the early variants of this idea was to use state-averaged trial wave functions in Diffusion Monte Carlo calculations [40, 41]. In this method, which follows from the state-averaged multi-configurational self consistent field method, instead of minimizing the energy evaluated with respect to the single ground state wave function, the state-averaged energy

$$E_{SA} = \sum_{i \in A} w_i \frac{\langle \Psi_i | \hat{H} | \Psi_i \rangle}{\langle \Psi_i | \Psi_i \rangle}, \tag{69}$$

is minimized, where i denotes the different states in the active space and w_i denotes each state's weight ($\sum_i w_i = 1$). A generalized variational theorem applies to this minimization so long as the state wave functions are kept orthogonal. Using this method, one obtains wave functions whose orbitals represent a compromise among all of the states included in the averaging. This method has been applied to the study of a variety of photoactive molecules including ethene [41], formaldimine [40], formaldehyde [40], green fluorescent protein [42], and oxirane [44]. These studies have shown that QMC calculations can help troubleshoot excited state energies provided by alternative methods such as time-dependent density functional theory.

A much more recently proposed cost function for targeting arbitrary excited states is

$$C(\Psi) = \frac{\langle \Psi | (\omega - \hat{H}) | \Psi \rangle}{\langle \Psi | (\omega - \hat{H})(\omega - \hat{H}) | \Psi \rangle}, \tag{70}$$

where $C(\Psi)$ represents the cost function and ω is an energy shift whose value is chosen to lie directly below the eigenvalue of the eigenstate whose energy is being targeted [50]. By optimizing this cost function, a wave function is obtained whose variational energy lies immediately above ω. This cost function may be sampled via correlated sampling. Using the resolution of the identity, Eq. 70 may be re-expressed as

$$C(\Psi) = \frac{\sum_m \langle \Psi | m \rangle \langle m | (\omega - \hat{H}) | \Psi \rangle}{\sum_m \langle \Psi | (\omega - \hat{H}) | m \rangle \langle m | (\omega - \hat{H}) | \Psi \rangle}, \tag{71}$$

which can in turn be written in a form hospitable to quantum Monte Carlo methods

$$C_{MC}(\Psi) = \frac{\sum_{m \in \xi} W_m}{\sum_{m \in \xi} W_m^2}, \tag{72}$$

with

$$W_m \equiv \frac{\langle m|(\omega - \hat{H})|\Psi\rangle}{\langle m|\Psi\rangle}. \tag{73}$$

Thus, the proposed cost function can be evaluated by sampling the elements ξ from $|\langle m|\Psi\rangle|^2$ and then evaluating W_m. This algorithm has been used to determine the excited state energies of C_2, CH_2, and H_6 and has yielded results with greater accuracy than those obtained using many standard methods, such as Equation of Motion-Coupled Cluster Singles and Doubles (EOM-CCSD) [50].

6.3 The Hubbard Model

The Hubbard Model, a model long studied by condensed matter physicists because it is thought by some to illustrate the same physics that is responsible for high-temperature superconductivity, has been the subject of many variational and projector quantum Monte Carlo explorations over the years [60–63, 102, 114]. Most VMC studies of the Hubbard model have focused on minimizing the energy of different proposed wave function forms, including Gutzwiller and modified Gutzwiller wave functions, to determine which terms in these wave functions best reproduce the model's physics [60–63]. Most projector quantum Monte Carlo studies have performed Diffusion or Auxiliary Field Quantum Monte Carlo calculations starting with relatively simple single-determinant wave functions [102, 115, 116]. Based upon the successes realized solving quantum chemistry problems using multideterminant Slater-Jastrow wave functions, Chang and coworkers have recently put forth a way to produce and use variationally-optimized multideterminant Slater-Jastrow wave functions as trial wave functions in Auxiliary Field Quantum Monte Carlo simulations of the Hubbard model [117]. Prior to their work, Jastrow factors only saw widespread application in real-space QMC methods, as described above. What they showed is that Jastrow factors can also be applied to wave functions constructed in Fock space by using the Hubbard-Stratonovich Transformation [118] to decouple the quadratic operators that often appear in Jastrow factors. In specific, they showed that they could apply a modified Gutzwiller projector onto a single determinant wave function by decoupling the terms in that Gutzwiller projector using the Hubbard-Stratonovich Transformation and then employing random sampling to obtain compact representations of the multideterminant expansion that stems from that projection. Even without variationally-optimizing the multideterminant expansions they obtained, they were able to reduce the constrained path approximation bias in the energies they computed for a variety of 2D Hubbard models by nearly an order of magnitude over the biases produced using just a Slater determinant trial wave function [117]. Variationally-optimizing their multideterminant expansions, they can reduce this bias nearly to zero, yielding highly accurate, almost approximation-free results for the Hubbard model.

7 Summary and Outlook

Variational Monte Carlo algorithms are a powerful set of algorithms that exploit random sampling in order to determine the lowest energy that can possibly be obtained given a specific wave function form. VMC methods constitute a cheap, relatively inexpensive way to estimate the ground state energy of a system without having to contend with the fermion sign problem that plagues more sophisticated projector quantum Monte Carlo methods. By giving scientists the ability to craft and test wave functions term by term by hand, VMC methods also enable researchers to get a sense of which interactions most affect the ground state physics of a given system. As discussed in this review, recent advances in VMC algorithms have brought the long-sought goal of achieving chemical accuracy using quantum Monte Carlo methods on molecules in reach and have substantially improved the predictive capability of QMC methods applied to the Hubbard model and strongly correlated materials. Recent steps have also been taken to extend ground state VMC methods to arbitrary excited states, which opens up a full range of opportunities to apply projector methods to excited states for the first time. Modern VMC methods are therefore paving the way toward using cheap, yet highly accurate quantum Monte Carlo methods on problems that have previously been inaccessible to more expensive quantum chemistry techniques.

Going forward, the successes realized developing VMC algorithms to optimize real space wave functions may be built upon by developing similarly robust methods for handling second-quantized wave functions and such complex wave function forms as matrix and tensor product states. Combining VMC optimization techniques with such powerful wave function ansatzes will provide the tools necessary to explore large, strongly-correlated molecules and materials using quantum Monte Carlo methods in the future.

Acknowledgements The author would like to acknowledge Drs. James Gubernatis and Miguel Morales-Silva for their continuous support and encouragement of my quantum Monte Carlo efforts over the many years. BMR would also like to thank the Lawrence Fellowship Program for funding and Mr. Raymond Clay and Dr. Cyrus Umrigar for their help editing this manuscript. Portions of this work were performed under the auspices of the U.S. Department of Energy by Lawrence Livermore National Laboratory under Contract DE-AC52-07NA27344.

References

1. J.B. Anderson, A random-walk simulation of the Schrodinger equation: H_3^+. J. Chem. Phys. **63**, 1499 (1975)
2. P.J. Reynolds, D.M. Ceperley, B.J. Alder, W.A. Lester, Fixed-node quantum Monte Carlo for molecules. J. Chem. Phys. **77**, 5593 (1982)
3. J.L. DuBois, H.R. Glyde, Natural orbitals and Bose-Einstein condensates in traps: a diffusion Monte Carlo analysis. Phys. Rev. A **68**, 033602 (2003)
4. W.M.C. Foulkes, L. Mitas, R.J. Needs, G. Rajagopal, Quantum Monte Carlo simulations of solids. Rev. Mod. Phys. **73**, 33 (2001)

5. M.D. Towler, Quantum Monte Carlo, or, how to solve the many-particle Schrodinger equation accurately whilst retaining favourable scaling with system size, in *Computational Methods for Large Systems: Electronic Structure Approaches for Biotechnology and Nanotechnology* (Wiley, 2011). http://onlinelibrary.wiley.com/doi/10.1002/9780470930779.ch4/summary
6. D.M. Ceperley, B.J. Alder, Ground state of the electron gas by a stochastic method. Phys. Rev. Lett. **45**, 566 (1980)
7. D.M. Ceperley, E.L. Pollock, Path integral computation of the low temperature properties of liquid He-4. Phys. Rev. Lett. **56**, 351 (1986)
8. D.M. Ceperley, Path integrals in the theory of condensed helium. Rev. Mod. Phys. **67**, 279 (1995)
9. D.M. Ceperley, Path integral Monte Carlo methods for fermions, in *Monte Carlo and Molecular Dynamics of Condensed Matter Systems*, ed. by K. Binder, G. Ciccotti (Editrice Compositori, Bologna, Italy, 1996)
10. E.W. Brown, B.K. Clark, J.L. Dubois, D.M. Ceperley, Path integral Monte Carlo simulation of the warm-dense homogenous electron gas. Phys. Rev. Lett. **110**, 146405 (2013)
11. G. Sugiyama, S.E. Koonin, Auxiliary field Monte Carlo for quantum many-body ground states. Ann. Phys. **168**, 1 (1986)
12. S.R. White, D.J. Scalapino, R.L. Sugar, E.Y. Loh, J.E. Gubernatis, R.T. Scalettar, Numerical study of the two-dimensional Hubbard model. Phys. Rev. B **40**, 506 (1989)
13. S. Zhang, Auxiliary-field quantum Monte Carlo for correlated electron systems, in *Emergent Phenomena in Correlated Matter: Modeling and Simulation*, vol. 3, ed. by E. Pavarini, E. Koch, U. Schollwock (2013)
14. B.M. Rubenstein, S. Zhang, D.R. Rubenstein, Finite-temperature auxiliary-field quantum Monte Carlo technique for Bose Fermi mixtures. Phys. Rev. B **86**, 053606 (2012)
15. S. Zhang, H. Krakauer, Quantum Monte Carlo method using phase-free random walkers with Slater determinants. Phys. Rev. Lett. **90**, 136401 (2003)
16. W.A. Al-Saidi, S. Zhang, H. Krakauer, Auxiliary-field quantum Monte Carlo calculations of molecular systems with a Gaussian basis. J. Chem. Phys. **124**, 224101 (2006)
17. F. Ma, W. Purwanto, S. Zhang, H. Krakauer, Quantum Monte Carlo calculations in solids with downfolded Hamiltonians. Phys. Rev. Lett. **114**, 226401 (2015)
18. P. Werner, A. Comanac, L. de' Medici, M. Troyer, A.J. Millis, Continuous-time solver for quantum impurity models. Phys. Rev. Lett. **97**, 076405 (2006)
19. E. Gull, A.J. Millis, A.I. Lichtenstein, A.N. Rubtsov, M. Troyer, P. Werner, Continuous-time Monte Carlo methods for quantum impurity models. Rev. Mod. Phys. **83**, 349 (2011)
20. L. Muhlbacher, E. Rabani, Real-time path integral approach to nonequilibrium many-body quantum systems. Phys. Rev. Lett. **100**, 176403 (2008)
21. P. Werner, T. Oka, A.J. Millis, Diagrammatic Monte Carlo simulation of nonequilibrium systems. Phys. Rev. B **79**, 035320 (2009)
22. E. Gull, D.R. Reichman, A.J. Millis, Numerically exact long-time behavior of nonequilibrium quantum impurity models. Phys. Rev. B **84**, 085134 (2011)
23. G. Cohen, E. Gull, D.R. Reichman, A.J. Millis, Taming the dynamical sign problem in real-time evolution of quantum many-body problems. Phys. Rev. Lett. **115**, 266802 (2015)
24. G.H. Booth, A.J.W. Thom, A. Alavi, Fermion Monte Carlo without fixed nodes: a game of life, death, and annihilation in Slater determinant space. J. Chem. Phys. **131**, 054106 (2009)
25. D. Cleland, G.H. Booth, A. Alavi, Communications: survival of the fittest: accelerating convergence in full configuration-interaction quantum Monte Carlo. J. Chem. Phys. **132**, 041103 (2010)
26. F.D. Malone, N.S. Blunt, J.J. Shepherd, D.K.K. Lee, J.S. Spencer, W.M.C. Foulkes, Interaction picture density matrix quantum Monte Carlo. J. Chem. Phys. **143**, 044116 (2015)
27. G.H. Booth, A. Alavi, Approaching chemical accuracy using full configuration-interaction quantum Monte Carlo: a study of ionization potentials. J. Chem. Phys. **132**, 174104 (2010)
28. D. Cleland, G.H. Booth, A. Alavi, A study of electron affinities using the initiator approach to full configuration interaction quantum Monte Carlo. J. Chem. Phys. **134**, 024112 (2011)

29. G.H. Booth, A. Gruneis, G. Kresse, A. Alavi, Towards an exact description of electronic wave functions in real solids. Nature **493**, 365 (2013)
30. N. Metropolis, A.W. Rosenbluth, M.N. Rosenbluth, A.H. Teller, E. Teller, Equation of state calculations by fast computing machines. J. Chem. Phys. **21**, 1087 (1953)
31. R.M. Noack, S.R. Manmana, Diagonalization- and numerical renormalization-group-based methods for interacting quantum systems. AIP Conf. Proc. **789**, 93 (2005)
32. A. Szabo, N.S. Ostlund, *Modern Quantum Chemistry: Introduction to Advanced Electronic Structure Theory* (Dover, New York, 1996)
33. I. Shavitt, R. Bartlett, *Many-Body Methods in Chemistry and Physics: MBPT and Coupled-Cluster Theory* (Cambridge, New York, 2009)
34. R.G. Parr, Density functional theory. Ann. Rev. Phys. Chem. **34**, 631 (1983)
35. S.F. Sousa, P.A. Fernandes, M.J. Ramos, General performance of density functionals. J. Phys. Chem. A **111**, 10439 (2007)
36. A. Georges, G. Kotliar, W. Krauth, M.J. Rozenberg, Dynamical mean-field theory of strongly correlated fermion systems and the limit of infinite dimensions. Rev. Mod. Phys. **68**, 13 (1996)
37. C.J. Umrigar, Variational Monte Carlo basics and applications to atoms and molecules, in *Quantum Monte Carlo Methods in Physics and Chemistry*, NATO Advanced Science Institutes, Series C, vol. 525, ed. by M.P. Nightingale, C.J. Umrigar (1999), p. 129
38. D. Frenkel, B. Smit, *Understanding Molecular Simulation: From Algorithms to Applications* (Elsevier, San Diego, 2002)
39. R.J. Needs, M.D. Towler, N.D. Drummond, P. Lopez Rios, Continuum variational and diffusion quantum Monte Carlo calculations. J. Phys.: Condens. Matter **22**, 023201 (2010)
40. F. Schautz, F. Buda, C. Filippi, Excitations in photoactive molecules from quantum Monte Carlo. J. Chem. Phys. **121**, 5836 (2004)
41. F. Schautz, C. Filippi, Optimized Jastrow-Slater wave functions for ground and excited states: application to the lowest states of ethene. J. Chem. Phys. **120**, 10931 (2004)
42. C. Filippi, M. Zaccheddu, F. Buda, Absorption spectrum of the green fluorescent protein chromophore: a difficult case for ab Initio methods? J. Chem. Theory Comput. **5**, 2074 (2009)
43. R. Guareschi, C. Filippi, Ground- and excited-state geometry optimization of small organic molecules with quantum Monte Carlo. J. Chem. Theory Comput. **9**, 5513 (2013)
44. F. Cordova, J. Doriol, A. Ipatov, M.E. Casida, C. Filippi, A. Vela, Troubleshooting time-dependent density-functional theory for photochemical applications: oxirane. J. Chem. Phys. **127**, 164111 (2007)
45. O. Valsson, C. Filippi, Photoisomerization of model retinal chromophores: insight from quantum Monte Carlo and multiconfigurational perturbation theory. J. Chem. Phys. **6**, 1275 (2010)
46. R. Send, O. Valsson, C. Filippi, Electronic excitations of simple cyanine dyes: reconciling density functional and wave function methods. J. Chem. Theory Comput. **7**, 444 (2011)
47. C. Filippi, F. Buda, L. Guidoni, A. Sinicropi, Bathochromic shift in green fluorescent protein: a puzzle for QM/MM approaches. J. Chem. Theory Comput. **8**, 112 (2012)
48. O. Valsson, C. Angeli, C. Filippi, Excitation energies of retinal chromophores: critical role of the structural model. Phys. Chem. Chem. Phys. **14**, 11015 (2012)
49. O. Valsson, P. Campomanes, I. Tavernelli, U. Rothlisberger, C. Filippi, Rhodopsin absorption from first principles: bypassing common pitfalls. J. Chem. Theory Comput. **9**, 2441 (2013)
50. L. Zhao, E. Neuscamman, An efficient variational principle for the direct optimization of excited states. arXiv:1508.06683
51. W. Krauth, *Statistical mechanics: Algorithms and computations* (Oxford University Press, New York, 2006)
52. N. Metropolis, A.W. Rosenbluth, M.N. Rosenbluth, A.H. Teller, E. Teller, Equations of state calculations by fast computing machines. J. Chem. Phys. **21**, 1087 (1953)
53. G.F. Lawler, *Introduction to Stochastic Processes* (Taylor and Francis, Boca Raton, FL, 2006)
54. J. Toulouse, R. Assaraf, C.J. Umrigar, Introduction to the variational and diffusion Monte Carlo methods, in *Advances in Quantum Chemistry, 2015, Electron Correlation in Molecules – ab initio Beyond Gaussian Quantum Chemistry*, pp. 0000

55. W.K. Hastings, Monte Carlo sampling methods using Markov chains and their applications. Biometrika **57**, 97 (1970)
56. C.J. Umrigar, Accelerated Metropolis method. Phys. Rev. Lett. **71**, 408 (1993)
57. M.P. Allen, D.J. Tildesley, *Computer Simulation of Liquids* (Oxford Science Publications, New York, 1987)
58. D. Bressanini, P.J. Reynolds, Between classical and quantum Monte Carlo methods: "Variational QMC". Adv. Chem. Phys. **105** (1998)
59. B. Edegger, V.N. Muthukumar, C. Gros, Gutzwiller-RVB theory of high-temperature superconductivity: results from renormalized mean-field theory and variational Monte Carlo calculations. Adv. Phys. **56**, 927 (2007)
60. R. Kaneko, S. Morita, I. Masatoshi, Improved multi-variable variational Monte Carlo method examined by high-precision calculations of one-dimensional Hubbard model. J. Phys. Conf. Ser. **454**, 012046 (2013)
61. H. Otsuka, Variational Monte Carlo studies of the Hubbard model in one- and two-dimensions—off-diagonal intersite correlation effects. J. Phys. Soc. Jpn. **61**, 1645 (1992)
62. T. Yanagisawa, S. Koike, Y. Kunihiko, D-Wave state with multiplicative correlation factors for the Hubbard model. J. Phys. Soc. Jpn. **68**, 3608 (1999)
63. D. Baeriswyl, D. Eichenberger, M. Menteshashvili, Variational ground states of the two-dimensional Hubbard model. New J. Phys. **11**, 075010 (2009)
64. C.J. Umrigar, J. Toulouse, C. Filippi, S. Sorella, R.G. Hennig, Alleviation of the fermion-sign problem by optimization of many-body wave functions. Phys. Rev. Lett. **98**, 110201 (2007)
65. J. Toulouse, C.J. Umrigar, Optimization of quantum Monte Carlo wave functions by energy minimization. J. Chem. Phys. **126**, 084102 (2007)
66. J. Toulouse, C.J. Umrigar, Full optimization of Jastrow-Slater wave functions with application to the first-row atoms and homonuclear diatomic molecules. J. Chem. Phys. **128**, 174101 (2008)
67. C. Filippi, C. Umrigar, Multiconfiguration wave functions for quantum Monte Carlo calculations of first-row diatomic molecules. J. Chem. Phys. **105**, 213 (1996)
68. E. Neuscamman, Size consistency error in the antisymmetric geminal power wave function can be completely removed. Phys. Rev. Lett. **109**, 203001 (2012)
69. E. Neuscamman, The Jastrow antisymmetric geminal power in Hilbert space: theory, benchmarking, and application to a novel transition state. J. Chem. Phys. **139**, 194105 (2013)
70. E. Neuscamman, Subtractive manufacturing with geminal powers: making good use of a bad wave function. Mol. Phys. **114**, 577–583 (2016)
71. M. Bajdich, L. Mitas, G. Drobny, L.K. Wagner, K.E. Schmidt, Pfaffian pairing wave functions in electronic-structure quantum Monte Carlo simulations. Phys. Rev. Lett. **96**, 130201 (2006)
72. M. Bajdich, L. Mitas, L.K. Wagner, K.E. Schmidt, Pfaffian pairing and backflow wavefunctions for electronic structure quantum Monte Carlo methods. Phys. Rev. B **77**, 115112 (2008)
73. S. Wouters, B. Verstichel, D. Van Neck, G.K.-L. Chang, Projector quantum Monte Carlo with matrix product states. Phys. Rev. B **90**, 045104 (2014)
74. C.J. Umrigar, Optimized trial wave functions for quantum Monte Carlo calculations. Phys. Rev Lett. **60**, 1719 (1988)
75. A.J. Williamson, S.D. Kenny, G. Rajagopal, A.J. James, R.J. Needs, L.M. Fraser, W.M.C. Foulkes, Optimized wave functions for quantum Monte Carlo studies of atoms and solids. Phys. Rev. B **55**, 9640 (1996)
76. H. Huang, Z. Cao, A novel method for optimizing quantum Monte Carlo wave functions. J. Chem. Phys. **104**, 200 (1996)
77. K.E. Schmidt, J.W. Moskowitz, Correlated Monte Carlo wave functions for the atoms He and Ne. J. Chem. Phys. **93**, 4172 (1990)
78. H. Bueckert, S.M. Rothstein, J. Vrbik, Optimization of quantum Monte Carlo wavefunctions using analytical derivatives. Can. J. Chem. **70**, 366 (1992)
79. C.J. Umrigar, K.G. Wilson, J.W. Wilkins, A method for determining many body wave functions. Comput. Simul. Stud. Condens. Matter Phys. **33**, 185 (1988)

80. P.R.C. Kent, R.J. Needs, G. Rajagopal, Monte Carlo energy and variance-minimization techniques for optimizing many-body wave functions. Phys. Rev. B **59**, 12344 (1999)
81. M. Snajdr, S.M. Rothstein, Are properties derived from variance-optimized wave functions generally more accurate? Monte Carlo study of non-energy-related properties of H_2, He, and LiH. J. Chem. Phys. **112**, 4935 (2000)
82. F.J. Galvez, E. Buendia, A. Sarsa, Atomic properties from energy-optimized wave functions. J. Chem. Phys. **115**, 1166 (2001)
83. B.H. Wells, The differential Green's function Monte Carlo method: the dipole moment of Li H. Chem. Phys. Lett. **115**, 89 (1985)
84. W.H. Press, S.A. Teukolsky, W.T. Vetterling, B.P. Flannery, *Numerical Recipes in C: The Art of Scientific Computing* (Cambridge University Press, New York, 1992)
85. S.-Y. Huang, Z. Sun, W.A. Lester, Optimized trial functions for quantum Monte Carlo. J. Chem. Phys. **92**, 597 (1990)
86. X. Lin, H. Zhang, A.M. Rappe, Optimization of quantum Monte Carlo wave functions using analytical energy derivatives. J. Chem. Phys. **112**, 2650 (2000)
87. M.W. Lee, M. Mella, A.M. Rappe, Electronic quantum Monte Carlo calculations of atomic forces, vibrations, and anharmonicities. J. Chem. Phys. **122**, 244103 (2005)
88. H. Huang, Q. Xie, Z. Cao, Z. Li, Z. Yue, L. Ming, A novel quantum Monte Carlo strategy: surplus function approach. J. Chem. Phys. **110**, 3703 (1999)
89. C.J. Umrigar, C. Filippi, Energy and variance optimization of many-body wave functions. Phys. Rev. Lett. **94**, 150201 (2005)
90. E. Neuscamman, C.J. Umrigar, G.K.-L. Chan, Optimizing large parameter sets in variational quantum Monte Carlo. Phys. Rev. B **85**, 04103 (2012)
91. M.P. Nightingale, V. Melik-Alaverdian, Optimization of ground- and excited-state wave functions and van der Waals clusters. Phys. Rev. Lett. **87**, 043401 (2001)
92. F.R. Petruzielo, Approaching chemical accuracy with quantum Monte Carlo. J. Chem. Phys. **136**, 124116 (2012)
93. M.D. Brown, J.R. Trail, P. Lopez Rios, R.J. Needs, Energies of the first row atoms from quantum Monte Carlo. J. Chem. Phys. **126**, 224110 (2007)
94. S. Sorella, C. Capriotti, Green function Monte Carlo with stochastic reconfiguration: an effective remedy for the sign problem. Phys. Rev. B **61**, 2599 (2000)
95. S. Sorella, Green function Monte Carlo with stochastic reconfiguration. Phys. Rev. Lett. **80**, 4558 (1998)
96. S. Sorella, Wave function optimization in the variational Monte Carlo method. Phys. Rev. B **71**, 241103 (2005)
97. S. Sorella, Generalized Lancos algorithm for variational quantum Monte Carlo. Phys. Rev. B **64**, 024512 (2001)
98. M. Casula, C. Attaccalite, S. Sorella, Correlated geminal wave function for molecules: an efficient resonating valence bond approach. J. Chem. Phys. **121**, 7110 (2004)
99. M. Casula, S. Sorella, Geminal wave functions with Jastrow correlation: a first application to atoms. J. Chem. Phys. **119**, 6500 (2003)
100. S. Sorella, M. Casula, D. Rocca, Weak binding between two aromatic rings: feeling the van der Waals attraction by quantum Monte Carlo methods. J. Chem. Phys. **127**, 014105 (2007)
101. S. Sorella, Variational Monte Carlo and Markov chains for computational physics, in *Strongly Correlated Systems*, ed. by A. Avella, F. Mancini. Springer Series in Solid-State Sciences (Springer-Verlag Berlin Heidelberg, 2013), p. 207
102. S. Zhang, J. Carlson, J.E. Gubernatis, Constrained path Monte Carlo method for fermion ground states. Phys. Rev. B **55**, 7464 (1997)
103. J.C. Grossman, Benchmark quantum Monte Carlo calculations. J. Chem. Phys. **117**, 1434 (2002)
104. N. Nemec, M.D. Towler, R.J. Needs, Benchmark all-electron ab initio quantum Monte Carlo calculations for small molecules. J. Chem. Phys. **132**, 034111 (2010)
105. J.A. Pople, M. Head-Gordon, D.J. Fox, K. Raghavachari, L.A. Curtiss, Gaussian-1 theory: a general procedure for prediction of molecular energies. J. Chem. Phys. **90**, 5622 (1989)

106. L.A. Curtiss, C. Jones, G.W. Trucks, K. Raghavachari, J.A. Pople, Gaussian-1 theory of molecular energies for second-row compounds. J. Chem. Phys. **93**, 2537 (1990)
107. L.A. Curtiss, K. Raghavachari, G.W. Trucks, J.A. Pople, Gaussian-2 theory for molecular energies of first- and second-row compounds. J. Chem. Phys. **94**, 7221 (1991)
108. M.A. Morales, J. McMinis, B.K. Clark, J. Kim, G.E. Scuseria, Multideterminant wave functions in quantum Monte Carlo. J. Chem. Theory Comput. **8**, 2181 (2012)
109. B.K. Clark, M.A. Morales, J. McMinis, J. Kim, G.E. Scuseria, Computing the energy of a water molecule using multideterminants: a simple, efficient algorithm. J. Chem. Phys. **135**, 244105 (2011)
110. C. Filippi, R. Assaraf, S. Moroni, Simple formalism for efficient derivatives and multideterminant expansions in quantum Monte Carlo. J. Chem. Phys. **144**, 194105 (2016)
111. L.K. Wagner, Ground state of doped cuprates from first-principles quantum Monte Carlo calculations. Phys. Rev. B **92**, 16116 (2015)
112. L.K. Wagner, P. Abbamonte, Effect of electron correlation on the electronic structure and spin-lattice coupling of high-T_c cuprates: quantum Monte Carlo calculations. Phys. Rev. B **90**, 125129 (2014)
113. B. Busemeyer, M. Dagrada, S. Sorella, M. Casula, L.K. Wagner, Competing collinear magnetic structures in superconducting FeSe by first principles quantum Monte Carlo calculations. arXiv:1602.02054
114. J.P.F. LeBlanc, A.E. Antipov, F. Becca, I.W. Bulik, G.K.-L. Chan, C.M. Chung, Y. Deng, M. Ferrero, T.M. Henderson, C.A. Jimenez-Hoyos, E. Kozik, X.W. Liu, A.J. Millis, N.V. Prokof'ev, M. Qin, G.E. Scuseria, H. Shi, B.V. Svistunov, L.F. Tocchio, I.S. Tupitsyn, S.R. White, S. Zhang, B.X. Zheng, Z. Zhu, E. Gull, Solutions of the two-dimensional Hubbard model: benchmarks and results from a wide range of numerical algorithms. Phys. Rev. X **5**, 041041 (2015)
115. H. Shi, C.A. Jimenez-Hoyos, R. Rodriguez-Guzman, G.E. Scuseria, S. Zhang, Symmetry-projected wave functions in quantum Monte Carlo calculations. Phys. Rev. B **89**, 125129 (2014)
116. H. Shi, S. Zhang, Symmetry in auxiliary-field quantum Monte Carlo calculations. Phys. Rev. B **88**, 125132 (2013)
117. C.-C. Chang, B.M. Rubenstein, M.A. Morales-Silva, Auxiliary-field based trial wave functions in quantum Monte Carlo calculations (2016). arXiv:1604.003545
118. J.E. Hirsch, Discrete Hubbard-Stratonovich transformation for fermion lattice models. Phys. Rev. B **28**, 4059(R) (1983)

Appendix: Calculus of Variations

Jianzhong Wu

This appendix provides a very brief, first-line introduction to calculus of variations, an extension of multivariable calculus that was first introduced by Leonhard Euler in 1733. The background material is expected to be sufficient for those who are mainly interested in application rather than mathematical development of variational methods for molecular modeling. To get a more comprehensive understanding of this fascinating subject, the reader is referred to standard texts of mathematical physics such as:

1. *Mathematical Methods of Physics*, J. Mathews and R. L. Walker, Addison-Wesley, 1970.
2. *Calculus of Variations*, I. M. Gelfand and S. V. Fomin, Dover Books on Mathematics, 2000.
3. *Variational Methods in Mathematical Physics*, P. Blanchard and E. Brüning, Springer- Verlag, 1992.

A.1 Functional

A functional is an extension of what we mean by a multivariable function. When we write a multivariable function $f(\mathbf{z})$, where $\mathbf{z}$ is an n-dimensional variable, we mean that for each set of numbers $\mathbf{z} = (z_1, \ z_2, \ldots, z_n)$, there is a number $f(\mathbf{z})$ associated with it. Simple examples of multivariable functions are $f(\mathbf{z}) = \mathbf{z}^2 = \sum_{i=1}^{n} z_i^2$ or $f(\mathbf{z}) = \mathbf{a} \cdot \mathbf{z}$, where $\mathbf{a}$ is an n-dimensional vector.
When we write a functional, $F[y]$, we mean that for each smooth (differentiable) function $y(x)$, there is a number $F[y]$ related to it. In other words, a functional maps a function into a number, or a functional is a function of functions. The integral $F[y] = \int_0^1 y(x)dx$ provides a simple example of functionals. For each smooth function $y(x)$, its integration from 0 to 1 yields a number. While the "input" of a

J. Wu (ed.), *Variational Methods in Molecular Modeling*,
Molecular Modeling and Simulation, DOI 10.1007/978-981-10-2502-0

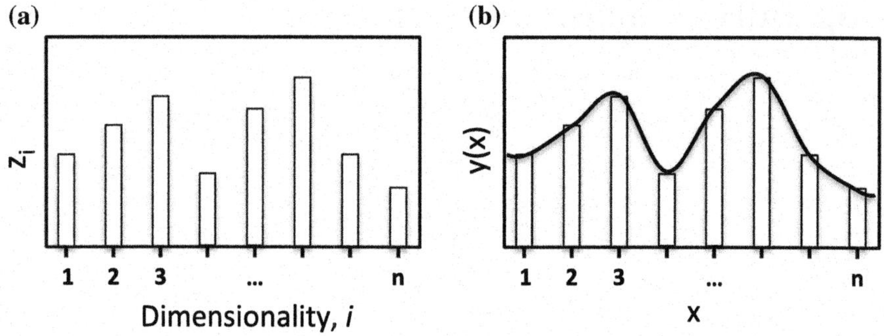

Fig. A.1 While the input for a multidimensional function is a vector, the input for a functional is a smooth function y(x). **a** An n-dimensional vector z contains a set of numbers affiliated with its dimensionality; **b** A one-dimensional function y(x) may be understood as a vector of infinite dimensionality

multivariable function is an n-dimensional vector, the "input" for a functional is a function. By comparing the similarity between a function and a vector, we see that a functional is a function of infinite dimensionality. Schematically, Fig. A.1 illustrates the difference between the inputs for a multi-dimensional function and a functional.

A.2 Variational Problem

To illustrate how a functional can be used to solve a realistic problem, consider the time required for a ball to fall along some frictionless path with two ends fixed at positions A and B, as indicated in Fig. A.2. For simplicity, assume that the path is two-dimensional and that it can be described by a smooth function $y = y(x)$. Let t denote the time required for the ball to go from A to B along a frictionless path. What path $y(x)$ should be chosen to make t a minimum?

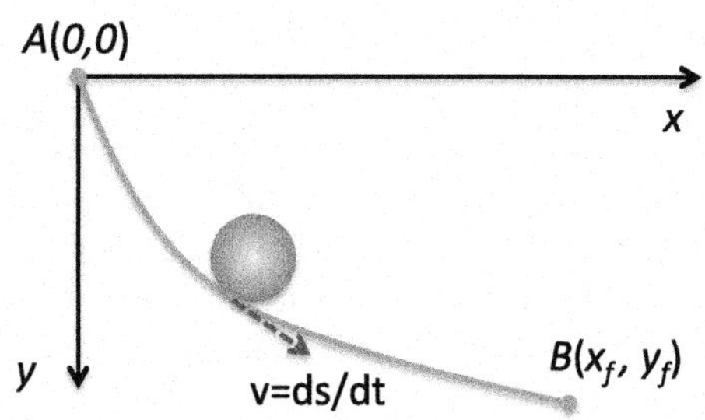

Fig. A.2 Calculus of variations can be used to identify a frictionless path that yields the shortest traveling time for a ball falling from point A to B

For convenience, we put point A at the origin of a coordinate system and measure y downward. At any instant, the ball speed is

$$v = \frac{ds}{dt}, \tag{A.1}$$

where v denotes the magnitude of speed, s represents the length along the path, and t is time. Rearrangement of Eq. (A.1) gives

$$dt = \frac{ds}{v}, \tag{A.2}$$

and thus the total traveling time is

$$t = \int_A^B \frac{ds}{v}. \tag{A.3}$$

The differential length of the path ds is

$$ds = \sqrt{1 + y'^2} dx, \tag{A.4}$$

where $y' = dy/dx$ represents the slope of the path. Because the ball starts at point A, conservation of energy requires that at any vertical distance y, the loss of potential energy per unit mass at y is equal to the gain in the kinetic energy per unit mass, i.e.,

$$gy = v^2/2 \tag{A.5}$$

where g stands for the gravity constant. Substituting Eqs. (A.4) and (A.5) into Eq. (A.3) gives

$$t = \int_0^{x_f} \sqrt{\frac{1 + y'^2}{2gy}} dx. \tag{A.6}$$

Equation (A.6) indicates that the total time t can be found if we know y as a function of x. For any path with ends fixed at $A(0, 0)$ and $B(x_f, y_f)$, there is a corresponding time for the ball to travel from A to B. Therefore, the total traveling time is a functional of path $y(x)$, that is, $t = F[y(x)]$.

The essential problem in calculus of variations is functional minimization,[1] i.e., to find a function that minimizes a given functional. In the above example, we want to know the path $y(x)$ with two ends fixed at A and B that gives the minimum descent time. To answer this question, we need to know how a functional responds to a change in its "input", where the "input" is not an ordinary variable, but a function.

[1]Functional maximization can be concerted to minimization by trivially adding a negative sign.

A.3 Functional Derivative

To obtain the unknown function that minimizes a functional, we use functional differentiation as discussed below. It is not much different from the partial derivative used in finding the minimum of a multidimensional function.

The variation of a functional with respect to its "input" is described by a functional derivative:

$$\begin{aligned}\frac{\delta F[y(x)]}{\delta y(x')} &\equiv \lim_{\varepsilon\to 0}\frac{F[y(x)+\varepsilon\delta(x-x')]-F[y(x)]}{\varepsilon}\\ &= \lim_{\varepsilon\to 0}\frac{F[y+\varepsilon\delta]-F[y]}{\varepsilon\delta}\delta \\ &= \frac{dF[y]}{dy}\delta(x-x')\end{aligned} \tag{A.7}$$

where ε is a real number, and $\delta(x-x')$ stands for the Dirac delta function. As shown in Fig. A.3, the Dirac function $\delta(x-x_0)$ represents a generalized probability density that is normalized and has a value of infinite at $x = x_0$. According to Eq. (A.7), the functional derivative $\delta F[y(x)]/\delta y(x')$ can be understood as the change in functional $F[y(x)]$ with respect to a change in the input function $y(x)$ at the point $x = x'$. Because the functional derivative is in general dependent on x', $\delta F[y(x)]/\delta y(x')$ is a function of x'.

The functional derivative defined above can be similarly applied to a function. Suppose $f(y)$ is a function of y, its functional derivative with respect to y is

$$\frac{\delta f(y)}{\delta y(x')} = f'(y)\delta(x-x'). \tag{A.8}$$

In a special case $f(y) = y$, we have

$$\frac{\delta y(x)}{\delta y(x')} = \delta(x-x'). \tag{A.9}$$

Equation (A.9) says that the functional derivative of a function with respect to itself is a Dirac delta function.

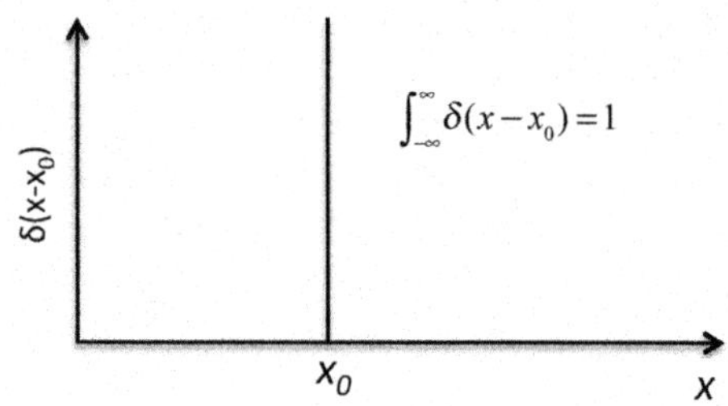

Fig. A.3 One-dimensional Dirac function $\delta(x-x_0)$ represents a probability density that is everywhere zero except at $x = x_0$ where it is infinite (∞)

Functional derivative may be considered as a natural extension of a partial derivative of a multivariable function to infinite dimensionality. To see this, consider again a multivariable function $f(\mathbf{z})$, where $\mathbf{z}$ stands for an n-dimensional vector. Partial derivative $\partial f/\partial z_i$ describes the change in $f(\mathbf{z})$ with respect to an infinitesimal change in the ith dimension of $\mathbf{z}$ while keeping all other dimensions unchanged, i.e.,

$$df = \sum_{j=1}^{n} \frac{\partial f}{\partial z_i} \delta_{ij} dz_i = \frac{\partial f}{\partial z_i} dz_i. \tag{A.10}$$

where δ_{ij} stands for the Kronecker delta function, i.e., $\delta_{ij} = 1$ for $i = j$ and zero otherwise. Similarly, the change of a functional with respect to its "input" (function) at a point x' can be written as

$$\delta F = \int dx \frac{dF}{dy} \delta(x - x')\delta y = \left. \frac{dF}{dy} \delta y \right|_{x'}. \tag{A.11}$$

Comparing Eqs. (A.10) and (A.11), we see that the variable x can be understood as a continuous index of function $y(x)$, similar to $i$as an index of vector $\mathbf{z}$.

As all partial derivatives of a multi-dimensional function vanish at the minimum point, a functional $F[y]$ reaches an minimum when

$$\frac{\delta F[y(x)]}{\delta y(x')} = 0 \tag{A.12}$$

for all values of x'.

A.4 Chain Rules of Functional Derivative

A functional derivative obeys chain rules similar to those for a partial derivative. For example, the chain rule of a partial derivative of a multivariable function $f(\mathbf{z})$ can be written as

$$\frac{\partial f\{\mathbf{g}(\mathbf{z})\}}{\partial z_i} = \sum_{j=1}^{n} \frac{\partial f}{\partial g_j} \frac{\partial g_j}{\partial z_i}, \tag{A.13}$$

where $\mathbf{g}(\mathbf{z})$ is an n-dimensional function of vector $\mathbf{z}$. The analogous chain rule for a functional derivative is

$$\frac{\delta F\{G[y(x)]\}}{\delta y(x')} = \int dx'' \frac{\delta F}{\delta G(x'')} \frac{\delta G(x'')}{\delta y(x')}, \tag{A.14}$$

where the summation of discrete indices in Eq. (A.13) is replaced by an integral over the continuous indices. In particular, if $F[y(x)] = y(x)$, we have

$$\delta(x-x') = \int dx'' \frac{\delta y(x)}{\delta G(x'')} \frac{\delta G(x'')}{\delta y(x')}. \tag{A.15}$$

Equation (A.15) represents a general relation between the reciprocals of functional derivatives.

It can be shown that the functional derivative of a function is commutable with a normal derivative, i.e.,

$$\frac{\delta(df/dx)}{\delta y} = \frac{d}{dx}\left(\frac{\delta f}{\delta y}\right) \tag{A.16}$$

where both f and g are functions of x. In a special case, the functional derivative of $y\prime(x)$ is

$$\frac{\delta}{\delta y(x')}\left[\frac{dy(x)}{dx}\right] = \frac{d}{dx}\left[\frac{\delta y(x)}{\delta y(x')}\right] = \frac{d\delta(x-x')}{dx}. \tag{A.17}$$

A.5 Higher-Order Functional Derivatives and Functional Taylor Expansion

Higher-order functional derivatives can be defined similar to the higher-order partial derivatives. In general, the mth-order functional derivative of $F[y(x)]$ is

$$\frac{\delta^{(m)} F[y(x)]}{\delta y(x_1)\delta y(x_2)\cdots\delta y(x_m)} = \frac{d^{(m)} F[y]}{dy^{(m)}} \delta(x-x_1)\delta(x-x_2)\cdots\delta(x-x_m) \tag{A.18}$$

These functional derivatives are used in the functional Taylor expansion. In parallel to a Taylor expansion of a multivariable function, $f(\mathbf{z})$,

$$f(\mathbf{z}+\Delta\mathbf{z}) = f(\mathbf{z}) + \sum_{i=1}^{n} \frac{\partial f}{\partial z_i}\Delta z_i + \frac{1}{2}\sum_{i=1}^{n}\sum_{j=1}^{n} \frac{\partial f}{\partial z_i}\frac{\partial f}{\partial z_j}\Delta z_i \Delta z_j + \cdots\ , \tag{A.19}$$

we can apply the Taylor expansion to a functional

$$F[y+\Delta y] = F[y] + \int dx \frac{\delta F}{\delta y(x)}\Delta y(x) + \frac{1}{2!}\int\int dx dx' \frac{\delta^2 F}{\delta y(x)\delta y(x')}\Delta y(x)\Delta y(x') + \cdots \tag{A.20}$$

Again, the difference between the multivariable and the functional Taylor expansions lies only on in the summation of the indices, i.e., the summation of integers and the integration of a continuous variable.

A.6 Functional Integration

Functional integration provides a general procedure to evaluate the change in functional at different input functions. It can also be used to calculate a functional from its derivative.

For a given function $y(x)$, $F[\lambda y(x)]$ represents a function of real variable λ. By the chain rule, the derivative of $F[\lambda y(x)]$ with respect to λ gives

$$\frac{dF[\lambda y(x)]}{d\lambda} = \frac{dF[\lambda y(x)]}{d(\lambda y)} \frac{\partial(\lambda y)}{\partial \lambda} = \int dx' \frac{\delta F[\lambda y(x)]}{\delta(\lambda y(x'))} y(x'). \tag{A.21}$$

The second equality in Eq. (A.21) can be verified by substituting the functional derivative by its definition (i.e., Eq. (A.7)). Equation (A.21) holds true when we replace $y(x)$ with $\Delta y(x) \equiv y(x) - y_0(x)$:

$$\frac{dF[y_0(x) + \lambda \Delta y(x)]}{d\lambda} = \int dx' \frac{\delta F[\lambda y(x)]}{\delta(\lambda y(x'))} \Delta y(x'). \tag{A.22}$$

where $y_0(x)$ is an arbitrary input function. Integrating the two sides of Eq. (A.22) with respect to λ from 0 to 1 gives

$$F[y] = F[y_0] + \int_0^1 d\lambda \int dx' \frac{\delta F[\lambda y(x)]}{\delta(\lambda y(x'))} \Delta y(x'). \tag{A.23}$$

Equation (A.23) indicates that the change of a functional with its input function is related to the integration of the functional derivative and a coupling parameter λ linking the input functions.

A.7 Functional of a Multidimensional Function

It is straightforward to extend the functional derivative, the functional Taylor expansion, and the functional integral when the input is a multidimensional function, i.e., $y = y(\mathbf{x})$ where $\mathbf{x}$ is a multidimensional vector. Following a procedure similar to that discussed for the one-dimensional case, we have a functional derivative

$$\frac{\delta F[y(\mathbf{x})]}{\delta y(\mathbf{x}\prime)} = \frac{dF[y]}{dy} \delta(\mathbf{x} - \mathbf{x}\prime) \tag{A.24}$$

where $\delta(\mathbf{x} - \mathbf{x}\prime)$ stands for a multidimensional Dirac delta function. The functional Taylor expansion of $F[y(\mathbf{x})]$ is

$$F[y + \Delta y] = F[y] + \int d\mathbf{x} \frac{\delta F}{\delta y(\mathbf{x})} \Delta y(\mathbf{x}) + \frac{1}{2!} \int \int d\mathbf{x} d\mathbf{x}\prime \frac{\delta^2 F}{\delta y(\mathbf{x}) \delta y(\mathbf{x}\prime)} \Delta y(\mathbf{x}) \Delta y(\mathbf{x}\prime) + \cdots, \tag{A.25}$$

and a functional integral is

$$F[y] = F[y_0] + \int_0^1 d\lambda \int d\mathbf{x'} \frac{\delta F[\lambda y(\mathbf{x})]}{\delta(\lambda y(\mathbf{x'}))} \Delta y(\mathbf{x'}). \tag{A.26}$$

A.8 An Illustrative Example

Now we return to the example shown in Fig. A.2. We want to find the functional derivative of

$$t[y(x)] = \int_0^{x_f} \sqrt{\frac{1+y'^2}{2gy}} dx. \tag{A.27}$$

and the path $y(x)$ that yields the shortest descent time from A to B.

For short notation, let

$$f(y, y') = \sqrt{\frac{1+y'^2}{2gy}}. \tag{A.28}$$

$f(y, y')$ can be understood as a normal two-dimensional function because y and y' are literately independent. Following the rules of ordinary calculus, we have

$$\frac{df}{dy} = \left(\frac{\partial f}{\partial y}\right)_{y'} + \left(\frac{\partial f}{\partial y'}\right)_y \frac{dy'}{dy}. \tag{A.29}$$

The functional derivative $t[y(x)]$ is thus given by

$$\begin{aligned}
\frac{\delta t[y(x)]}{\delta y(x')} &= \int_0^{x_f} \frac{df}{dy} \delta(x - x') dx \\
&= \int_0^{x_f} \left[\left(\frac{\partial f}{\partial y}\right)_{y'} + \left(\frac{\partial f}{\partial y'}\right)_y \frac{dy'}{dy}\right] \delta(x - x') dx \\
&= \left(\frac{\partial f}{\partial y}\right)_{x=x'} + \int_0^{x_f} \left(\frac{\partial f}{\partial y'}\right)_y \frac{\delta y'}{\delta y} dx \\
&= \left(\frac{\partial f}{\partial y}\right)_{x=x'} + \int_0^{x_f} \left(\frac{\partial f}{\partial y'}\right)_y d\delta(x - x') \\
&= \left(\frac{\partial f}{\partial y}\right)_{x=x'} - \frac{d}{dx}\left(\frac{\partial f}{\partial y'}\right)_{x=x'}
\end{aligned} \tag{A.30}$$

where $0 < x' < x_f$. The last equality is obtained by integrating by parts. The path of shortest descent satisfies

$$\frac{\delta t[y(x)]}{\delta y(x')} = 0. \tag{A.31}$$

From Eq. (A.29), we have

$$\left(\frac{\partial f}{\partial y}\right) - \frac{d}{dx}\left(\frac{\partial f}{\partial y'}\right) = 0. \tag{A.32}$$

Equation (A.31) is known as the Euler-Lagrange equation. Solution to this ordinary differential equation gives the path of shortest descent.

Equation (A.31) can be solved most conveniently by using an indirect method. First, we notice that

$$\frac{df}{dx} = \frac{\partial f}{\partial y'}\frac{dy'}{dx} + y'\frac{\partial f}{\partial y} \tag{A.33}$$

$$\frac{d}{dx}\left[y'\frac{\partial f}{\partial y'}\right] = \frac{\partial f}{\partial y'}\frac{dy'}{dx} + y'\frac{d}{dx}\left[\frac{\partial f}{\partial y'}\right] \tag{A.34}$$

Subtracting Eq. (A.32) by Eq. (A.33), and utilizing Eq. (A.31), we find

$$\frac{d}{dx}\left[f - y'\frac{\partial f}{\partial y'}\right] = y'\left[\frac{\partial f}{\partial y} - \frac{d}{dx}\left(\frac{\partial f}{\partial y'}\right)\right] = 0. \tag{A.35}$$

Thus

$$f - y'\frac{\partial f}{\partial y'} = k \tag{A.36}$$

where k is a constant. From $f = \sqrt{(1+y'^2)/(2gy)}$, we find

$$\frac{\partial f}{\partial y'} = \frac{1}{\sqrt{2gy}}\frac{y'}{\sqrt{1+y'^2}} = f\frac{y'}{1+y'^2} \tag{A.37}$$

and

$$f - y'\frac{\partial f}{\partial y'} = f(1 - \frac{y'^2}{1+y'^2}) = \frac{f}{1+y'^2} = \sqrt{\frac{1}{2gy(1+y'^2)}} = k. \tag{A.38}$$

Rearranging Eq. (A.37) gives

$$y(1+y'^2) = 2a, \tag{A.39}$$

where $a = gk^2$. Without loss of generality, we may assume $y' = \tan(\theta/2)$ where θ is a parameter. Then we have from Eq. (A.38)

$$y = \frac{2a}{1+y'^2} = 2a\sin^2(\theta/2) = a(1-\cos\theta) \tag{A.40}$$

and

$$\frac{dx}{d\theta} = \frac{1}{y'}\frac{dy}{d\theta} = a\tan(\theta/2)\sin\theta = a(1-\cos\theta) \qquad \text{(A.41)}$$

Integrating Eq. (A.40) with respect to θ equation, subject to the boundary condition $x = 0$ at $\theta = 0$, yields

$$x = a(\theta - \sin\theta) \quad . \qquad \text{(A.42)}$$

From Eqs. (A.39) and (A.41), we obtain the parametric equation for path $y(x)$ that yields shortest time for the ball to descend from A to B:

$$\begin{cases} x = a(\theta - \sin\theta) \\ y = a(1 - \cos\theta) \end{cases} \qquad \text{(A.43)}$$

Parameters a and θ_f can be found by the condition that this path must end at point $B(x_f, y_f)$:

$$\begin{cases} x_f = a(\theta_f - \sin\theta_f) \\ y_f = a(1 - \cos\theta_f) \end{cases} . \qquad \text{(A.44)}$$

Printed by Printforce, the Netherlands